Bo Ebenman · Lennart Persson (Eds.)

Size-Structured Populations

Ecology and Evolution

With 91 Figures

Springer-Verlag
Berlin Heidelberg New York London Paris Tokyo

Dr. Bo Ebenman
Department of Biology
University of Linköping
S-581 83 Linköping
Sweden

Dr. Lennart Persson
Department of Ecology, Limnology
P. O. Box 65
S-221 00 Lund
Sweden

ISBN 3-540-50188-6 Springer-Verlag Berlin Heidelberg New York
ISBN 0-387-50188-6 Springer-Verlag New York Berlin Heidelberg

Library of Congress Cataloging-in-Publication Data. Size-structured populations: ecology and evolution / Bo Ebenman. Lennart Persson (Eds.). p. cm. Includes index. 1. Population biology. 2. Ecology. 3. Evolution. I. Ebenman, Bo, 1953– . II. Persson, Lennart, 1948– . QH352.S59 1988 574.5'248–dc19 88-26561

Printed in Germany

Typesetting: K+V Fotosatz GmbH, Beerfelden
Printing and binding: Konrad Triltsch, Graphischer Betrieb, Würzburg
2131/3130-543210

Contents

Contributors

ARNOLD, S. J., Department of Biology, The University of Chicago, 940 East 57th Street, Chicago, IL 60637, USA

CASWELL, H., Biology Department, Woods Hole Oceanographic Institution, Woods Hole, MA 02543, USA

EBENMAN, B., Department of Biology, University of Linköping, 581 83 Linköping, Sweden

FRASER, D. F., Department of Biology, Siena College, Loudonville, NY 12211, USA

GILLIAM, J. F., Behavioral Ecology Group, Department of Biological Sciences, State University of New York at Albany, Albany, NY 12222, USA

GREY, D. R., Department of Probability and Statistics, the University, Sheffield, S3 7RH, England

KIRKPATRICK, M., Department of Zoology, University of Texas, Austin, Texas 78712, USA

LAW, R., Biology Department, York University, Heslington, York, YO1 5DD, England

LEIBOLD, M. A., Kellogg Biological Station and Department of Zoology, Michigan State University, Hickory Corners, MI 49060, USA

LYNCH, M., Department of Ecology, Ethology and Evolution, Shelford Vivarium, University of Illinois, 606 E. Healey St., Champaign, IL 61820, USA

MAY, R. M., Biology Department, Princeton University, Princeton, NJ 08544, USA; present address: Zoology Department, Oxford University, Oxford, OX1 3PS, England

METZ, J. A. J., Institute of Theoretical Biology, University of Leiden, Groenhovenstratt 5, 2311 BT Leiden, The Netherlands

MITTELBACH, G. G., Kellogg Biological Station and Department of Zoology, University of British Columbia, Vancouver, B.C., Canada V6T 2A9

NEILL, W. E., The Ecology Group, Department of Zoology, University of British Columbia, Vancouver, B.C., Canada V6T 2A9

NORBERG, R. Å., Department of Zoology, University of Gothenburg, Box 25059, 400 31 Gothenburg, Sweden

OSENBERG, C.W., Kellogg Biological Station and Department of Zoology, Michigan State University, Hickory Corners, MI 49060, USA

PERSSON, L., Department of Animal Ecology, University of Umeå, 90187 Umeå, Sweden

POLIS, G.A., Department of Biology, Vanderbilt University, Box 93, Station B, Nashville, Tennessee 37235, USA

DE ROOS, A.M., Institute of Theoretical Biology, University of Leiden, Groenhovenstratt 5, 2311 BT Leiden, The Netherlands

VAN DEN BOSCH, F., Institute of Theoretical Biology, University of Leiden, Groenhovenstratt 5, 2311 BT Leiden, The Netherlands

WERNER, E.E., Department of Biology, University of Michigan, Ann Arbour, Michigan 48109, USA

WILBUR, H.M., Department of Zoology, Duke University, Durham, North Carolina 27706, USA

Prolog

R. M. May

Most elementary textbooks in population biology and population genetics deal with homogeneous collections of individuals, studying the way population density or gene frequency changes over time in response to ecological and evolutionary forces. More detailed investigations recognize the age-structure within a population, often using techniques borrowed from actuarial studies of humans, to explore the demographic consequences of age-specific birth and death rates.

As Ebenman and Persson emphasize in the opening chapter of this book, the interactions of an individual organism with others of its own species, or with other species, or with the physical environment, are likely to depend on its size and condition (and in some cases on the organism's stage in a complex life cycle), rather than age as such. Age and size may, of course, often be highly correlated. The various studies in the book make it plain, however, that there are many situations where fundamental understanding or practical management is best attained by dealing directly with the population's size structure (and not using age-structure as a shadowy surrogate). Indeed – given the observed correlations between overweight and mortality factors such as heart ailments – it could even be that insurance agencies might do well to have their actuarial tables take account of size along with age, at least in the intermediate age-range around 20–60 years!

Conventional models for fisheries management, for example, usually deal with age-classes, using a variety of policy instruments (quotas, net size, etc.) in order to optimize the sustainable catch, according to criterias that all too often convolve biological and political considerations in messy ways. It is likely, however, that the mortality resulting from harvesting of adult fish or marine mammals creates strong selective pressures favouring earlier reproduction. The ensuing changes in the size and stage-structure of the target population will usually not be described well by dealing simply with the age-structure (May 1984). By the same token, plant growth typically depends more on weather patterns, nutrient availability, presence or absence of shading by competitors, and other such factors, than upon age as such; again, size-structured or stage-structured models are likely to be more appropriate management tools than simple age-structured models will be.

Ebenman and Persson's opening chapter outlines how the different chapters fit together to provide a formal framework for analyzing the ecological and evolutionary effects of size-structure in populations, and to apply this formal apparatus to empirical studies of particular populations in the field and laboratory.

Biology Department, Princeton University, Princeton, NJ 08544, USA
Present address: Zoology Department, Oxford University, Oxford, OX1 3PS, England

Size-Structured Populations
B. Ebenman and L. Persson (Eds.)

There is a consistent theme of reaching toward detailed understanding of particular examples, based on general principles.

In populations of fish, mammals, trees and many other organisms, the individual's basic body-plan remains essentially unchanged, even though size may change greatly over time. Many arthropods, helminths, and other organisms, however, possess complicated life cycles in which successive stages can differ more dramatically – both in physical appearance and in ecological roles – than do different species of vertebrates. The consequent stage-structured models in many ways are closer to models for a multispecies community than to age-structured models; the one organism functions for many purposes as if it were a succession of different species. Indeed, much of the taxonomy of tropical lepidopterans and other arthropods is made difficult by our current ignorance about which larval organisms develop into which adults. For organisms with complex life cycles, we therefore need to develop stage-structured models that blend appropriate elements of size-structured models with elements of multispecies population models. Some of the chapters in the present book represent a good beginning to this task. They remind us that the vertebrate-centered approach to population biology, with its traditional emphasis on age-structured equations, is not adequate to deal with the evolution and ecology of the vast number of species possessing complex life histories.

Organisms with complex life histories present many interesting empirical patterns that currently are ill-understood. One among many such enigmatic patterns is Dyar's Rule, which notes that successive instars of many insect larvae differ in mass by a factor close to two (Hutchinson 1959). Ultimately, a fuller understanding of the evolution of complex life histories will probably require widening our focus to consider developmental constraints, and even the compromises and trade-offs that may have been struck to reconcile the interests of individual cell lineages with the interests of multicellular aggregates when the first multi-cellular organisms appeared. Viewed in this light, the work on stage-structured models for organisms with complex life cycles that is recorded here is a part of a larger enterprise, currently gathering momentum, that weaves strands ranging from developmental biology to population biology, toward understanding the evolution of individuality (Buss 1988). These are problems that must be resolved if we are ever to obtain a basic understanding of why there are as many species as there are on earth; that is, a basic understanding of how diversity is generated and maintained.

In summary, the interesting and important collection of papers gathered together in this book is dedicated to systematic exploration of the proposition that how you look, and what "life stage" you have attained, is usually more important than your chronological age. Californians have known this for some time, but the ecological community has been a bit slow in giving it explicit recognition!

References

Buss LW (1988) The Evolution of Individuality. Princeton University Press, Princeton

Hutchinson GE (1959) Homage to Santa Rosalia, or why are there so many kinds of animals? Amer Natur 93:145–159

May RM (ed) (1984) Exploitation of Marine Communities. Springer, New York

Introduction
Dynamics of Size-Structured Populations: An Overview

B. EBENMAN[1] and L. PERSSON[2]

The size range of organisms is enormous, spanning over 21 orders of magnitude, with the blue whale and giant sequoia being 10^{21} heavier than the smallest microbe (McMahon and Bonner 1983). An organism is often identified as an individual in a given developmental stage, usually as a full-grown adult, which is, in many cases, a questionable point of view. As J. T. Bonner (1965) has argued, the ultimate description of an individual includes the whole life cycle. In many organisms the individuals pass through a wide spectrum of sizes, spanning more than four orders of magnitude, during the independent part of their life cycles; well-known examples are plants, fishes and reptiles. Thus, also within a species, individuals often vary greatly in size. Such large variation in size will have profound evolutionary and population dynamic consequences, which is the topic of this book.

In classical demography, mortality and fecundity have been considered as functions of age. However, in many organisms, size is more important than age in determining the fate of an individual and is, no doubt, one of the most important characteristics of any organism. Size will directly affect the type of prey an organism can eat, what predators will attack it, its physiology and energetics, and its reproductive success (Peters 1983; Calder 1984; Werner and Gilliam 1984; Sauer and Slade 1987; Sebens 1987; many contributions in this Volume). Size and morphology determine the ecological niche and hence influence competition within and between species. In some species, size even determines the sex of an individual (Policansky 1982; Charnov 1982).

The fact that fecundity and mortality are often directly connected with the size of an individual and only indirectly with its age suggests that it is more appropriate to discuss life histories in terms of size than in terms of age. Present life history theory is almost exclusively based on age-classified demographic models (e.g., Charlesworth 1980). In only a few cases has size itself been a component in life-history optimization models (Roff 1986; Kozlovski and Wiegert 1986, 1987; Pugliese 1987). Including size makes it possible to undertake a more mechanistic analysis of the constraints and trade-offs in life-history evolution that are caused by genetic correlations between traits expressed in different stages of a life cycle. For example, in traditional life history theory the concept of trade-off between current reproductive effort and residual reproductive value is applied to explain a negative correlation between early reproduction and later survival and/or reproduction. However, as pointed out by Kozlovski and Uchmanski (1987), it

[1] Department of Biology, Linköping University, 581 83 Linköping, Sweden
[2] Department of Ecology, Limnology, P.O. Box 65, 221 00 Lund, Sweden

Size-Structured Populations
B. Ebenman and L. Persson (Eds.)

seems that taking growth into account and considering the relation between size and survival/birth rate allow a more straightforward and mechanistic explanation. Understanding the evolution of growth and size is, thus, critical to an understanding of the life history and population biology of many organisms (Kirkpatrick; Lynch; Lynch and Arnold; Werner this Vol.).

Natural selection can act either on size itself (Kirkpatrick this Vol.) or on the degree of plasticity by which organisms vary growth and size in response to the environment (Via and Lande 1985; Stearns and Koella 1986; Schlichting 1986). Kirkpatrick outlines a predictive theory for the evolution of growth trajectories based on quantitative genetics and argues that only a relatively restricted set of growth trajectories can be realized because patterns of genetic covariation between sizes at different ages constrain evolution. Genetic covariance between sizes/morphologies at different ages is thus an important constraining factor in the evolution of growth patterns, and it is therefore essential to develop methods to be able to measure such constraints. Path analysis is one method that can be used to gain insight into the genetic and environmental factors that ar responsible for variation and constraints on a growth process (Lynch this Vol.). Lynch and Arnold (this Vol.) outline a method for reconstructing the phenotypic covariance matrix for components of growth and size before selection, and then use this reconstruction to estimate measures of directional, stabilizing and correlational selection.

The genetic covariance structure is itself subject to evolutionary modification (Clark 1987). One way of breaking up the genetic covariance structure may be the evolution of metamorphosis and complex life cycles (Werner this Vol.; Ebenman unpubl.). In a sense, complex life cycles represent an ultimate expression of the problems confronted when considering the ecology and evolution of size-structured populations. In species with complex life cycles, individuals of different ages differ not only in size, but usually also in form, function and in their ecological niche (Werner this Vol.). Development from newborn to adult for many groups that today exhibit sharp and abrupt metamorphic events (holometabolous insects, many marine invertebrates and amphibians) was, at the origin of their phylogeny, of a more continuous nature. The ancestors of frogs closely resembled the metamorphic stage, not the larvae or the adults, of present-day species (Szarski 1957; Wassersug and Hoff 1982; Alberch 1987); they passed through a long metamorphosis without undergoing any dramatic morphological changes. The same pattern was shown by the ancestors of the holometabolous insects (Sharov 1957; Kukalova-Peck 1978) as well as by the forerunners of many marine invertebrates living today (Strathmann 1985). Thus, there has been a general evolutionary trend towards a morphological diversification of the life cycle in several widely unrelated taxa.

Werner (this Vol.) advances the thesis that the increase in body size over the life cycle of an individual often necessitates a shift in the ecological niche, and that, to be effectively utilized, the different niches may require different body designs. Thus, the different ecological niches of the life stages result in different selection pressures on each stage. However, genetic correlations between traits expressed during a live cycle (e.g., body shape, structure of mouth parts as young and adult) constrain the evolutionary adaptation to the different ecological niches. There should, therefore, be a strong selection pressure to reduce the genetic correlation between homologous characters expressed in the different stages of a life cycle. Evolution of complex life cycles and metamorphosis may be a way of breaking up the genetic covariance structure during development, so that relatively independent evolution becomes possible in the different stages (Freeman 1982; Horn et al. 1982; Ebenman unpubl.).

The growth rates of individuals in many species are often so variable that age provides little information about their demographic status (Caswell this Vol.). To deal with such organisms, it is important to formulate demographic models that take size into account. Such models provide a means of investigating the consequences of growth, survival and reproduction on population dynamic behavior in size-structured populations (Caswell, Ebenman, Law and Grey, Metz et al. this Vol.; VanSickle 1977; Botsford 1981; Murphy 1983; Law 1983; Hughes 1984; Kirkpatrick 1984; Caswell 1985; Metz and Diekmann 1986; Solbrig et al. 1988).

Individuals of many species change their use of resources as they develop from newborns to adults. Such niche changes during life cycles will affect intraspecific competition and hence population dynamics. The strength of competition between individuals of different ages will often depend on their sizes and hence on their growth patterns. However, the growth patterns of individuals may vary considerably because of differences in the abiotic and/or biotic environment. It is well known, from many studies on invertebrates, amphibians, fishes and plants, that competition within and among age classes may result in retarded individual growth rates (e.g., Sebens 1987; Wilbur and Collins 1973; Jones 1987; Weiner 1985; Uchmanski 1985; many contributions in this Vol.). Ebenman (this Vol.) discusses, from a theoretical point of view, the dynamic interplay between growth of individuals and growth of populations that results when growth of individuals is plastic and density-dependent and when competition between age classes is size-dependent. The great advantage of models that take size into account, as compared to those purely based on age, is that they provide a means of explicitly including ecologically important characteristics of individuals in explanations of population dynamic behavior (Metz et al. this Vol.). Many biologically interesting questions – like, for example, the effect of plastic growth of individuals on population dynamic behavior and coexistence of competing species – can simply not be dealt with from a purely age-specific point of view (Ebenman this Vol.).

The analysis of models that account for the interplay between growth of individuals and growth of populations may also shed new light on some of ecology's most general empirical patterns, like the relation between number of species and their sizes (e.g., May 1978; Griffiths 1986) and the relations between size and population density in animals (e.g., Damuth 1981, 1987; Peters 1983; Brown and Maurer 1986; Begon et al. 1986) and plants (Norberg 1988 and this Vol.; Weller 1987). Of all organisms, plants are perhaps those whose populations show the most pronounced size structure. When individual plants grow in an uncrowded population, there is no reduction in growth rate and no mortality caused by competition. But as plants begin to fill up the available space, crowding commences, and competition causes mortality to increase and growth rate of individuals to decrease (Norberg this Vol.). This leads to the well-known $-3/2$ power law (or self-thinning rule), one of the most general empirical rules in plant population biology (see Westoby 1984 for review). It relates plant density to the mean biomass per plant in monospecific stands undergoing density-dependent and size-dependent mortality. The self-thinning law has, thus far, been demonstrated by empirical data, but not satisfactorily derived from theory (see Weller 1987). According to the work of Norberg (this Vol. and 1988) the causal basis of the thinning rule is to be sought in general, geometric principles of growth and packing.

The field experimental studies and field studies presented in this Volume cover a variety of organisms such as rotifers, crustacean zooplankton, dipterans, scorpions, fish, and amphibians. These studies unravel a vast number of complex interactions that are introduced by the presence of size structure. In recent years there has been a controversy over the importance of competition versus predation as forces structuring ecological communities (see reviews by Schoener 1982, 1983; Connell 1983; Wilbur this Vol.). However, the circumstance that individuals grow in size or radically alter morphology during their life cycles gives rise to a situation in which interactions may alter between competition and predation during the course of ontogeny. As a consequence, studies focusing on size-structured populations generally emphasize how competition and predation interact to shape the dynamic behavior of populations. This is, for example, illustrated by the concept of mixed competition/predation, commonly used by researchers in the field of size-structured populations (Werner and Gilliam 1984). Mixed competition/predation includes both interspecific (competition and intraguild predation) and intraspecific interactions (competition and cannibalism) (Polis this Vol.). Plastic growth exhibited by many organisms also makes the effects of predation much more complex. For example, in some cases the prey population may actually benefit from the thinning effect of the predator if the quality of the survivors is sufficiently enhanced (Wilbur this Vol.). Due to different individual growth rates, species pairs may also reverse their interactions over time, so that one species that preys on another species during one point in its life will itself later become the prey and the preyed-upon species will become the predator (Polis, Wilbur this Vol.). Another example of the complexity of the interactions in size-structured populations is that, depending on whether intraspecific predation (cannibalism) – an extreme form of interference – is strong or weak, populations with similar total biomasses can be composed of either a few large or many small individuals (Botsford 1981).

The interaction between competition and predation, resulting from an increase in body size or a change in morphology during the life cycle, is not restricted to an alternation between the two. At any instant of time, the performance of an individual may be framed as a trade-off between maximizing foraging (growth) rates and minimizing mortality rates due to predation (Gilliam and Fraser, Werner this Vol.). The inclusion of behaviorally based active habitat selection leads to a situation in which the effects of the predator are much more subtle than a simple harvesting of the prey. As a consequence of predation-induced shifts in the habitats of juveniles, the presence of predators may decrease competition between juveniles and adults of the same species and competition among adults of different species that are not susceptible to the predator. At the same time, the intensity of interspecific competition among juveniles may increase

(Mittelbach et al. this Vol.). Such complexities in interactions are further enhanced when the resources in different habitats are depleted (either by the consumers or due to other factors) (Gilliam and Fraser this Vol.).

One interesting phenomenon in interactions among size-structured populations is that although interspecific competition is restricted to only one stage in the life cycle, this may have profound effects on population size and population size structure. Such effects of interspecific competitors generally affect the juveniles of a species, and are hence referred to as competitive juvenile bottlenecks (Neill, Persson this Vol.). One mechanism that has been advanced as a cause of competitive juvenile bottlenecks is that juveniles of a species with a larger adult size will have to compete for resources utilized by smaller species. Consequently, these latter species may competitively limit the recruitment of smaller size classes of the larger species to adult size classes (Werner 1986). Competitive juvenile bottlenecks usually extend to even more intricate interactions due to the fact that the adults of the larger species may prey on the species causing the juvenile bottleneck (Neill, Persson this Vol.). It is also evident that other mechanisms may lead to competitive juvenile bottlenecks; for example, when a species outcompetes the prey species utilized by the smallest stages of the larger species (Neill 1988, this Vol.).

Thus, in size-structured populations both competition and predation are important structuring forces. But is there any evidence that the relative strength of these two biotic forces may vary predictably between systems? Two of the contributions (Neill, Persson) in this Volume suggest that this is the case. Both studies imply that the interaction between a species pair changes from being mainly competitive to being mainly predatory, along a productivity (nutrient) gradient of the system. These and other contributions in this Volume also point to a possibility of developing general theories regarding community regulation among size-structured populations; theories similar to those proposed for, for example, intertidal systems (albeit the implications of size structure have largely been ignored for these latter systems) (Menge and Sutherland 1976, 1987; Lubschenko 1986).

Finally, during recent years evolutionary ecologists have shown growing interest in processes extending to the ecosystem level. Theoretical developments and models in this area have constituted a major step toward an understanding of community and ecosystem regulation (e.g., Oksanen et al. 1981). The contributions in this Volume demonstrate how including size structure will profoundly affect the predictions of such models, due to the fact that individuals change their position in the food web during their life cycles. For example, theoretical considerations suggest that positive correlations between consumers and their resources may be found under conditions when such correlations would not be expected for models excluding stage structure (Mittelbach et al. this Vol.).

Acknowledgments. The organization of the symposium Dynamics of Size-Structured Populations held in Falsterbo, Sweden, October 26–30, 1987, from which this Volume emanated, was made possible by a grant from the Swedish Natural Science Research Council. Additional financial support was provided by the Department of Ecology, University of Lund, Sweden; this support is gratefully acknowledged. We are grateful to Börje Ekstam and Uno Wennergren, Linköping University, for valuable help with reviewing some of the chapters in this book. We are also grateful to Patricia Ödman, Linköping University, for checking the English of this introduction. Finally, we express our appreciation to Gunilla Lindqvist for helping solve the practical problems in connection with the symposium.

References

Alberch P (1987) Evolution of a developmental process: Irreversibility and redundancy in amphibian metamorphosis. In: Raff R, Raff E (eds) Development as an evolutionary process. Liss, New York, pp 23–46
Begon M, Firbank L, Wall R (1986) Is there a self-thinning rule for animal populations? Oikos 46:122–124
Bonner JT (1965) Size and cycle. Univ Press, Princeton, NJ
Botsford LW (1981) The effects of increased individual growth rates on depressed population size. Am Nat 117:38–63
Brown JH, Maurer BA (1986) Body size, ecological dominance and Cope's rule. Nature (London) 324:248–250
Calder WA (1984) Size, function and life history. Harvard Univ Press, Cambridge
Caswell H (1985) The evolutionary demography of clonal reproduction. In: Jackson JBC, Buss LW, Cook RE (eds) Population biology and evolution of clonal organisms. Yale Univ Press, New Haven, pp 187–224
Charlesworth B (1980) Evolution in age-structured populations. Univ Press, Cambridge
Charnov EL (1982) The theory of sex allocation. Univ Press, Princeton, NJ
Clark AG (1987) Genetic correlations: The quantitative genetics of evolutionary constraints. In: Loeschcke V (ed) Genetic constraints on adaptive evolution. Springer, Berlin Heidelberg New York, pp 25–45
Connell JH (1983) On the prevalence and relative importance of interspecific competition: evidence from field experiments. Am Nat 122:661–696
Damuth J (1981) Population density and body size in mammals. Nature (London) 290:699–700
Damuth J (1987) Interspecific allometry of population density in mammals and other animals: the independence of body mass and population energy use. Biol J Linn Soc 31:193–246
Freeman GL (1982) What does the comparative study of development tell us about evolution? In: Bonner JT (ed) Evolution and development. Springer, Berlin Heidelberg New York, pp 155–167
Griffiths D (1986) Size-abundance relations in communities. Am Nat 127:140–166
Horn HS (1982) Adaptive aspects of development. In: Bonner JT (ed) Evolution and development. Springer, Berlin Heidelberg New York, pp 215–235
Hughes TP (1984) Population dynamics based on individual size rather than age: A general model with a reef coral example. Am Nat 123:778–795
Jones GP (1987) Competitive interactions among adults and juveniles in a coral reef fish. Ecology 68:1534–1547
Kirkpatrick M (1984) Demographic models based on size, not age, for organisms with indeterminate growth. Ecology 65:1874–1884
Kozlowski J, Uchmanski J (1987) Optimal individual growth and reproduction in perrenial species with indeterminate growth. Evol Ecol 1:214–230
Kozlowski J, Wiegert RG (1986) Optimal allocation of energy to growth and reproduction. Theor Popul Biol 29:16–37
Kozlowski J, Wiegert RG (1987) Optimal age and size at maturity in annuals and perrenials with determinate growth. Evol Ecol 1:231–244
Kukalova-Peck J (1978) Origin and evolution of insect wings and their relationship to metamorphosis, as documented by the fossil record. J Morphol 156:53–126
Law R (1983) A model for the dynamics of a plant population containing individuals classified by age and size. Ecology 64:224–230
Lubchenko J (1986) Relative importance of competition and predation: Early colonization by seaweeds in New England. In: Diamond J, Case T (eds) Community ecology. Harper & Row, New York, pp 344–358
May RM (1978) The dynamics and diversity of insect faunas. In: Mound LA, Waloff N (eds) Diversity of insect faunas. Blackwell, Oxford, pp 188–204

McMahon TA, Bonner JT (1983) On size and life. Freeman, New York
Menge BA, Sutherland JP (1976) Species diversity gradients: synthesis of the roles of predation, competition and temporal heterogeneity. Am Nat 110:351–369
Menge BA, Sutherland JP (1987) Community regulation: variation in disturbance, competition, and predation in relation to environmental stress and recruitment. Am Nat 130:730–757
Metz JAJ, Diekmann O (1986) The dynamics of physiologically structured populations. Springer, Berlin Heidelberg New York (Springer, lecture notes in biomathematics, vol 68)
Murphy LF (1983) A nonlinear growth mechanism in size structured population dynamics. J Theor Biol 104:493–506
Neill WE (1988) Complex interactions in oligotrophic lake food webs: responses to nutrient enrichment. In: Carpenter SC (ed) Complex interactions in lake communities. Springer, Berlin Heidelberg New York (in press)
Norberg RÅ (1988) Theory of growth geometry of plants and self-thinning of plant populations: geometric similarity, elastic similarity, and different growth modes of plant parts. Am Nat 131:220–256
Oksanen L, Fretwell SD, Arruda J, Niemela P (1981) Exploitation ecosystems in gradients of primary productivity. Am Nat 118:240–261
Peters RH (1983) The ecological implications of body size. Univ Press, Cambridge
Policansky D (1982) Sex change in plants and animals. Annu Rev Ecol Syst 13:471–495
Pugliese A (1987) Optimal resource allocation and optimal size in perrenial herbs. J Theor Biol 126:33–49
Roff DA (1986) Predicting body size with life history models. BioScience 36:316–323
Sauer JR, Slade NA (1987) Size-based demography of vertebrates. Annu Rev Ecol Syst 18:71–90
Schlichting CD (1986) The evolution of phenotypic plasticity in plants. Annu Rev Ecol Syst 17:667–693
Schoener TW (1982) The controversy over interspecific competition. Am Sci 70:586–595
Schoener TW (1983) Field experiments on interspecific competition. Am Nat 122:240–285
Sebens KP (1987) The ecology of indeterminate growth in animals. Annu Rev Ecol Syst 18:371–407
Sharov AG (1957) Types of insect metamorphosis and their relationship. Rev Entomol USSR 36:569–576
Solbrig OT, Sarandon R, Bossert W (1988) A density-dependent growth model of a perrenial herb, *Viola fimbriatula.* Am Nat 131:385–400
Stearns SC, Koella JC (1986) The evolution of phenotypic plasticity in life-history traits: Predictions of reaction norms for age and size at maturity. Evolution 40:893–913
Strathmann RR (1985) Feeding and nonfeeding larval development and life-history evolution in marine invertebrates. Annu Rev Ecol Syst 16:339–361
Szarski H (1957) The origin of the larva and metamorphosis in Amphibia. Am Nat 91:283–301
Uchmanski J (1985) Differentiation and frequency distributions of body weights in plants and animals. Philos Trans R Soc London Ser B 310:1–75
VanSickle L (1977) Analysis of a distributed parameter population model based on physiological age. J Theor Biol 64:571–586
Via S, Lande R (1985) Genotype-environment interaction and the evolution of phenotypic plasticity. Evolution 39:505–522
Wassersug R, Hoff K (1982) Developmental changes in the orientation of the anuran jaw suspension: A preliminary exploration into the evolution of anuran metamorphosis. Evol Biol 15:223–245
Weiner J (1985) Size hierarchies in experimental populations of annual plants. Ecology 66:743–752
Weller DE (1987) A reevaluation of the $-3/2$ power rule of plant self-thinning. Ecol Monogr 57:23–43
Werner EE (1986) Species interactions in freshwater fish communities. In: Diamond J, Case T (eds) Community ecology. Harper & Row, New York, pp 344–357
Werner EE, Gilliam JF (1984) The ontogenetic niche and species interactions in size-structured populations. Annu Rev Ecol Syst 15:393–425
Westoby M (1984) The self-thinning rule. Adv Ecol Res 14:167–225
Wilbur HM, Collins JP (1973) Ecological aspects of amphibian metamorphosis. Science 182:1305–1314

The Evolution of Growth Patterns and Size

The Evolution of Size in Size-Structured Populations

M. KIRKPATRICK

1 Introduction

In many populations of animals and plants, particularly those with indeterminate growth, size is the dominant ecological attribute of individuals (see Caswell 1982 and this Vol.; Peters 1983; Werner and Gilliam 1984; Sauer and Slade 1987; Sebens 1987). Understanding the evolution of size is therefore critical to understanding the population biology of these species. Because individuals of different ages are often of very different size, a full treatment must consider the ontogeny of size. This paper outlines a predictive theory for the evolution of growth trajectories based on quantitative genetics.

The major conclusion that emerges from this analysis is that growth trajectories seen in nature are most likely determined jointly by what selection favors and what evolutionary constraints allow. This suggests that knowledge of the ecological circumstances alone is not sufficient to determine the evolutionary fate of size and growth – the evolutionary constraints must be considered also. These constraints can be identified and measured using the methods of quantitative genetics. The details of how constraints operate has an impact on how size and growth respond to changes in the regime of selection, including those changes caused by human harvesting.

Since many biologists interested in size-structured populations may not be familiar with quantitative genetics, this paper begins with a review of the elements of that field that are important to the evolution of growth trajectories. An extension of standard quantitative genetics is then introduced that allows the evolution of characters like growth trajectories to be studied. Next, I describe a method for quantifying the forces of selection acting on growth trajectories. The evolution of growth trajectories is then considered in light of the results on inheritance and selection.

2 An Introduction to Evolutionary Quantitative Genetics

The field of quantitative genetics has been extensively developed over the last 50 years, primarily by animal and plant breeders working with domesticated species. More recently, it has become widely used in evolutionary biology. Excellent introductions to the field have been written by Falconer (1981) and Bulmer (1985).

Dept. of Zoology, University of Texas, Austin, Texas 78712, USA

Size-Structured Populations
B. Ebenman and L. Persson (Eds.)

What follows is a concise description of some basic results important to the evolution of growth trajectories.

2.1 Evolution of the Phenotypic Mean

The central problem of interest for many evolutionary studies is to understand how the phenotypic mean of one or more characters in a population evolves under the influence of some form of selection. Quantitative genetics provides the answer if a set of assumptions about the hereditary basis for the characters are met (Bulmer 1985; Falconer 1981). Most important among these is that the additive genetic values of individuals in the population are normally distributed. The additive genetic value is simply that component of an individual's phenotype that is transmitted hereditarily to its offspring. The results from a long history of artificial selection in many species of animals and plants leads us to believe that this is generally a reasonable assumption. The theory for the evolution of characters that meet these assumptions is well established (see Lande 1979; Bulmer 1985; Falconer 1981).

Lande and Arnold (1983) provided a very useful notation for describing the evolution of the phenotypic means of a set of quantitative characters under selection. In that notation, following a single generation of selection, the population mean for character i at the start of the next generation is

$$\bar{z}_i' = \bar{z}_i + G_{i1}\beta_1 + G_{i2}\beta_2 + \ldots + G_{in}\beta_n = \bar{z}_i + \sum_{j=1}^{n} G_{ij}\beta_j \ , \tag{1}$$

where $\bar{z}_i$ is the phenotypic mean at the start of the current generation. Equation (1) shows that the evolutionary change in the mean of character i results from the sum of a series of contributions that represent the effects of selection acting on each of the traits. Each term in this sum consists of two components. The first, G_{ij}, is the *additive genetic covariance* between character i and character j. The additive genetic covariance represents the contribution that heredity makes to the phenotypic resemblance between the two characters. Many morphological characters, for example, show positive genetic covariances caused by genetic variation in growth hormone levels. The additive genetic covariance of trait i with itself, G_{jj}, is the additive genetic variance for that trait. This is a measure of the genetic variation for trait i that is transmitted from parents of offspring. These variances and covariances are often organized in a square table that is referred to as the additive genetic covariance matrix.

The second component in each term of Eq. (1) is β_j, the *selection gradient* acting on trait j. A selection gradient is a measure of the force of directional selection acting directly on a trait. It is defined as the regression of relative fitness on the phenotypic value for that trait, holding the values of all other traits constant (Lande and Arnold 1983). Selection gradients are often organized in a list that is referred to as the vector of selection gradients.

We see from Eq. (1) that the evolution of the mean of any particular trait depends not only on selection acting on that trait, but also on selection acting on the others characters. The characters are evolutionarily coupled by the additive genetic covariances. The evolutionary effect on one character of selection on another is referred to as an indirect or correlated evolutionary response to selection.

Although simple, Eq. (1) provides a fundamental insight. The evolution of any character depends on both inheritance (reflected in the additive genetic variances and covariances) and selection (reflected in the selection gradients). Quantitative prediction of the evolutionary trajectory of a character requires knowledge of both of these components. A simple qualitative argument, however, shows that a set of characters can reach an evolutionary equilibrium in either of two ways. The first is that the traits evolve to a point at which all the selection gradients acting on them are zero, so that directional selection vanishes. This occurs when the traits are optimized so that their means lie at a fitness peak. There is, however, a second way in which an evolutionary equilibrium can be reached. Even if the directional selection persists, a pattern of genetic covariances can cause the net evolutionary effects of selection to cancel so that there is no change in the phenotypic means. The importance of this possibility for growth trajectories will be a major theme in this paper.

2.2 Measuring the Parameters of Selection and Inheritance

A major strength of the quantitative-genetic approach is that its parameters can be measured directly. Animal and plant breeders have developed an extensive inventory of experimental and statistical techniques to measure additive genetic variances and covariances. While these methods have traditionally been based on controlled breeding experiments, evolutionary biologists have been increasingly successful in extending this approach to natural populations (e.g., Dingle and Hegmann 1982; Via 1984a, b; Shaffer 1986; Boag and van Noordwijk 1987). In general, the data needed are measurements of the characters in a series of related individuals, such as half-sibs or parents and their offspring. The genetic variances and covariances can then be calculated using standard statistical methods (see Falconer 1981).

Selection gradients can also be measured directly. The discovery that the selection gradients can be interpreted as the partial regressions of relative fitness on phenotypic value provides the basis for statistical methods that allow one to directly measure selection gradients in natural populations (Lande and Arnold 1983; see also Arnold and Wade 1984; Price et al. 1984). Alternatively, selection gradients can also be calculated from a theoretical description of how selection acts on a population. If relative fitnesses are constant in time and phenotypes are normally distributed, then the selection gradient on character i is equal to the rate at which the population's mean fitness changes as a function of the mean of that character:

$$\beta_i = \frac{d}{d\bar{z}_i} \ln (\overline{W}) , \qquad (2)$$

where $\overline{W}$ is the mean fitness of the population (Lande 1979). Thus if we know (or are willing to assume) how an individual's fitness varies as a function of its phenotype and how the character is distributed in the population, the selection gradient can be calculated from this relationship.

3 Growth Trajectories Are Infinite-Dimensional Characters

The machinery of quantitative genetics is clearly appropriate to characters like size that are continuous and very often (after suitable transformation of the data) normally distributed. But how can quantitative genetics be applied to a character as complex as a growth trajectory? Previous approaches to this problem involve two different strategies.

The first of these has been to assume that all individuals in the population follow a growth "law" (Cock 1966). That is, the growth trajectories fall along curves that are described by some family of curves, such as those defined by the von Bertalanffy growth equation:

$$z(a) = l_\infty (1 - e^{-ka}) \ . \tag{3}$$

Here, $z(a)$ is the size of an individual at age a, k is a parameter that measures how rapidly individuals grow, and l_∞ is the final size that is approached asymptotically by individuals as they age. (Throughout this paper z rather than s denotes size in order to be consistent with how these variables are used in quantitative genetics.) Variation among individuals in the population is assumed to result from variation in the parameters of the equation (l_∞ and k, in the case of the von Bertalanffy equation), as illustrated in Fig. 1. These parameters can be treated as quantitative characters in the framework described above by measuring genetic variation in them and the force of selection acting on them, then projecting their evolution using Eq. (1).

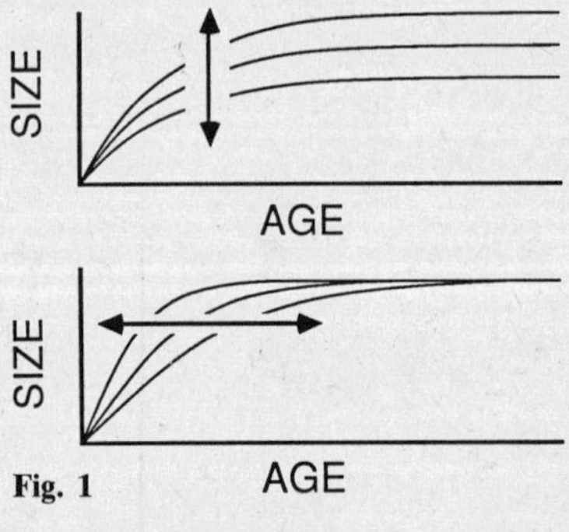

Fig. 1

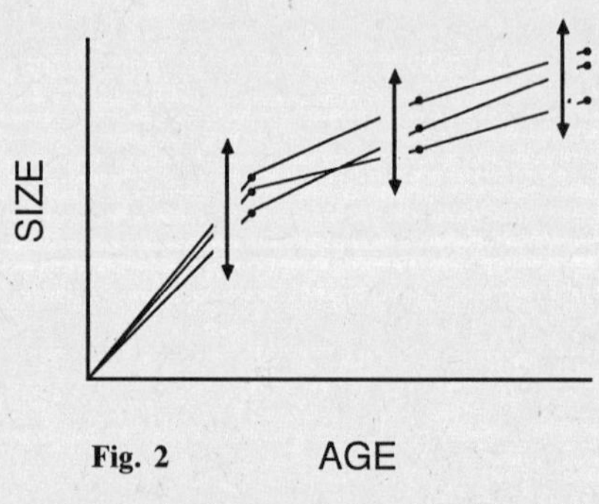

Fig. 2

Fig. 1. Examples of growth trajectories described by the von Bertalanffy growth law [Eq. (3)]. This growth law permits only two forms of variation in the trajectories: variation in l_∞ (*above*), and variation in k (*below*)

Fig. 2. Examples of growth trajectories described by treating the size of individuals at each of three measurement ages as three separate characters. No a priori restriction is made concerning the relationships of the sizes of individuals at these three ages

The second approach has been to summarize the growth trajectory of each individual by its size at each of a fixed set of ages (e.g., Riska et al. 1984). The size of each age can be conceptualized as distinct character, as sketched in Fig. 2. Thus by measuring the size of an individual at three ages we are, in effect, taking measurements of three different traits. These three traits are coupled by genetic correlations that arise, for example, from the pleiotropic effects of growth hormone (individuals with a high titer of growth hormone are likely to be large at all ages). The genetic covariances, selection gradients, and other parameters for size at these three ages can in principle be measured like those of any other traits (see Lynch and Arnold this Vol.), which allows their evolution to be determined from Eq. (1).

Each of these two approaches has strengths and weaknesses. By fitting a parametric growth law we obtain a complete description of the growth trajectory of each individual (that is, its size for every moment in its life). A severe restriction of this approach is that it rules outs, a priori, any departure from the family of curves that is chosen. If a von Bertalanffy growth curve is postulated, for example, it is implicitly assumed that it is impossible for the population to evolve a growth pattern in which size increases without asymptote. This precludes the possibility that a bush might ultimately evolve the growth form of a tree, or a minnow the growth pattern of a cod. The second approach to growth trajectories, on the other hand, makes no such restriction. Its drawback is that it produces a summary description of the full growth trajectory, since it gives information only about the discrete set of ages at which the measurements are made. An ideal method would combine the best attributes of both approaches by giving a complete, continuous description of the growth trajectories without making any restrictive assumptions about the possible forms.

A natural way to approach this goal is to extend the second method by including measurements at more and more ages in order to fill in the growth trajectory. A brief consideration of this suggestion, however, shows that the trajectory cannot be fully described by any finite number of measurements. Since an individual has a size at every age during its life, and since time is continuous in nature, there are an infinite number of size measures that make up its full growth trajectory. Growth trajectories are referred to as *infinite-dimensional characters* for this reason. Recent theoretical work on the quantitative genetics of infinite dimensional characters provides a framework for studying the evolution of growth trajectories that meets the objectives of being continuous and parameter-free. In order to develop this approach, I will summarize the basic results of the quantitative genetics of infinite-dimensional characters. Details can be found in Kirkpatrick and Heckman (1989).

3.1 Means and Covariances of Growth Trajectories

Any predictive theory for evolution requires a way to quantify the character in an individual and its mean and variation in the population. In the case of growth trajectories, we will denote the size of an individual at age a as $\mathfrak{z}(a)$. The *mean growth trajectory,* $\bar{\mathfrak{z}}(a)$, is determined by simply taking the average size of all individuals in the population at each age a. In practice, size is actually measured

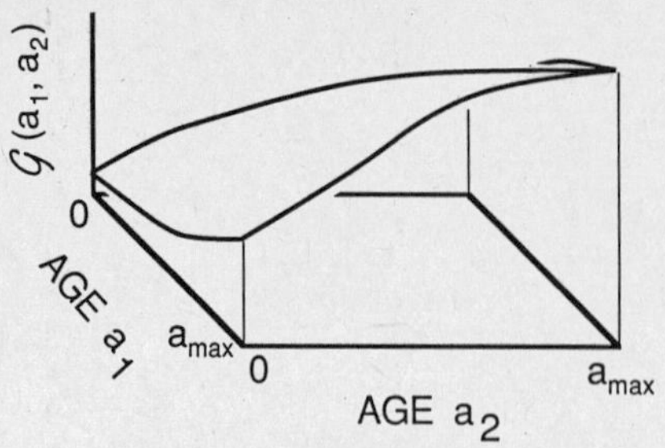

Fig. 3. A visualization in three dimensions of a hypothetical additive genetic covariance function. The x and y coordinates specify two ages, while the altitude of the surface over that point corresponds to the additive genetic covariance for size between those two ages. The additive genetic variances for size at different ages is given by the altitude of the ridge in the covariance function that runs along the diagonal from *far left* to the *near right corners* of the surface

at a finite set of ages. The mean size at ages between those that actually have been measured can be estimated by interpolating between the measured points using some sort of smooth curve such as a polynomial or a spline.

In standard quantitative genetics, the patterns of variation and covariation for a finite set of characters is described by a covariance matrix. The analog of a covariance matrix for infinite-dimensional characters is a *covariance function.* The additive genetic covariance function for growth trajectories, for example, is written $\mathcal{G}(a_1, a_2)$ and is a continuous function whose value is the additive genetic covariance between the sizes of individuals at any two ages a_1 and a_2. Whereas a covariance matrix has entries only at a finite number of discrete points, a covariance function is a function of two continuous variables. A covariance function can be visualized as a surface whose horizontal coordinates are age and whose altitude corresponds to the covariance between size at those two ages, as shown in Fig. 3. The continuous covariance function can be estimated from a covariance matrix by interpolation, just as the mean function was. The covariance function is best estimated using orthogonal functions, such as orthogonal polynomials, because the coefficients from these functions provide a natural method to carry out the analyses described below (Kirkpatrick and Heckman 1989). The additive genetic covariance function $\mathcal{G}$ is estimated from the additive covariance matrix, while the phenotypic covariance function $\mathcal{P}$ is estimated from the phenotypic covariance matrix. These matrices can be obtained using the established methods discussed above.

3.2 A Genetic Model for the Evolution of Growth Trajectories

The discussion thus far has provided only a way to describe means and variances in growth trajectories. To gain evolutionary insights we need to move from description to prediction. This requires an evolutionary model based on explicit assumptions.

The assumptions of the evolutionary model for growth trajectories are direct extensions of the standard assumptions of quantitative genetics (Bulmer 1985; Falconer 1981). Two of these are fundamental. The first is that growth trajectories can be statistically decomposed into additive and nonadditive parts. That is, we can think of the growth trajectory of an individual as a sum of an additive component that it transmits hereditarily to its offspring, and a nonadditive component caused by genetic interaction (dominance and epistasis) and environmental effects. The second basic assumption is that these additive and nonadditive components of growth trajectories are multivariate-normally distributed in the population. This implies, among other things, that the phenotypic size of individuals of the same age will be normally distributed in the population (on some scale of measurement, such as logarithmic).

Here I will make the additional assumption that generations are non-overlapping. Although individuals grow continuously in time, the birth of new individuals is assumed to occur after the death of the last member of the previous generation. This life history is found in many species of annual plants and invertebrates. The effect of this assumption is to eliminate the evolutionary consequences of population growth (Fisher 1958). This last assumption is made for the sake of simplicity; the model can be extended to accommodate overlapping generations.

These assumptions lead to an equation which predicts the evolutionary change in the mean growth trajectory. Following one generation of selection, the mean growth trajectory in the population will be

$$\bar{\mathscr{z}}'(a) = \bar{\mathscr{z}}(a) + \int_0^{a_{max}} \mathscr{G}(a,x)\ \beta(x)\ dx\ , \tag{4}$$

where the function β is the selection gradient acting on size at age a, and a_{max} is the maximum age that individuals reach. A positive value of $\beta(x)$ implies that selection favors an increase in size at age x, while a negative value implies selection favors a decrease in size. Equation (4) is clearly the infinite-dimensional analog of Eq. (1). As in the finite-dimensional case, we see that the evolutionary response of the population's mean growth trajectory to selection depends both on the forces of selection, summarized by β, and the patterns of inheritance, summarized by $\mathscr{G}$.

4 Selection Gradients Acting on Growth Trajectories

How can we determine the selection gradients acting on growth trajectories? Selection gradients can be empirically determined in natural populations by applying regression analysis to several different kinds of data (see Lynch and Arnold, this Vol. and the references therein). Considerable insight can be gained, however, by using theoretical analysis.

4.1 The Selection Gradient on Size

The relationship between the selection gradient and mean population fitness [Eq. (2)], can be extended to infinite dimensional characters, which shows that the selection gradient acting on size at age a is

$$\beta_z(a)\ da = \frac{d}{d\bar{\mathscr{z}}(a)} \ln(\overline{W}) \tag{5}$$

(Kirkpatrick and Heckman 1989; Kirkpatrick and Lofsvold 1989). (The subscript on β_z indicates we are referring to selection on size, and will distinguish this selection gradient from another that will be introduced below.) The problem is now cast in terms of finding the relationship between mean size, $\bar{z}(a)$, and mean population fitness, $\overline{W}$.

A large number of ecological studies have shown that size is a primary determinant of life history characteristics in a wide variety of indeterminately growing organisms (reviewed in Werner and Gilliam 1984; Sauer and Slade 1987; Sebens 1987). Fecundity is often found to be directly related to some measure of size (e.g., Bagenal 1978), while mortality rates are often strongly determined by size through size-dependent predation (e.g., Brooks and Dodson 1965). This suggests that a reasonable starting assumption for size-structured populations is that these life history parameters are determined by size alone, and are independent of other factors such as age. A different approach is taken by Gilliam and Fraser (this Vol.) and others, who implicitly assume that selection acts on growth rate rather than size per se at some life history stages. (In the case of Gilliam and Fraser's model, this follows from their assumption that habitat choice of juvenile fishes is evolving, and that associated with each habitat are a characteristic mortality rate and growth rate which are independent of size.) In the following I will assume that birth and death rates depend on size and age, but not on growth rate or other variables. The models can easily be recast under the assumption that mortality depends on growth rate.

Assuming that size and age determine birth and death rates, the population mean fitness can be written in general as

$$\overline{W} = \int_0^{a_{max}} \int_{-\infty}^{\infty} l(a)\, b(z,a)\, f(z,a)\, dz\, da \ , \tag{6}$$

where $l(a)$ is the probability that a newborn survives to reach age a, $f(z,a)$ is the probability that an individual is of size z among those that have survived to age a, and $b(z,a)$ is the birth rate (fecundity) of individuals of size z at age a. The integration extends over negative as well as positive values of size to allow for the possibility that size is measured on a logarithmic or other scale in which it can take on values less than zero.

The effects of size on mortality are hidden in the survival term $l(a)$, which is related to the size- and age-dependent mortality rate $\mu(z,a)$ by the equation

$$l(a) = \exp\left\{-\int_0^a \bar{\mu}(x)\, dx\right\} \ , \tag{7}$$

where $\bar{\mu}(x)$ is the average mortality rate among individuals that are alive at age x:

$$\bar{\mu}(x) = \int_{-\infty}^{\infty} f(z,x)\, \mu(z,x)\, dz \ . \tag{8}$$

By substituting Eqs. (6–8) into (5), we obtain an explicit formula for the selection gradient on size:

$$\beta_z(a) = \left[l(a) \frac{d}{d\bar{z}(a)} \bar{b}(a) - u(a) \frac{d}{d\bar{z}(a)} \bar{\mu}(a) \right] / \overline{W} \ , \tag{9}$$

where $\bar{b}(a)$ is the average birth rate of individuals surviving to age a, and u(a) is the expected reproductive contribution of individuals of age a to the next generation:

$$u(a) = \int_{a}^{a_{max}} \int_{-\infty}^{\infty} l(a)\,\bar{b}(a)\,f(z,a)\,dz\,da \quad . \tag{10}$$

This shows that the force of selection acting on size at any given age a results from a compromise between two effects. The first term inside the brackets of Eq. (9) represents the effect of size on birth rate, since $\frac{d}{d\bar{z}(a)}\,\bar{b}(a)$ is the change in the mean birth rate of individuals of age a per unit change in their mean size. The factor l(a) accounts for the fact that the impact of selection is mitigated by the probability that an individual survives to reach age a. The second term inside the brackets of Eq. (9) represents the effect of size on mortality, since $\frac{d}{d\bar{z}(a)}\,\bar{\mu}(a)$ is the change in the mean mortality rate per unit change in mean size. The factor u(a) weights this derivative by the expected reproductive contribution of individuals of age a to the next generation.

4.2 Evaluating the Force of Selection Acting on the Growth Trajectory

To actually evaluate the selection gradient on size [Eq. (9)] we need information on cohort survivorship [the function l(a)], the distribution of sizes among individuals alive at each age [the function f(z, a)], the effects of size and age on mortality [the function $\mu(z,a)$], and the effects of size and age on birth rate [the function b(z, a)]. These data can be gathered empirically from any particular population of interest. Alternatively, a general picture of how selection acts can be deduced using a demographic model that provides this information, given certain assumptions about growth, birth, and mortality. One such model is that of Kirkpatrick (1984). The assumptions of that model will now be briefly outlined, and then its results used to determine some general characteristics of selection gradients acting on growth trajectories.

The demographic model of Kirkpatrick (1984) assumes that birth and mortality rates depend on size only, and are independent of age and other variables (such as growth rate). Growth rates vary so that individuals of the same age typically differ in size; this variation can be introduced into the model in two alternative ways. The mortality rate is assumed to be a quadratic function of size, so that there is a single size at which mortality is minimized. The three parameters that determine the mortality function are the optimal size at which mortality is minimized, the minimum mortality rate, and the sensitivity of the mortality rate to departures from the optimal size. The model allows for an arbitrary relationship between size and birth rate.

Given these assumptions, expressions for f(z, a) and l(a) can be derived (Kirkpatrick 1984). Those results then can be used to determine the selection gradient β_z using Eq. (9). I investigated a range of parameter values numerically and found (not surprisingly) that a wide variety of outcomes is possible.

Two kinds of outcome, however, illustrate an important point about the role of selection in determining a population's growth trajectory. Under some sets of conditions, the selection gradient is positive for all ages. This implies that selection favors an increase in the average body size of individuals at all points in their growth. Conditions favorable to this outcome are when mortality decreases and birth rate increases with size. A second outcome, seen under different ecological conditions, produces selection gradients that are negative up to some age, and then are positive for all subsequent ages. This means that the growth trajectory is selected towards smaller sizes up to a threshold age, at which point size is selected to increase. This outcome is favored when the birth rate is zero for individuals up to some critical size but then rapidly increases with size thereafter, while mortality increases gradually with increasing size. While neither of these outcomes is particularly surprising, they do lead to interesting insights about how growth trajectories are shaped by evolution.

5 The Evolution of Size and Growth in Size-Structured Populations

Determining how the force of selection acts on size is an important first step in determining how growth trajectories will evolve. But is it sufficient? That is, can the evolutionary fate of the mean growth trajectory be deduced from knowledge of selection alone? The implications of the results from the last section show that the answer is clearly no.

5.1 Evolution Without Constraints

Consider a hypothetical population in which the mean growth trajectory is free to evolve with no genetic constraint. By this I mean that there is additive genetic variation for size at every age and there are no genetic correlations between the sizes at different ages. The only restriction will be that the size of individuals at conception is fixed. These conditions imply that the mean size of individuals at any given age is free to evolve independently of the mean size at other ages, the most constraint-free situation imaginable. If the forms of selection described in the last section were applied to such an imaginary population, biologically absurd outcomes would result.

In the first example, selection favors an increase in size at every age. The force of this selection will change over time because of evolutionary changes in the mean growth trajectories. Nevertheless, under the conditions described above, the selection gradient is guaranteed to be positive for all ages regardless of the form of the mean growth trajectory. The imaginary constraint-free population will respond to this selection by evolving ever larger size at every age. The ultimate evolutionary equilibrium can be determined from Eq. (4) (or, in these two examples, directly from intuition). The equilibrium, shown in Fig. 4, is a mean growth trajectory that begins at the size of the egg and then instantaneously jumps to in-

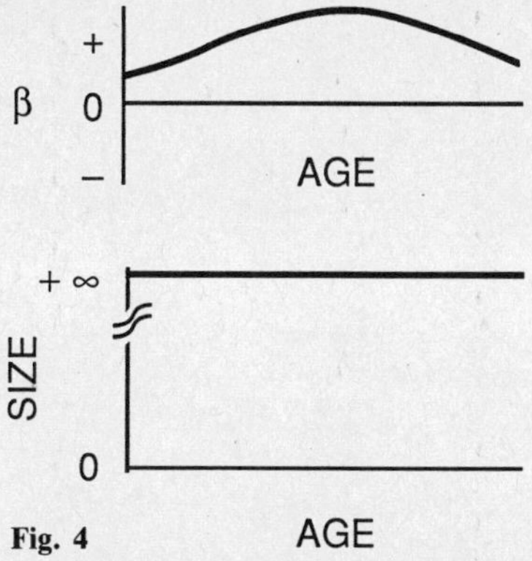

Fig. 4

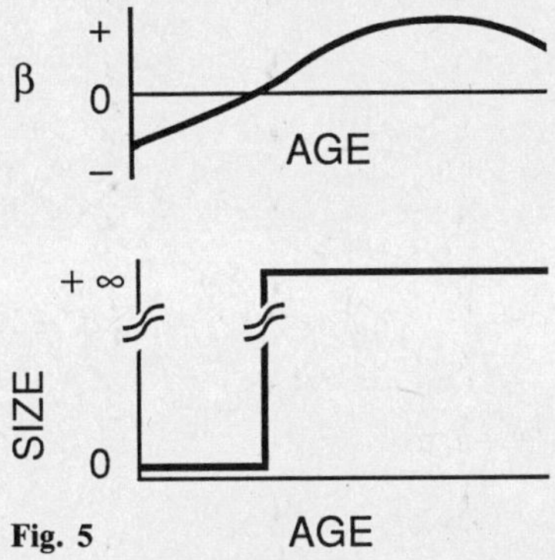

Fig. 5

Fig. 4. *Above* qualitative sketch of the selection gradient on size that results when mortality decreases and birth rate increases will size. *Below* the equilibrium growth trajectory which results from the above selection gradient in the absence of genetic constraints

Fig. 5. *Above* qualitative sketch of the selection gradient on size that results when mortality increases gradually with size, and birth rate is zero up to the critical size and rapidly increases with size thereafter. *Below* the equilibrium growth trajectory which results from the above selection gradient in the absence of genetic constraints

finite size. In the second example, smaller size is favored up to a certain age, after which larger size is favored for the rest of the life of the organism. As in the first example, the conditions described above guarantee that the sign (but not the magnitude) of the selection gradient at each age will stay the same as the growth trajectory evolves. The evolutionary equilibrium for the hypothetical constraint-free population, shown in Fig. 5, is a mean growth trajectory in which individuals remain at zero size from birth until the threshold age, at which point they grow instantly to infinite size.

Obviously, these two outcomes are absurd. Where did the logic break down? The problem must lie either in the assumptions about selection or the assumptions about inheritance. Although the demographic model is highly simplified, it seems plausible that at least some organisms experience ecological conditions that produce selection gradients qualitatively similar to those described above. The flaw apparently lies in the assumption that a population can have an infinitely flexible genetic architecture. This returns us to the earlier point that an evolutionary equilibrium can be achieved in two ways. By considering only how the forces of selection might mold the mean growth trajectory, we have neglected the role of genetic constraints imposed by patterns of genetic variation and covariation. Those patterns are very different in actual populations than in the hypothetical example just considered, and are apparently what prevents evolution from producing the kinds of fanciful growth trajectories just described. This motivates a search for a method than can identify the evolutionary constraints acting on growth trajectories.

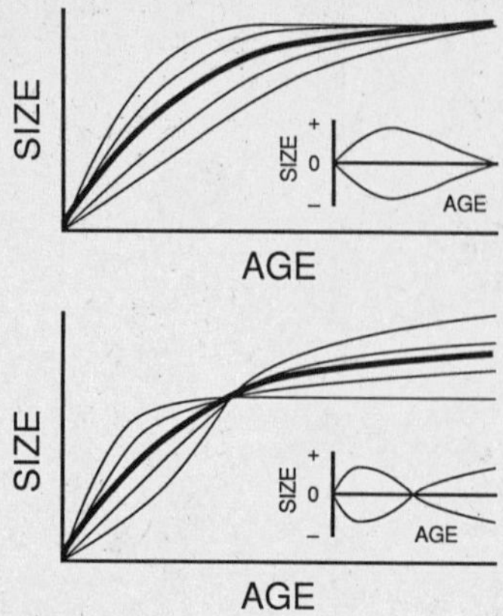

Fig. 6. Mean growth trajectory (*bold curves*) and two sets of evolutionary deformations corresponding to two hypothetical eigenfunctions. The eigenfunctions themselves are shown in the *small inset panels.* Given genetic variation for these eigenfunctions, any combination of the two families of deformations (*top and bottom panels*) could be produced by selection

5.2 Genetic Constraints on the Evolution of Growth Trajectories

Patterns of genetic variation in the growth trajectory can be visualized using a technique called spectral analysis (Kirkpatrick and Heckman 1989; Kirkpatrick and Lofsvold 1989). Spectral analysis is the infinite-dimensional analog of principal components analysis. That is, it reveals the axes along which there is variation. A spectral analysis of the additive genetic covariance function yields a series of curves called eigenfunctions. Each eigenfunction represents one family of evolutionary deformations of the population's mean growth trajectory. The eigenfunctions are statistically independent of one another, and any conceivable growth trajectory can be constructed from the sum of the population's current mean growth trajectory and some combination of the eigenfunctions. Spectral analysis thus extracts from the covariance data the natural families of evolutionary deformations of the mean growth trajectory without imposing a priori some specific function such as the von Bertalanffy growth law [as in Eq. (3) and Fig. 1]. The concept is illustrated by hypothetical eigenfunctions in Fig. 6.

In general, however, there will not be genetic variation for all conceivable deformations (eigenfunctions) of the mean growth trajectory that selection might favor. Each eigenfunction has a number associated with it, called an eigenvalue, that measures the additive genetic variation present in the population for that particular deformation. The eigenvalues are determined by spectral analysis at the same time that the eigenfunctions are. In general, all but a few of the eigenvalues will be zero; that is, there will be no genetic variation for many of the eigenfunctions. If selection favors changes in the mean growth trajectory that correspond to these eigenfunctions, the population will not be able to respond evolutionarily.

This point is illustrated by the data of Riska et al. (1984), who performed an analysis of the genetics of growth in mice using an extensive breeding experiment in which 2693 individuals were measured at each of nine ages. Spectral analysis shows that, of the nine possible eigenfunctions that can be extracted from their data, only two or three have significant amounts of genetic variation associated with them (Riska et al. 1984; Kirkpatrick and Lofsvold 1989).

The conclusion from this result is that the mean growth trajectory in this mouse population has effectively only two or three evolutionary degrees of freedom which correspond to the eigenfunctions that have genetic variation. The eigenfunctions for which there is genetic variation describe the families of possible evolutionary deformations of the mean growth trajectory which are evolutionarily accessible. Other growth trajectories which lie outside these families (described by the remaining eigenfunctions) are evolutionarily precluded by a lack of genetic variation. They would only become accessible to evolution if new mutations appeared which changed the genetic covariance structure in the population.

The mean growth trajectory will respond to selection with evolutionary deformations that involve the eigenfunctions for which there is genetic variation. If, however, selection favors a growth trajectory that lies outside the range of what is evolutionarily accessible, the population will approach but not attain the optimal trajectory. This accounts for the fact that we do not see the extreme growth patterns considered in the previous section (for example, individuals growing to infinite size the moment after birth). In general, the ultimate evolutionary equilibrium is determined by a compromise between selection and heredity. Of the evolutionarily accessible growth trajectories, it will be the one that lies closest to that which is favored by selection (i.e., that maximizes the mean population fitness). The equilibrium can be determined using Eq. (4) from knowledge of the selection gradient [Eq. (9)] and the genetic covariance function (Kirkpatrick and Lofsvold 1989).

The mouse data lead to the conclusion that only a relatively restricted set of growth trajectories can be realized by evolution because patterns of genetic covariation constrain evolution. While this conclusion is intriguing, two limitations should be noted. First, this is a purely descriptive analysis and does not give us insight into the mechanisms that generate the constraints. A different kind of investigation is needed to determine what proximate physiological mechanism (e.g., limitations of energy intake or resource allocation) is responsible. Second, it is not known how broadly the conclusion from the mouse data can be generalized, since it is at present the only population whose growth trajectories have been analyzed. It is of great interest to see of similar results hold not just for determinately-growing species, such as mice, but also indeterminately-growing (and size-structured) populations, such as the Arcto-Norwegian cod.

5.3 Other Sources of Genetic Constraints

Spectral analysis provides a picture of the evolutionary constraints imposed by the genetic covariances between the sizes at different ages. Does this then fully describe the evolutionary limits of the growth trajectory? Whenever there are additional characters under selection, the answer is likely to be no.

The evolution of the mean growth trajectory is determined not only by selection acting on that trait but also by selection acting on all other traits genetically correlated with it, as indicated by Eq. (1). Correlated characters that are neglected can be an important source of error in the conclusions made about the evolution of a trait (Lande and Arnold 1983). The best answer to this difficulty is to try to anticipate what characters are likely to be strongly genetically correlated and also under selection. An obvious candidate in the context of growth trajectories is reproduction.

Physiologists often consider the allocation of resources between growth and reproduction to be the primary partition of energy in indeterminately-growing organisms. Given limited energetic intake, diverting energy to one of these functions must result in a decreased allocation to the other. In the framework of quantitative genetics, this tradeoff would appear as a strong negative genetic correlation between growth and reproduction. On these grounds, it seems likely that selection acting on reproduction will play an important role in the evolution of growth trajectories.

In the evolutionary model developed above, birth rate was considered to be a fixed function of the size and age of the individual. But the birth function, also known as reproductive effort, itself evolves. The evolutionary consequences for the growth trajectory can be determined by explicitly considering selection and evolution of the birth function. For simplicity, I will assume now that birth rate depends only on size and not on age. Extending Eq. (4) to this end, the mean growth trajectory and mean birth functions at the start of the next generation are

$$\bar{\mathscr{z}}'(a) = \bar{\mathscr{z}}(a) + \int_0^{a_{max}} \mathscr{G}_z(a,x)\beta_z(x)\,dx + \int_{-\infty}^{\infty} \mathscr{G}_{zb}(a,x)\beta_b(x)\,dx \ , \tag{11}$$

$$\bar{b}'(z) = \bar{b}(z) + \int_{-\infty}^{\infty} \mathscr{G}_b(z,x)\beta_b(x)\,dx + \int_0^{a_{max}} \mathscr{G}_{zb}(x,z)\beta_z(x)\,dx \ , \tag{12}$$

where $\beta_z(a)$ is the selection gradient on size for individuals of age a, $\beta_b(z)$ is the selection gradient on birth rate for individuals at size z, $\mathscr{G}_z$ is the additive genetic covariance function for size, and $\mathscr{G}_b$ is the covariance function for birth rate. The function $\mathscr{G}_{zb}$ is defined such that $\mathscr{G}_{zb}(a,z)$ is the additive genetic covariance between size at age a and birth rate at size z. Equation (11) shows that the additional character, birth rate, enters into the evolutionary dynamics for the growth trajectory simply as an additional term appended to Eq. (4), tantamount to adding an additional term to the sum of Eq. (1). The two new genetic covariance functions that appear, $\mathscr{G}_b$ and $\mathscr{G}_{zb}$, can be experimentally determined using the methods mentioned earlier.

The selection gradient on birth rate, β_b, which appears in Eqs. (11) and (12), is new to this discussion. It is evaluated by applying Eq. (5) to Eq. (6), which shows that

$$\beta_b(z) = \frac{1}{\bar{W}} \int_0^{\infty} l(a) f(z,a)\,da \ . \tag{13}$$

This result shows that the force of selection favoring an increase in the birth rate at size z is proportional to the fraction of its life that an average individual spends at size z.

6 Conclusions

The major theme from this discussion is that the evolution of size and growth can only be understood through knowledge of the selection pressures acting on the population and the patterns of genetic variation that are present. A description of selection alone is not sufficient. In general, the evolution of size is likely to depend on a compromise between what is favored by selection and what is allowed by genetic constraints. Quantitative genetics provides methods for measuring both selection and genetic variation. This allows hypotheses concerning the evolution of growth and size to be tested quantitatively.

Two alternative equations for the evolution of the mean growth trajectory have been presented. In the first [Eq. (4)] selection is assumed to act only on size, while in the second [Eq. (11)] the impact on size of selection on birth rate is also considered. It is appropriate to ask what advantage is gained by including the additional complication of selection on birth rate. The evolutionary change in the growth rate predicted by the two models will generally differ. This is because the second model includes the correlated response of the growth trajectory to selection on birth rate.

A second and perhaps more generally interesting consequence of including selection on birth rate is that the evolutionary equilibrium predicted for the growth trajectory will be different. This is important, since many biologists regard populations to be in approximate evolutionary equilibrium. Both basic and applied biology are interested in identifying the forces and constraints that are responsible for the equilibria of growth trajectories in natural populations. Ecologists, for example, might like to know how well the body size of a species matches the optimal size determined by its niche. If tradeoffs between size and reproduction are unimportant, then the expectation might be that the fit would be good, limited only by the genetic constraints on growth described by Eq. (4). If, on the other hand, tradeoffs between size and reproduction are important [as considered in Eq. (11)] then selection on reproduction might well cause body size to deviate substantially from the survival optimum. For an example in applied science, consider the problem improving growth in an indeterminately-growing population by artificial selection. Imagine the species has experienced a long history of natural selection favoring increased size at all life stages (as in Fig. 4). If the species has reached its current growth pattern primarily through a compromise between the forces of selection and the genetic constraints operating on size alone, then the evolution of the growth trajectory may reasonably be described by Eq. (4). In this case, strong artificial selection favoring increased growth rates will lead to little or no progress, since the population long since reached the genetic constraints on maximizing growth. On the other hand, if the population reached its current growth trajectory through a compromise between selection on size and selection on birth rate, as described by Eq. (11), then expectations about the outcome of artificial selection will be very different. Since the artificial selection will be applied only to size, the population can respond evolutionarily by increasing its size at the expense of reproduction. Whether this tradeoff of size against reproductive output is acceptable depends on the goals of the particular selection program – for example, whether the population will have to sustain itself or will be maintained by artificial means.

It is clear that understanding the evolution of growth trajectories demands an understanding of the genetic correlations that constrain the evolution of growth trajectories in addition to information on how selection acts. Considering this situation, it is surprising that these quantities are virtually unknown for any size-structured population in nature. Hopefully, motivated by the growing interest in size-structured populations, biologists will soon begin amassing information on the selection and inheritance of size in these species.

Acknowledgments. I thank H. Caswell for instructing me on life history theory, S. Arnold and M. Lynch for discussions on the quantitative genetics of growth, and the participants of the Symposium on the Dynamics of Size-Structured Populations for numerous helpful comments. I am grateful to S. Adolph, B. Ebenman, D. Lofsvold, and L. Persson for comments on the manuscript. The research was supported by N.S.F. grants BSR-8604743 and BSR-8657521.

7 References

Arnold SJ, Wade MJ (1984) On the measurement of natural and sexual selection: applications. Evolution 38:720–734

Bagenal TB (1978) Aspects of fish fecundity. In: Gerking SD (ed) Ecology of freshwater fish production. John Wiley & Sons, New York, pp 75–101

Boag PT, van Noordwijk AJ (1987) Quantitative genetics. In: Cooke F, Buckley PA (eds) Avian genetics: A population and ecological approach. Academic Press, New York London, pp 45–78

Brooks JL, Dodson SI (1965) Predation, body size, and composition of plankton. Science 150:28–35

Bulmer M (1985) The mathematical theory of quantitative genetics. Univ Press, Oxford

Caswell H (1982) Stable population structure and reproductive value for populations with complex life cycles. Ecology 63:1223–1231

Cock AG (1966) Genetical aspects of metrical growth and form in animals. Q Rev Biol 41:131–190

Dingle H, Hegmann JP (eds) (1982) Evolution and genetics of life histories. Springer, Berlin Heidelberg New York

Falconer DS (1981) Introduction to quantitative genetics, 2nd edn. Longman, London

Fisher RA (1958) The genetical theory of natural selection, 2nd edn. Dover, New York

Kirkpatrick M (1984) Demographic models based on size, not age, for organisms with indeterminate growth. Ecology 65:1874–1884

Kirkpatrick M, Heckman N (1989) A quantitative genetic model for growth, shape, and other infinite-dimensional characters. (submitted)

Kirkpatrick M, Lofsvold D (1988) The evolution of complex quantitative characters. (submitted)

Lande R (1979) Quantitative genetic analysis of multivariate evolution, applied to brain: body size evolution. Evolution 33:402–416

Lande R, Arnold SJ (1983) The measurement of selection on correlated characters. Evolution 37:1210–1226

Peters RH (1983) The ecological implications of body size. Univ Press, Cambridge

Price TD, Grant PR, Gibbs HL, Boag PT (1984) Recurrent patterns of natural selection in a population of Darwin's finches. Nature (London) 309:787–789

Riska R, Atchley WR, Rutledge JJ (1984) A genetic analysis of targeted growth in mice. Genetics 107:79–101

Sauer JR, Slade NA (1987) Size-based demography of vertebrates. Annu Rev Ecol Syst 18:71–90

Sebens KP (1987) The ecology of indeterminate growth in animals. Annu Rev Ecol Syst 18:371–407

Shaffer HB (1986) The utility of quantitative genetic parameters in character weighting. Syst Zool 35:124–134

Via S (1984a) The quantitative genetics of polyphagy in an insect herbivore. I. Genotype-environment interaction in larval performance on different host plant species. Evolution 38:881–895

Via S (1984b) The quantitative genetics of polyphagy in an insect herbivore. II. Genetic correlations in larval performance within and among host plants. Evolution 38:896–905

Werner EE, Gilliam JF (1984) The ontogenetic niche and species interactions in size-structured populations. Annu Rev Ecol Syst 15:393–425

Path Analysis of Ontogenetic Data

M. LYNCH

1 Introduction

Ecologists and evolutionary biologists have long been interested in the quantitative description of growth processes (Huxley 1932; Thompson 1942; Cock 1966; Gould 1977). Mathematical expressions for individual growth trajectories are required for the analysis of a diversity of problems in biology including: (1) optimal harvesting strategies for species of economic importance, (2) evolutionary theory of life history strategies, and (3) development of gross morphology and shape. Thus, considerable effort has been expended in the search for relatively simple growth formulations and in the development of suitable curve-fitting procedures. General references on these subjects include Nair (1954), Bertalanffy (1957), Beverton and Holt (1957), Richards (1959), Fabens (1965), Paloheimo and Dickie (1965), Ricklefs (1967), Parks (1970), Pruitt et al. (1979), Majowski and Uchmanski (1980), and Lande (1985).

Outside of the animal breeding literature, the vast majority of quantitative growth analyses (all cross-sectional studies) involve the fitting of general formulae to the mean sizes of individuals at different ages. Such treatment may be entirely adequate for many economic applications, but studies in evolution and developmental biology, which are concerned with variation, require alternative approaches. In the case of nonlinear growth, the function obtained by fitting to mean age-specific sizes will generally be different than the mean of the functions fitted to each individual. Thus, any attempt to relate the former to physiological, developmental, or genetic mechanisms, all of which are properties of individuals, can be misleading.

Longitudinal information on individual size is a key to understanding growth as a process (Cock 1966). Such data allow the investigator to evaluate the extent to which growth components, such as age-specific increments, are jointly influenced by beneficial or antagonistic factors. This point has not gone unnoticed by quantitative geneticists, who for many years have attempted to ascertain the causal sources of variance and covariance of size-related traits (Hazel et al. 1943; Young and Legates 1965; Monteiro and Falconer 1966; Thomas et al. 1968; Moore et al. 1970; Rutledge et al. 1972; Falconer et al. 1978; Herbert et al. 1979). The recent work of Atchley and his associates (Atchley and Rutledge 1980; Cheverud et al. 1983 a, b; Atchley 1984; Leamy and Cheverud 1984; Riska et al. 1984; Riska and Atchley 1985) on laboratory mice and rats provides an excellent example of

Dept. of Ecology, Ethology, and Evolution, Shelford Vivarium, University of Illinois, 606 E. Healey St., Champaign, Illinois 61820, USA

Size-Structured Populations
B. Ebenman and L. Persson (Eds.)

the insights into developmental and evolutionary constraints that can be gained from a multivariate analysis of growth components.

It is often argued that the fundamental determinants of a growth process may be evaluated by fitting the data with various functions (e.g., logistic, Gompertz, Bertalanffy, Winberg, Parks) for which the biological underpinnings have been explicitly stated. However, since many data sets can be fitted quite well with several growth functions, the statistical power of this approach is not very great. The purpose of this paper is to illustrate how information on individual size variation can be exploited, without reference to any specific growth function, to yield insight into the proximal causes of growth trajectories. Path analysis, a simple extension of multiple regression theory, will be shown to be capable of partitioning the variation in size at any age into the direct and indirect contributions of an arbitrary number of sources.

Wright (1921a, b) introduced the method of path analysis as a means of assessing the contributions of various causal factors to phenotypic variances and covariances. Although he heavily relied on the technique throughout his career, only a few other evolutionary biologists (e.g., Doyle and Myers 1982; Arnold 1983; Cheverud 1984) have given it much attention. Path analysis is extensively employed in the applied fields of animal breeding, genetic epidemiology (Cloninger et al. 1983), and the social sciences (Li 1975).

One possible reason that path analysis has not gained wider attention from evolutionary biologists is that the structure of a path diagram is often highly subjective, particularly in nongenetic analyses. The rules for setting up a causal framework for a path analysis are not always explicit. The major criticism of the technique is that it attempts to deduce causality from systems of correlated variables (Niles 1922; Karlin et al. 1983). However, Wright (1921a, b, 1923, 1934, 1968, 1983, 1984) has repeatedly emphasized that the method was developed as a means of quantifying causal sources of variation after the causal connections between variables have been ascertained through supplemental empirical and/or theoretical investigation. In the analysis of growth, final size is necessarily the sum of growth increments or parts. The assignment of causality at this level is unambiguous.

2 General Theory

2.1 Path Analysis

Consider the linear model

$$Y = a + b_1X_1 + b_2X_2 + \dots + b_nX_n + e \ , \tag{1}$$

where Y, $X_1, \dots, X_n$ are observed variables, $b_1, \dots, b_n$ are the partial regression coefficients obtained by the method of least-squares, a is the fitted intercept, and e is the residual deviation of the observed Y from the regression prediction. If all of the observed values are standardized by subtracting the mean and dividing by the standard deviation, then

$$Y' = p_1 X_1' + p_1 X_2' + \ldots + p_n X_n' + p_e e' , \tag{2}$$

where the primes denote standardized variables and

$$p_i = b_i [\mathrm{Var}(X_i)/\mathrm{Var}(Y)]^{1/2} , \tag{3a}$$

$$p_e = [\mathrm{Var}(e)/\mathrm{Var}(Y)]^{1/2} . \tag{3b}$$

These new fitted variables, standardized partial regression coefficients, are called path coefficients. Since the variance of a standardized variable is always equal to one, the variance of Y' is simply

$$1 = \sum_{i=1}^{n} p_i^2 + \sum_{i=1}^{n} \sum_{\substack{j=1 \\ i \neq j}}^{n} p_i p_j r_{ij} + p_e^2 , \tag{4}$$

where r_{ij} is the correlation between X_i and X_j. This is one of the fundamental expressions of path analysis, known as the equation of complete determination.

Equation (4) decomposes the variance of Y into the proportional contributions of various explanatory pathways. p_i^2 represents the direct contribution of variation in X_i to the variance in Y, while $p_i p_j r_{ij}$ is the additional, indirect influence that X_i has on Y because of correlation between X_i and X_j. It should be noted that any of the p_i^2 may exceed one and that the contribution of an indirect path may be negative. The only constraint is that the contributions of all paths sum to one.

Equations (1)–(4) have a direct bearing on the analysis of incremental growth. Consider a cohort of individuals, each measured on t occasions. The size at time t may be written as

$$\begin{aligned} z_t &= (z_1 - 0) + (z_2 - z_1) + (z_3 - z_2) + \ldots + (z_t - z_{t-1}) \\ &= \Delta z_0 + \Delta z_1 + \Delta z_2 + \ldots + \Delta z_{t-1} , \end{aligned} \tag{5}$$

where Δz_0 is the initial size and $\Delta z_1, \ldots, \Delta z_{t-1}$ are subsequent growth increments. The form of Eq. (5) is fundamentally the same as Eq. (1) except that: (1) there is no residual error term, (2) the intercept is zero, and (3) the partial regression coefficients are all equal to one. Because or (3), the path coefficients are simply ratios of standard deviations, $p_i = [\mathrm{Var}(\Delta z_i)/\mathrm{Var}(z_t)]^{1/2}$, and the equation of complete determination reduces to

$$1 = \frac{\sum_{i=0}^{t-1} \mathrm{Var}(\Delta z_i) + \sum_{i=0}^{t-1} \sum_{\substack{j=0 \\ i \neq j}}^{t-1} \mathrm{Cov}(\Delta z_i, \Delta z_j)}{\mathrm{Var}(z_t)} , \tag{6}$$

where Cov refers to covariance. Equation (6) partitions the variance of the size of a part or an organism into contributions from t^2 causal paths, t of which are direct and $t(t-1)$ of which are indirect (Fig. 1). Each indirect path is actually counted twice. For example, the contribution of path $\Delta z_0 \leftrightarrow \Delta z_1 \rightarrow z_t$ is identical to that of path $\Delta z_1 \leftrightarrow \Delta z_0 \rightarrow z_t$; both are $p_0 p_1 r_{01}$. In applying Eq. (6), individuals with missing data should not be utilized, since their inclusion will normally prevent the components of variance and covariance from summing to $\mathrm{Var}(z_t)$.

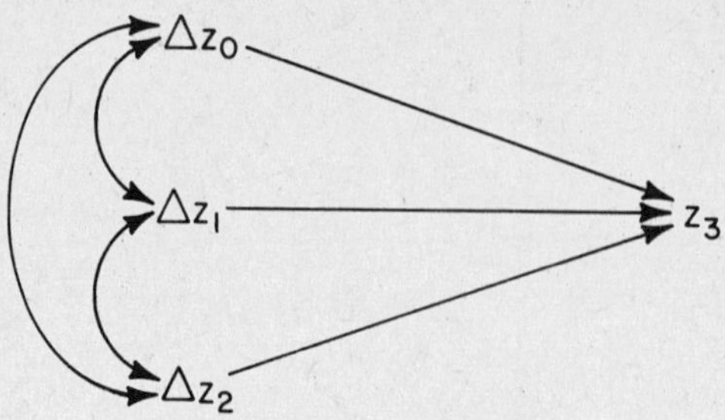

Fig. 1. Path diagram for final size for a system in which all individuals have been measured at three ages

2.2 Sampling Error and Measurement Error

A natural concern that arises in any attempt to partition a variance into components is the degree of accuracy of the estimated components. A general expression for the sampling variance of a path contribution is given in the Appendix. If the data are multivariate-normally distributed and the sample size (N) is on the order of 100 or greater, the expression for the standard error of a direct path contribution simplifies to

$$SE(p_i^2) \simeq 2p_i^2[(1-r_{it}^2)/N]^{1/2} , \tag{7}$$

where r_{it} is the estimated correlation between Δz_i and z_t. This is consistent with Wright (1934), who stated the sampling variance of p_i to be $p_i^2(1-r_{it}^2)/N$ for a causal system such as Fig. 1. Under the same conditions, the standard error of an indirect path contribution (sum of both directions) is

$$SE(2p_ip_jr_{ij}) \simeq 2p_ip_j[(1+3r_{ij}^2-4r_{it}r_{jt}r_{ij})/N]^{1/2} . \tag{8}$$

These expressions may be employed to obtain a general impression of the accuracy of path contribution estimates, but the construction of exact confidence intervals would require a more formal understanding of the sampling distribution of p_i^2 and $p_ip_jr_{ij}$.

Measurement error causes an upward bias in the estimates of the path coefficients and a downward bias in the estimated correlation between adjacent growth increments. However, if multiple measurements are taken on individuals, an unbiased analysis is possible by isolating the variance of z_t and its components resulting from measurement error from that caused by biological factors. The observed variances can be written as

$$Var(z_t) = Var_a(z_t) + Var_e(z_t) , \tag{7a}$$

$$Var(\Delta z_0) = Var_a(\Delta z_0) + Var_e(z_1) , \tag{7b}$$

$$Var(\Delta z_i) = Var_a(\Delta z_i) + Var_e(z_i) + Var_e(z_{i+1}) \quad \text{for} \quad i>0 , \tag{7c}$$

$$Cov(\Delta z_i, \Delta z_{i+1}) = Cov_a(\Delta z_i, \Delta z_{i+1}) - Var_e(z_{i+1}) , \tag{7d}$$

$$Cov(\Delta z_i, \Delta z_j) = Cov_a(\Delta z_i, \Delta z_j) \quad \text{for} \quad j>i+1 \quad \text{or} \quad j<i-1 , \tag{7e}$$

where the sample variances and covariances exclusive of measurement error are subscripted with an a and the error variances are subscripted with an e. The error variances are equal to the variance of multiple measures on the same individual at the same stage divided by the number of measures. Use of the corrected variances and covariances, $\mathrm{Var}_a(..)$ and $\mathrm{Cov}_a(..,..)$, does not alter the integrity of Eq. (6) and is preferable whenever possible.

3 Applications

To illustrate the utility of the techniques, growth data from two experiments with *Daphnia* will be analyzed. A major advantage of working with an asexual organism such as *Daphnia* is that genotypes can be replicated in large numbers. This makes it possible to examine genotype-specific variance-covariance structures of the components of a growth process and to evaluate how the developmental pathway is altered by a change in environment. It also allows for the possibility of estimating the variance-covariance structure among clones and providing some insight into the genetic architecture of the growth process.

3.1 Environmental Modification of the Developmental Pathway

An obligately parthenogenetic clone of *Daphnia pulex* (Lynch 1985) has been grown on several food concentrations and measured daily for growth, survival, and reproduction. Here I will focus entirely on the growth trajectory and the factors that shape it. In all of the experiments reported in this paper, the animals were maintained in a growth chamber at 20 °C on a 12 : 12 light-dark cycle. The food supply consisted of pure *Scenedesmus* or a *Scenedesmus-Chlamydomonas* mixture grown and prepared as described in Lynch et al. (1986). The animals were always acclimated to the experimental food level for at least two generations prior to data collection.

The lengths of individuals are based on the distance from the top of the head to the base of the tail spine. Because the carapace of a daphnid does not grow within an instar and the instar duration ranges from 1 to 3 days under the experimental conditions, estimates of the measurement variance of instar-specific sizes could be obtained from multiple measures. First and second instar individuals generally lie flat on the microscope stage and the base of their tail spine is located unambiguously. Despite the fact that they are only rarely measured more than once per instar (because of short instar durations), their measurement variance is relatively low (Fig. 2). Although older individuals are measured two or three times per instar, their measurement variance is much greater and increases slightly with instar number. The data in Fig. 2, are used throughout this paper in conjunction with Eqs. (7a–e) to eliminate bias caused by measurement error.

A large experiment performed by Toline (1986) will first be considered in some detail. Although the animals employed in this experiment were taken from three lines that had been isolated from a single stem mother 4 years in the past, their growth properties had not diverged at all (Toline 1986). Thus, the entire data set has been pooled in the following analysis. The food level in this experiment was

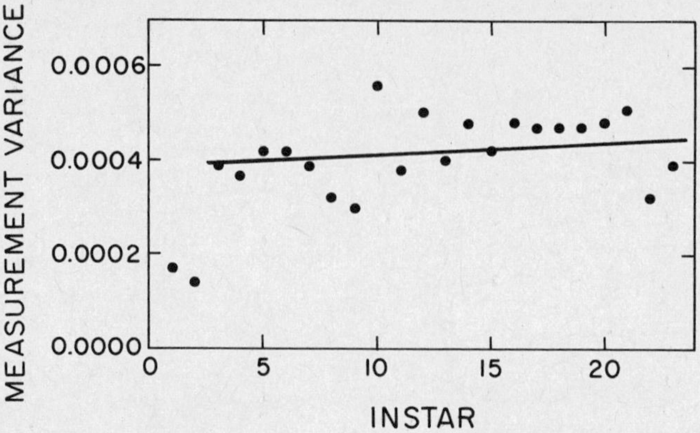

Fig. 2. Average measurement variance for instars 1–23 pooled over several experiments. Individual points are based on 70–320 estimates. The variances for instars 1 and 2 are not significantly different and have a mean of 0.000156. For instars ≥ 3, the variance is given by the weighted least-squares regression, 0.00000262 (instar) + 0.000390, $p < 0.01$

very high ($\sim 4.2 \times 10^5$ *Scenedesmus* cell $ml^{-1} \simeq 3.61$ μg C/ml, not including bacteria which undoubtedly colonized the medium) and was replenished each day. The animals were housed individually in beakers containing 200 ml of medium.

The mean growth curve is very nearly linear for the first seven instars, at which point a shift to a second, less pronounced phase of linear growth takes place (Fig. 3). The variance in length also increases in a roughly linear fashion for the first seven instars, but then levels off abruptly. Since all of the instar-specific growth rates are variable, approximate constancy of the variance of length beyond the seventh instar implies that the growth in different instars must have become compensatory, i.e., instars in which an individual grows at a higher than average rate tend to be followed by instars with low growth rates, and vice versa. This was examined by performing a path analysis up to the tenth instar, treating the initial size and the instar-specific growth increments as the explanatory variables. The analysis included 282 individuals, all born on the same date.

The simple correlations between initial size (Δz_0) and the nine growth increments (Δz_1 to Δz_9) leading to size in instar 10 are given below the diagonal in Table 1. Most of the components of early growth (Δz_0 to Δz_4) are significantly positively correlated. However, the majority of the remaining correlations are negative, several significantly so. Thus, for this particular clone under a very enriched environment, the components of growth up until instar 5 (reproductive maturity) are jointly influenced in the same direction by variable effects. Thereafter, tradeoffs begin to appear, i.e., a growth-enhancing effect in one instar tends to have an inhibitory effect on growth in a later instar.

The proportional contributions of the 55 paths to size in instar 10 are given in the upper half of Table 1. The vast majority of the variance of final size is determined by the direct paths (diagonal elements). Their contributions sum to 1.19, while those of all of the indirect paths (off-diagonal elements) sum to -0.19.

The pattern of growth described above is substantially modified when the food supply is depressed. The ontogenetic trajectories of mean length and

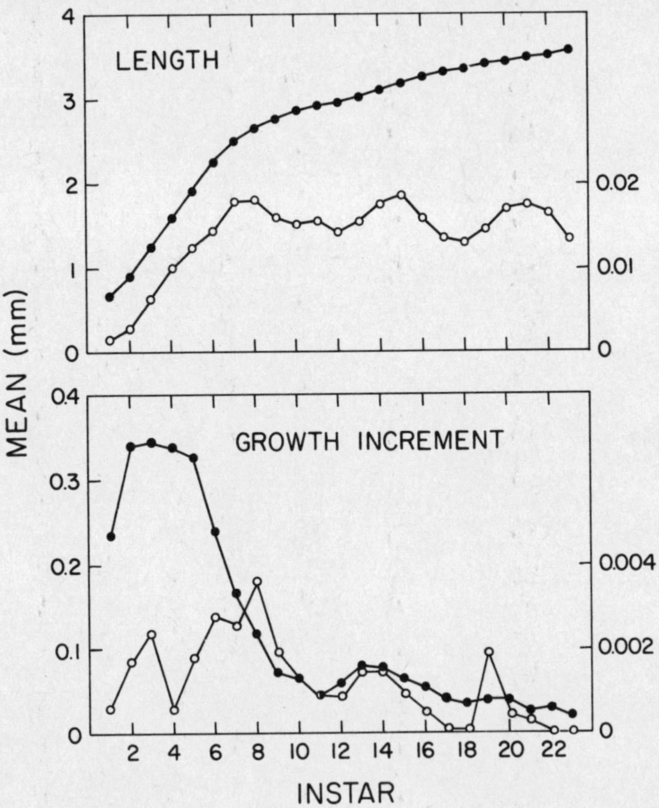

Fig. 3. The mean (*closed circles*) and variance (*open circles*) for instar-specific lengths and growth increments for a cohort of a clone of *Daphnia pulex.* Sample sizes decline from 306 for instar 1 to 56 for instar 23

Table 1. Simple correlations (below diagonal) and path contributions (diagonal and above) of components of growth to size in the 10th instar of a clone of *Daphnia pulex*

	Δz_0	Δz_1	Δz_2	Δz_3	Δz_4	Δz_5	Δz_6	Δz_7	Δz_8	Δz_9
Δz_0	0.08	0.04	0.07	0.04	0.03	−0.01	0.04	−0.02	−0.03	−0.02
Δz_1	0.41**	0.04	0.07	0.04	−0.00	−0.01	0.01	−0.02	0.01	−0.02
Δz_2	0.39**	0.50**	0.12	−0.01	−0.00	−0.04	−0.03	−0.01	−0.01	−0.04
Δz_3	0.22**	0.12*	−0.04	0.12	0.10	0.04	0.03	−0.06	−0.05	0.02
Δz_4	0.30**	−0.08	0.00	0.62**	0.04	0.05	0.02	−0.00	−0.06	−0.00
Δz_5	−0.07	−0.02	−0.14*	0.01	0.45**	0.12	−0.01	−0.04	−0.03	−0.02
Δz_6	0.13	0.07	−0.08	−0.03	0.15	0.01	0.15	0.01	−0.05	−0.01
Δz_7	−0.02	−0.13*	−0.05	−0.20**	0.03	−0.09	0.00	0.14	−0.09	−0.02
Δz_8	−0.11	0.05	−0.03	−0.16*	−0.31**	−0.09	−0.16*	−0.22**	0.24	−0.10
Δz_9	−0.10	−0.11	−0.15*	0.07	−0.03	−0.08	−0.02	−0.06	−0.27**	0.13

Δz_0 = the length at birth (instar 1); Δz_i for $i>0$ = the growth from instar i to i+1. * and ** denote significant correlations at the 0.05 and 0.01 levels.

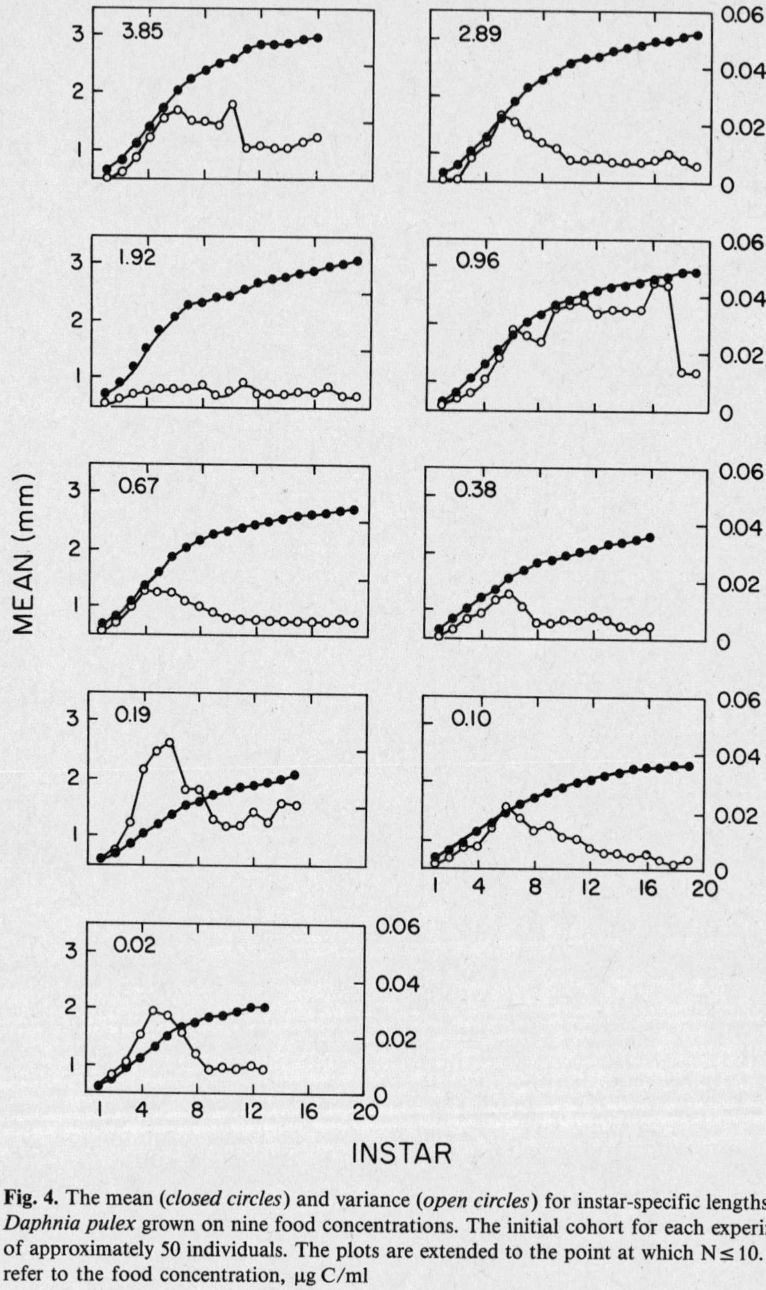

Fig. 4. The mean (*closed circles*) and variance (*open circles*) for instar-specific lengths for a clone of *Daphnia pulex* grown on nine food concentrations. The initial cohort for each experiment consisted of approximately 50 individuals. The plots are extended to the point at which N ≤ 10. Inset numbers refer to the food concentration, μg C/ml

variance for the same clone grown on nine food concentrations are presented in Fig. 4. The treatment in these experiments was somewhat different than in the previously described experiment. Many fewer animals (~50/treatment) were monitored, and although the same temperature and light conditions were employed, the animals were maintained in only 40 ml of medium and the food supply was renewed every other day. The food consisted of a mixture of *Scenedesmus* und *Chlamydomonas*, but since both algal species appear to be nutritionally equivalent for this clone (Lynch et al. 1986), the carbon equivalent allows a comparison with the previous experiment.

As in the previous experiment, there is a reduction in the average growth rate at approximately the seventh instar at all food concentrations (Fig. 4). The variance in length increases until the fifth or sixth instar, then levels off and eventually declines. This reduction in variance, which is particularly striking at the lower food concentrations, again implies that through compensatory growth the entire cohort is being focused into the same growth trajectory.

The path analyses for length at instar 10 are provided for the four largest data sets in these experiments in Table 2. At this point, the variance in length is near its maximum for the two highest food concentrations but strongly reduced at the two lowest food levels. An indication of the degree of food limitation in these four cases is provided by the mean ($\pm$SE) lengths in instar 10 (2.62 ± 0.02, 2.41 ± 0.03, 2.30 ± 0.01, and 1.95 ± 0.01 mm), in comparison to 2.84 ± 0.01 mm in the previous experiment. The variance-covariance structures for growth components at the highest food levels, 3.85 and 0.96 μg C/ml, are qualitatively similar to those described in Table 1. In these cases, the vast majority of the variance in length is accounted for by the direct paths (Table 3) with almost all of the indirect paths having absolute values less than 0.1.

On the other hand, at the two lowest food concentrations, 0.67 and 0.38 μg C/ml, the negative contributions of the indirect paths are very large, summing to about -2.7 in comparison to an average of 0.0 in the three experiments at higher food levels (Table 3). Many of the indirect paths have values more negative than -0.10, indicating strong compensatory mechanisms regulating the growth in different instars. Nevertheless, at 0.67 μg C/ml, growth in the first three instars was mutually favored (disfavored) by the same factors.

The path analyses for this clone may be interpreted as follows. While long-term growth is indeterminate, homeostatic (negative-feedback) mechanisms prevent the divergence of sizes of individuals later in life. The growth of adult individuals is either roughly parallel (high food levels) or convergent (low food levels). Growth in the first five instars is divergent at all food levels. The onset of compensatory growth mechanisms around the fifth instar may be related to the initiation of investment into reproduction at this stage. Individuals that grow at relatively high rates in one instar may also invest a greater amount of energy in reproduction at the expense of growth in the following instar. Such a tradeoff would be expected to be accentuated when resources are in short supply.

Also provided in Table 3 are composite results that are obtained when the variances and covariances of growth increments are not corrected for measurement error. Rather substantial numerical differences are apparent, although the conclusions reached above would not have been altered qualitatively had measurement error been ignored.

Table 2. Simple correlations (below diagonal) and path contributions (diagonal and above) of

3.85 μg C/ml

	Δz_0	Δz_1	Δz_2	Δz_3	Δz_4	Δz_5	Δz_6	Δz_7	Δz_8	Δz_9
Δz_0	0.00	0.02	0.03	0.03	0.02	0.01	−0.01	−0.00	−0.01	0.02
Δz_1	0.60**	0.05	0.06	0.03	0.01	−0.02	−0.06	−0.03	−0.04	0.00
Δz_2	0.79**	0.50**	0.07	0.11	0.07	0.01	−0.03	−0.00	−0.08	0.12
Δz_3	0.69**	0.21	0.67**	0.11	0.11	0.01	−0.10	−0.08	−0.08	0.09
Δz_4	0.53**	0.05	0.46**	0.55**	0.10	0.04	−0.06	−0.04	−0.05	0.05
Δz_5	0.50**	−0.28	0.10	0.17	0.47**	0.03	0.07	0.02	−0.00	−0.01
Δz_6	−0.32*	−0.63**	−0.38**	−0.75**	−0.44**	0.94**	0.04	0.09	0.07	−0.06
Δz_7	−0.09	−0.32*	−0.13	−0.54**	−0.29*	0.24	0.97**	0.06	0.07	−0.03
Δz_8	−0.19	−0.29*	−0.59**	−0.44**	−0.33*	−0.00	0.63**	0.54**	0.07	−0.08
Δz_9	0.29	0.01	0.59**	0.35*	0.19	−0.09	−0.38**	−0.14	−0.37**	0.16

0.67 μg C/ml

	Δz_0	Δz_1	Δz_2	Δz_3	Δz_4	Δz_5	Δz_6	Δz_7	Δz_8	Δz_9
Δz_0	0.34	0.16	0.46	0.01	−0.03	0.16	−0.31	−0.05	−0.17	−0.15
Δz_1	0.41**	0.15	0.27	0.32	−0.04	−0.16	−0.49	−0.01	−0.16	0.09
Δz_2	0.55**	0.53**	0.51	0.47	−0.10	−0.18	−0.74	−0.35	−0.14	−0.17
Δz_3	0.02	0.61**	0.51**	0.47	−0.10	−0.46	−0.50	−0.17	−0.24	−0.02
Δz_4	−0.14	−0.26	−0.34*	−0.36*	0.04	0.29	0.05	−0.06	0.00	0.06
Δz_5	0.23	−0.36*	−0.24	−0.62**	1.00**	0.28	0.47	−0.22	0.11	−0.20
Δz_6	−0.31*	−0.71**	−0.60**	−0.42**	0.20	0.54**	0.78	−0.24	0.14	−0.26
Δz_7	−0.06	−0.02	−0.32*	−0.16	−0.17	−0.23	−0.14	0.67	−0.10	0.25
Δz_8	−0.28	−0.43**	−0.22	−0.35*	−0.04	0.15	0.12	−0.14	0.22	−0.21
Δz_9	−0.26	0.24	−0.25	−0.03	0.31*	−0.39**	−0.30*	0.31*	−0.45**	0.24

Δz_0 = the length in instar 1, while Δz_i for $i > 0$ = the growth from instar i to i + 1. * and ** denote

Table 3. Total contributions of the direct and indirect paths to length at instar 10, obtained from Tables 1 and 2

	Total path contributions		
Food conc.	Direct	Indirect	N
3.85	0.70 (1.00)	0.30 (0.00)	46
3.61	1.19 (1.59)	−0.19 (−0.59)	282
0.96	1.01 (1.18)	−0.01 (−0.18)	43
0.67	3.72 (4.62)	−2.72 (−3.62)	45
0.38	3.72 (4.47)	−2.72 (−3.47)	37

N = the number of individuals used in the path analysis. The food level is expressed as μg C/ml. The numbers in parentheses are the results obtained when measurement error is not accounted for.

components of growth to size in the 10th instar of a clone of *Daphia pulex* grown at four food levels

0.96 µg C/ml

Δz_0	Δz_1	Δz_2	Δz_3	Δz_4	Δz_5	Δz_6	Δz_7	Δz_8	Δz_9
0.05	0.02	0.01	−0.03	0.06	0.01	−0.01	0.03	0.03	−0.01
0.23	0.04	−0.03	0.04	0.03	0.02	−0.06	0.00	−0.03	−0.03
0.13	−0.30*	0.06	−0.02	0.03	−0.02	0.03	−0.03	−0.02	−0.02
−0.15	0.22	−0.10	0.15	−0.05	0.09	−0.11	−0.02	0.11	−0.01
0.42**	0.24	0.21	−0.20	0.11	0.02	−0.03	−0.04	0.01	−0.02
0.06	0.13	−0.09	0.31*	0.09	0.14	−0.02	−0.08	0.16	0.01
−0.06	−0.39*	0.18	−0.38*	−0.13	−0.06	0.14	−0.00	−0.00	−0.02
0.20	0.03	−0.24	−0.08	−0.19	−0.32*	−0.02	0.10	−0.07	−0.02
0.16	−0.18	−0.09	0.37*	0.03	0.56**	−0.00	−0.30*	0.15	0.05
−0.07	−0.26	−0.18	−0.03	−0.09	0.07	−0.10	−0.09	0.24	0.08

0.38 µg C/ml

Δz_0	Δz_1	Δz_2	Δz_3	Δz_4	Δz_5	Δz_6	Δz_7	Δz_8	Δz_9
0.13	0.08	0.12	0.02	−0.01	0.07	−0.17	0.01	−0.03	−0.01
0.32*	0.22	0.08	−0.02	−0.04	0.02	−0.26	−0.11	0.02	−0.14
0.24	0.03	0.33	−0.09	−0.10	−0.17	0.09	−0.19	−0.22	0.05
0.06	−0.02	−0.09	0.23	−0.05	0.37	−0.03	−0.25	−0.07	0.05
−0.01	−0.04	−0.10	−0.05	1.10	−0.43	−0.59	−0.50	0.04	0.08
0.11	0.03	−0.19	0.42**	−0.25	0.73	−0.24	−0.49	−0.06	−0.19
−0.36*	−0.44**	0.14	−0.00	−0.42**	−0.25	0.40	0.30	−0.04	0.11
0.02	−0.19	−0.30*	−0.46**	−0.41**	−0.45*	0.35*	0.35	0.24	0.05
−0.20	0.10	−0.84**	−0.31	0.07	−0.12	−0.17	0.86**	0.06	−0.00
−0.03	−0.35*	0.10	0.13	0.09	−0.26	0.20	0.09	−0.02	0.18

significant correlations at the 0.05 and 0.01 levels.

3.2 Partitioning the Genetic and Environmental Sources of Variance

The previous examples dealt only with the environmental/developmental influences on growth dynamics. Since all individuals of an experimental cohort were members of the same clone, none of the variation in size between individuals was attributable to genetic causes. If the same type of analysis was applied to a genetically undefined population, the paths would be jointly influenced by both genetic and environmental factors. Insight into the evolutionary flexibility of a growth trajectory would be gained if these two sources of variation could be discriminated. Numerous techniques of quantitative genetics, all of which rely on analyses of resemblance between relatives, can accomplish this task to various degrees of precision (Bulmer 1980; Falconer 1981; Hallauer and Miranda 1981).

The example provided here, an analysis of the growth trajectories of a multiclonal population of *Daphnia pulex*, is conceptually simple. Forty-one clones were isolated from a cyclically parthenogenetic population inhabiting a temporary pond at Portland Arch, Indiana, and raised at 20 °C on a food supply of *Scenedesmus* (2.58 µg C/ml), 200 ml replenished daily. Each clone was replicated into two daughter sublines and maintained for two generations under the defined laboratory conditions. Two offspring from the second clutch of each subline were then established and measured daily. Thus, 164 individuals (41 clones, 4 replicates/clone) were available for analysis. The reason for partitioning each clone into sublines has been described elsewhere (Lynch 1985). Briefly, maternal environmental effects can be important sources of phenotypic variation in *Daphnia.* The creation of sublines insures that the maternal effects, and in this case grandmaternal effects, contribute to the within-clone rather than the between-clone component of variance.

The final analysis is a multivariate analysis of variance, yielding two matrices of error and clone mean squares and cross-products. Letting the phenotype of each individual be the sum of a genetic value common to all members of the same clone, and an independently distributed environmental deviation, the error mean squares and cross-products provide estimates of the environmental variances and covariances of the growth components. The genetic variances (covariances) are obtained by subtracting the error mean squares (cross-products) from the respective clone mean squares (cross-products) and dividing by the number of replicates. The path analysis is identical to Eq. (6) except that each of the previous paths at the phenotypic level is partitioned into a genetic and an environmental path.

The path analysis was performed on the length in the fifth instar, which in this experiment was equivalent to the size at first reproduction. Summing over both genetic and environmental effects, there were therefore ten direct paths and 20 indirect paths (Fig. 5). It is clear that both genetic and environmental sources of variance influence growth in this population. Each of the growth components, with the possible exception of growth in instar 4, exhibited significant genetic variation (Table 4). The environmental path coefficients ranged from 0.16 to 0.26, while the genetic path coefficients ranged from 0.10 to 0.20 (Fig. 5). Genetic correlations between the growth components were uniformly high, implying that clones tend to have uniformly high or low growth rates in all instars (Fig. 5). Environmental correlations between adjacent growth increments also tend to be high. (The correlations in excess of 1 are a result of sampling error in the estimation

Table 4. Mean phenotypic value, environmental and genetic components of variance and broad-sense heritability, $H^2 = \mathrm{Var(G)/[Var(G) + Var(E)]}$, for five growth components for a multi-clonal population of *Daphnia pulex*

Trait	Mean	Var(E)	Var(G)	H^2	P
Δz_0	0.675	0.000340	0.000450	0.57	0.0001
Δz_1	0.227	0.000321	0.000181	0.36	0.0008
Δz_2	0.308	0.000593	0.000326	0.35	0.0008
Δz_3	0.382	0.000271	0.000259	0.49	0.0025
Δz_4	0.383	0.000729	0.000114	0.14	0.1417

P = the significance attached to the F ratio of the analysis of variance. The computations for H^2 are corrected for measurement error but the F ratios are not, so the estimated significance levels are conservative.

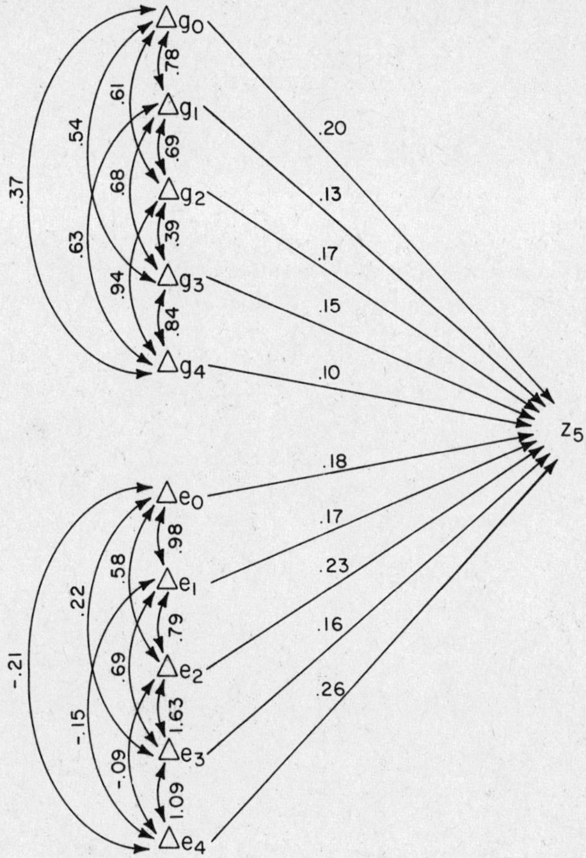

Fig. 5. Path diagram for length in instar 5 (z_5) for a mixed cohort of clones of *Daphnia pulex*. *g* refers to a genetic value and *e* to an environmental deviation. Δg_0 and Δe_0 refer to size at birth, while all other explanatory variables are growth increments. *Numbers on the single-headed arrows* are path coefficients. *Numbers on double-headed arrows* are genetic and environmental correlations

of environmental variances and covariances.) Consistent with the results in the previous section, the environmental correlations involving early growth components were fairly large and positive. However, there are substantial differences in the genetic and environmental correlations involving growth in the fourth instar. For example, the correlation between Δg_2 and Δg_4 is 0.94 while that between Δe_2 and Δe_4 is -0.09.

The contributions of the various pathways to length in instar 5 are given in Table 5. As usual, the direct path contributions are simply the squares of the path coefficients, while the indirect path contributions are the products of two path coefficients and a connecting correlation coefficient. There is no compelling evidence for compensatory growth at either the genetic or environmental level. In light of the previous results, this is perhaps to be expected since the analysis did not go beyond the fifth instar and was performed at a high food concentration. Twenty-one percent of the variance is referable to the direct environmental paths, but only 12% to the direct genetic paths. The indirect genetic paths account for 28% of the variance, and the indirect environmental paths for the remaining 39%.

Table 5. Path contributions of the genetic (above diagonal) and environmental (below diagonal) components of growth to length in instar 5 in a multi-clonal population of *Daphnia pulex*

	Δz_0	Δz_1	Δz_2	Δz_3	Δz_4
Δz_0	0.04 0.03	0.04	0.04	0.03	0.01
Δz_1	0.06	0.02 0.03	0.03	0.03	0.02
Δz_2	0.05	0.06	0.03 0.05	0.02	0.03
Δz_3	0.01	0.04	0.12	0.02 0.03	0.03
Δz_4	−0.02	−0.01	−0.01	0.09	0.01 0.07

Δz_0 = the size in instar 1; Δz_i for $i>0$ = the growth from instar i to i+1.

4 Discussion

This paper has focused on the use of path analysis to gain insight into the proximal environmental and genetic factors that are responsible for variation and/or constraints on a growth process. However, the statistical procedures that have been presented are not restricted to growth analysis. They apply to any situation in which a dependent variable can be decomposed into the sum of two or more parts. Many problems exist in ecology and behavior in which this is the case: the partitioning of total lifetime fitness into contributions from various age, the partitioning of an individual's diet into contributions from different prey types, or the partitioning of offspring production by hermaphrodites into contributions through male and female function.

For statistical purposes, investigators routinely transform their data to conform with normality as closely as possible. In some cases, transformation to a less easily measured scale may give more biological insight. Provided the same transformation is used for all size measurements, a path analysis can be applied to the transformed data since this will not alter the additive relationship implicit in Eq. (5). For instance, in the examples presented above, the lengths could have been converted to dry weights prior to analysis. This would change the partitioning of sources of variation since dry weight is a nonlinear function of length.

The examples provided above were concerned with the analysis of size at one particular stage. However, the procedures are readily generalized to the simultaneous analysis of size at all stages. Figure 6 illustrates the complete path diagram for a cohort of individuals that have been measured at four stages, z_1 to z_4. The path coefficient leading from a growth increment, Δz_i, to a size, z_j, is now a product of intervening path coefficients. For example, the coefficient for the path leading from Δz_0 to z_2 is $p(\Delta z_0, z_2) = p(\Delta z_0, z_1)p(z_1, z_2)$ where each individual path coefficient is simply the ratio of the variance of the first variable to the second. The equations of complete determination for the first three measures are:

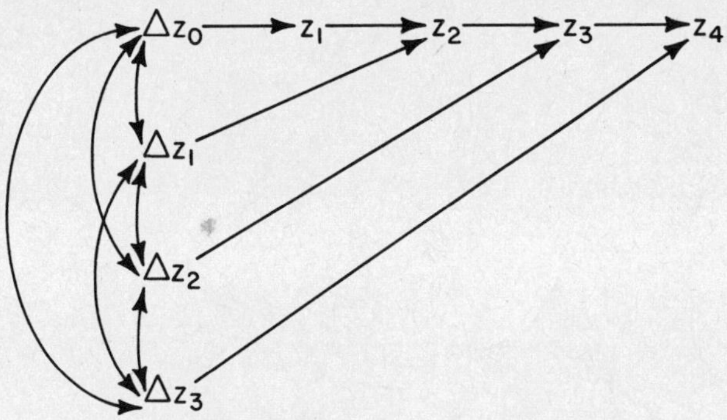

Fig. 6. Hierarchical path diagram for a growth trajectory when measurements (z_1, z_2, z_3, z_4) have been taken at four ages. $\Delta z_i = \Delta z_{i+1} - \Delta z_i$, and $\Delta z_0 = z_1$

$$z_1{:}\quad 1 = p^2(\Delta z_0, z_1) \ ,$$

$$z_2{:}\quad 1 = [p(\Delta z_0, z_1)\,p(z_1, z_2)]^2 + p^2(\Delta z_1, z_2) + 2r(\Delta z_0, \Delta z_1)\,p(\Delta z_0, z_1)\,p(z_1, z_2)\,p(\Delta z_1, z_2) \ ,$$

$$\begin{aligned} z_3{:}\quad 1 = {} & [p(\Delta z_0, z_1)\,p(z_1, z_2)\,p(z_2, z_3)]^2 + [p(\Delta z_1, z_2)\,p(z_2, z_3)]^2 + p^2(\Delta z_2, z_3) \\ & + 2r(\Delta z_0, \Delta z_1)\,p(\Delta z_0, z_1)\,p(z_1, z_2)\,p^2(z_2, z_3)\,p(\Delta z_1, z_2) \\ & + 2r(\Delta z_1, \Delta z_2)\,p(\Delta z_1, z_2)\,p(z_2, z_3)\,p(\Delta z_2, z_3) \\ & + 2r(\Delta z_0, \Delta z_2)\,p(\Delta z_0, z_1)\,p(z_1, z_2)\,p(z_2, z_3)\,p(\Delta z_2, z_3) \ . \end{aligned}$$

Finally, it should be noted that although the emphasis of this paper has been on the proximal causes of variation in the growth of individuals, the variances and covariances of the growth components are likely to be products of natural selection. Compensatory growth, for example, may in some cases be an evolutionary consequence of stabilizing selection on adult size. If size-selective mortality were much less intense on juveniles than adults, then genes for juvenile development that caused rapid growth early and slow growth later would be selectively equivalent to those with the reciprocal effects provided the same size at maturity was attained. The failure of selection to discriminate between such genes would result in the maintenance of a polymorphism which would be revealed as a negative genetic correlation between adjacent juvenile growth increments (Lande 1980, 1982). This appears not to be true for *Daphnia*, which exhibit positive genetic correlations between growth components (Fig. 5).

Elsewhere in this Volume (Lynch and Arnold this Vol.), we present a methodology that can be used to measure the intensities of directional and stabilizing selection operating on size and growth components in natural populations. Those methods, in combination with the procedures presented in this paper, should assist us in arriving at an evolutionary interpretation of within- and between-species variation in growth trajectories.

Acknowledgments. I thank K. Spitze and A. Toline for providing the data that form the basis of some of the analyses presented in this paper, and S. Arnold and M. Kirkpatrick for helpful comments. The work was supported by National Science Foundation grants BSR 83-06072, 86-00476 and SUB U MICHX98674.

5 References

Anderson TW (1984) An introduction to multivariate statistical analysis, 2nd edn. John Wiley & Sons, New York

Arnold SJ (1983) Morphology, performance and fitness. Am Zool 23:347–361

Atchley WR (1984) Ontogeny, timing of development, and genetic variance-covariance structure. Am Nat 123:519–540

Atchley WR, Rutledge JJ (1980) Genetic components of size and shape. I. Dynamics of components of phenotypic variability and covariability during ontogeny in the laboratory rat. Evolution 34:1161–1173

Bertalanffy L von (1957) Quantitative laws in metabolism and growth. Q Rev Biol 32:217–231

Beverton RJH, Holt SJ (1957) On the dynamics of exploited fish populations. Min Agric Fish Food, Fish Invest Ser 2, 19:1–533

Bullmer MG (1980) The mathematical theory of quantitative genetics. Univ Press, Oxford

Cheverud JM (1984) Evolution by kin selection: a quantitative genetic model illustrated by maternal performance in mice. Evolution 38:766–777

Cheverud JM, Leamy LJ, Atchley WR, Rutledge JJ (1983a) Quantitative genetics and the evolution of ontogeny. I. Ontogenetic changes in quantitative genetic variance components in randombred mice. Genet Res 42:65–75

Cheverud JM, Rutledge JJ, Atchley WR (1983b) Quantitative genetics of development: genetic correlations among age-specific trait values and the evolution of ontogeny. Evolution 37:895–905

Cloninger CR, Rao DC, Rice J, Reich T, Morton NE (1983) A defense of path analysis in genetic epidemiology. Am J Human Genet 35:733–756

Cock AG (1966) Genetical aspects of metrical growth and form in animals. Q Rev Biol 41:131–190

Doyle RW, Myers RA (1982) The measurement of the direct and indirect intensities of natural selection. In: Dingle H, Hegmann JP (eds) Evolution and genetics of life histories. Springer, Berlin Heidelberg New York, pp 157–176

Fabens AJ (1965) Properties and fitting of the von Bertalanffy growth curve. Growth 29:265–289

Falconer DS (1981) Introduction to quantitative genetics, 2nd edn. Longman, London New York

Falconer DS, Gauld IK, Roberts RC (1978) Cell numbers and cell sizes in organs of mice selected for large and small body size. Genet Res 31:287–301

Gould SJ (1977) Ontogeny and phylogeny. Harvard Univ Press, Cambridge, Mass

Hallauer AR, Miranda JB (1981) Quantitative genetics in maize breeding. Iowa State Univ Press, Ames

Hazel LN, Baker ML, Reinmiller CF (1943) Genetic and environmental correlations between the growth rate of pigs at different ages. J Anim Sci 2:118–128

Herbert JG, Kidwell JF, Chase HB (1979) The inheritance of growth and form in the mouse. IV. Changes in the variance components of weight, tail length and tail width during growth. Growth 43:36–46

Huxley JS (1932) Problems of relative growth. Methuen, London

Karlin S, Cameron EC, Chakraborty R (1983) Path analysis in genetic epidemiology: a critique. Am J Human Genet 35:695–732

Lande R (1980) The genetic covariance between characters maintained by pleiotropic mutations. Genetics 94:203–215

Lande R (1982) A quantitative genetic theory of life history evolution. Ecology 63:607–615

Lande R (1985) Genetic and evolutionary aspects of allometry. In: Jungers WL (ed) Size and scaling in primate biology. Plenum, New York, pp 21–32

Leamy L, Cheverud JM (1984) Quantitative genetics and the evolution of ontogeny. II. Genetic and environmental correlations among age-specific characters in randombred house mice. Growth 48:339–353

Li CC (1975) Path analysis: a primer. Boxwood, Pacific Grove, Cal

Lynch M (1985) Spontaneous mutations for life history characters in an obligate parthenogen. Evolution 39:804–818

Lynch M, Weider LJ, Lampert W (1986) Measurement of the carbon balance in *Daphnia.* Limnol Oceanogr 31:17–33

Majkowski J, Uchmanski J (1980) Theoretical foundations of individual growth equations in animals. Pol Ecol Stud 6:7–31

Monteiro LS, Falconer DS (1966) Compensatory growth and sexual maturity in mice. Anim Prod 8:179–192
Moore RW, Eisen EJ, Ulberg LC (1970) Prenatal and postnatal maternal influences on growth in mice selected for body weight. Genetics 64:59–68
Nair KR (1954) The fitting of growth curves. In: Kempthorne O, Bancroft TA, Gowen JW, Lush JL (eds) Statistics and mathematics in biology. Iowa State Univ Press, Ames, pp 119–132
Niles HE (1922) Correlation, causation and Wright's theory of path coefficients. Genetics 7:258–273
Paloheimo JE, Dickie LM (1965) Food and growth of fishes. I. A growth curve derived from experimental data. J Fish Res Board Can 22:521–542
Parks JR (1970) Growth curves and the physiology of growth. I. Animals. Am J Physiol 219:833–836
Pruitt KM, DeMuth RE, Turner ME Jr (1979) Practical application of generic growth theory and the signficance of the growth curve parameters. Growth 43:19–35
Richards FJ (1959) A flexible growth function for empirical use. J Exp Bot 10:290–300
Ricklefs RE (1967) A graphical method of fitting equations to growth curves. Ecology 48:978–983
Riska B, Atchley WR (1985) Genetics of growth predict patterns of brain-size evolution. Science 229:668–671
Riska B, Atchley WR, Rutledge JJ (1984) A genetic analysis of targeted growth in mice. Genetics 107:79–101
Rutledge JJ, Robinson OW, Eisen EJ, Legates JE (1972) Dynamics of genetic and maternal effects in mice. J Anim Sci 35:911–918
Thiele TN (1931) Theory of observations. Ann Math Stat 2:165–308
Thomas CH, Blow WL, Cockerham CC, Glazener EW (1968) The heritability of body weight, gain, and feed conversion in broilers. Poult Sci 37:862–869
Thompson D'A (1942) On growth and form, 2nd edn. Cambridge Univ Press, London
Toline CA (1986) Mutational divergence of life-history traits in an obligate parthenogen. MS Thesis Biol, Univ Ill, Champaign
Wright S (1921a) Systems of mating. I. The biometric relations between parents and offspring. Genetics 6:111–123
Wright S (1921b) Correlation and causation. J Agric Res 20:557–585
Wright S (1923) The theory of path coefficients: a reply to Niles' criticism. Genetics 8:239–255
Wright S (1934) The method of path coefficients. Ann Math Stat 5:161–215
Wright S (1968) Evolution and the genetics of populations, vol 1. Genetic and biometric foundations. Univ Press, Chicago, Ill
Wright S (1983) On "Path analysis in genetic epidemiology: a critique." Am J Human Genetics 35:757–768
Wright S (1984) Diverse uses of path analysis. In: Chakravarti A (ed) Human population genetics: the Pittsburgh symposium. VanNostrand Reinhold, New York, pp 1–34
Young CW, Legates JE (1965) Genotypic, phenotypic, and maternal interrelationships of growth in mice. Genetics 52:563–576

6 Appendix

When a dependent variable z_t can be expressed as the sum of t parts $(\Delta z_0, \ldots, \Delta z_{t-1})$, the contribution of each of the t^2 paths of causation can be represented by the ratio V_{ij}/V_t where V_{ij} is the measured covariance between components i and j (if $i = j, V_{ii}$ is the variance of component i) and V_t is the measured variance of the dependent variable. By Taylor expansion, the sampling variance of the path contribution is found to be

$$\mathrm{Var}(V_{ij}/V_t) \simeq \sigma_t^{-4}\{\mathrm{Var}(V_{ij}) + (\sigma_{ij}/\sigma_t^2)^2 \mathrm{Var}(V_t) - 2(\sigma_{ij}/\sigma_t^2)\mathrm{Cov}(V_{ij}, V_t)\} \;, \quad \text{(A 1)}$$

where σ_t^2 and σ_{ij} represent expected variances and covariances. Exact expressions for the variance terms are:

$$\mathrm{Var}(V_t) = \frac{1}{N}\,E[(z_t - \bar{z}_t)^4] - \frac{N-3}{N(N-1)}\,\sigma_t^4\ ,$$

$$\mathrm{Var}(V_{ij}) = \frac{1}{N}\,E[(\Delta z_i - \bar{\Delta} z_i)^2(\overline{\Delta z}_j - \bar{\Delta} z_j)^2] - \frac{N-3}{N(N-1)}\,\sigma_{ij}^2\ ,$$

where N is the sample size and $\bar{z}_t$ is the sample mean. Similarly, for the covariance term,

$$\mathrm{Cov}(V_{ij}, V_t) = \frac{1}{N}\,E[(\Delta z_i - \bar{\Delta} z_i)(\overline{\Delta z}_j - \overline{\Delta z}_j)(z_t - \bar{z}_t)^2] - \frac{N-3}{N(N-1)}\,\sigma_{ij}\sigma_t^2\ .$$

If the components are multivariate normally distributed, then

$$E[(z_t - \bar{z}_t)^4] = 3\eta\sigma_t^4\ ,$$

$$E[(\Delta z_i - \overline{\Delta z}_i)^2(\Delta z_j - \overline{\Delta z}_j)^2] = \eta(\sigma_{ii}\sigma_{jj} + 2\sigma_{ij}^2)$$

$$E[(\Delta z_i - \overline{\Delta z}_i(\Delta z_j - \overline{\Delta z}_j)(z_t - \bar{z}_t)^2] = \eta(\sigma_{ij}\sigma_t^2 + 2\sigma_{it}\sigma_{jt})\ ,$$

where $\eta = (N-1)(N^2 - 8N + 6)/N^3$ (Thiele 1931; Anderson 1984). If N is large, then $\eta \simeq 1$ ($N = 50$, $\eta = 0.83$; $N = 200$, $\eta = 0.96$) and the previous expressions simplify to

$$\mathrm{Var}(V_t) \simeq 2\sigma_t^4/(N-1)\ ,$$

$$\mathrm{Var}(V_{ij}) \simeq \frac{\sigma_{ii}\sigma_{jj}}{N} + \frac{N+1}{N(N-1)}\,\sigma_{ij}^2\ ,$$

$$\mathrm{Cov}(V_{ij}, V_t) \simeq \frac{2\sigma_{it}\sigma_{jt}}{N} + \frac{2\sigma_{ij}\sigma_t^2}{N(N-1)}\ .$$

Approximate sampling variances for the path contributions are obtained by inserting the appropriate expressions into Eq. (A1) and then substituting the observerd variances and covariances for the expectations σ_t^2, σ_{ij}, and σ_{it}.

The Measurement of Selection on Size and Growth

M. LYNCH[1] and S. J. ARNOLD[2]

1 Introduction

A good deal of interest in natural selection is focused on the size of individuals or individual parts. For example, the significance of size-selective predation has long been a dominant theme in research on zooplankton ecology (Brooks and Dodson 1965; Kerfoot 1980; Lynch 1980). Much of the theory on the evolution of complex life histories is focused on the interaction of size-specific growth and mortality rates (Wilbur 1980; Werner and Gilliam 1984; Werner 1986 and this Vol.). Size-dependent competition and reproductive performance are central issues in plant population biology (Harper 1977; Dirzo and Sarukhán 1984; see also Ebenman this Vol.). In order to couch studies of these phenomena in an evolutionary framework, techniques are required for analyzing the intensity of natural selection on components of size and growth. A methodology for the measurement of selection on age-invariant characters (e.g., size at age x) has been outlined in Lande and Arnold (1983) and Mitchell-Olds and Shaw (1987), but selection on ontogenetic patterns raises some practical difficulties that were not addressed in those papers.

A conceptual difficulty arises in the analysis of selection on size. Because the size at any age is a sum of growth increments during previous age intervals, the question can arise as to whether selection is operating on size per se or on growth rate. From the standpoint of evolutionary interpretation, this is not a trivial circularity, since different agents of selection may operate on size and growth rate. For example, given a population of available prey, a predator will often select prey items solely on the basis of current size regardless of their previous rates of growth. On the other hand, individuals of identical size but different growth rates may vary in their vulnerability to predation if variation in growth rate is a consequence of selection for habitats that vary in resource level as well as in predation risk. Since selection may often act simultaneously on size and growth rate, a method is needed for measuring the direct forces of selection on both types of characters.

As an example of an evolutionary phenomenon whose analysis requires a quantification of the forces of selection operating on size and growth rate, consider compensatory growth in which individuals tend to converge on the same mature size despite differences in size at birth and early growth rate (Monteiro and

[1] Department of Ecology, Ethology, and Evolution, Shelford Vivarium, University of Illinois, 606 E. Healey St., Champaign, IL 61820, USA
[2] Department of Biology, The University of Chicago, 940 East 57th St., Chicago, IL 60637, USA

Size-Structured Populations
B. Ebenman and L. Persson (Eds.)

Falconer 1966; Atchley 1984; Riska et al. 1984). Such a growth pattern requires that weight gains during some successive growth intervals exhibit negative phenotypic correlations (Lynch this Vol.). Perhaps the simplest hypothesis that can account for the evolution of compensatory growth is strong stabilizing selection on size at sexual maturity with weaker selection on size at earlier ages. Alternatively, correlational selection may favor a negative tradeoff between growth in adjacent age intervals with little selection operating directly on size. Multivariate selection gradients, a subject of this chapter, provide the relevant parameter estimates for testing these and other hypotheses.

Growth components are fundamentally different from sets of morphological traits that are expressed simultaneously. An individual reveals each growth component only during a single interval of time and only if it lives through that interval. Thus, mortality during the period of study will result in individuals with incomplete data. The data set becomes progressively smaller for traits expressed later in life. This raises a technical problem. To properly correct for the effects of correlation among traits when measuring selection, we need to know the phenotypic covariance matrix for the traits before selection (Pearson 1903; Lande and Arnold 1983). However, we can only estimate a complete matrix for the subset of individuals that survived until the last measurement. If selection has operated on growth rate, size, or any other correlated trait, the latter matrix will be a biased estimate of the former.

In the following sections, we show how estimates of the phenotypic covariance matrices that prevailed before selection can be obtained. This is accomplished in a stepwise fashion by applying information on the strength and direction of selection for individual growth intervals. The age-specific parameter estimates can then be used to estimate the total forces of directional, stabilizing, and correlational selection operating on size and growth characters. Some simple algebraic relationships between selection on size and growth will be pointed out, and path analysis will be shown to provide a means of partitioning the variance of fitness into components attributable to variation in age-specific size and/or growth rate.

2 Conditional Selection Differentials and Gradients

As in Lynch (this Vol.), we express the size of an individual at time t, z_t, as the sum of an initial size at birth and a series of growth increments between various measurements,

$$z_t = z_0 + (z_1 - z_0) + (z_2 - z_1) + \ldots + (z_t - z_{t-1}) = z_0 + \sum_{i=1}^{t} \Delta z_i \,, \tag{1}$$

where z_0 is the size at the first census. We start with a cohort of same-aged individuals. At each of a series of censuses, we record whether each individual is alive or dead and measure the live ones. Figure 1 provides an example of the type of growth data that may arise from such a longitudinal study on a hypothetical cohort of 15 individuals. Size is measured on all survivors at three points (i = 0, 1, 2) in time. A fourth census is necessary to evaluate the survival of individuals

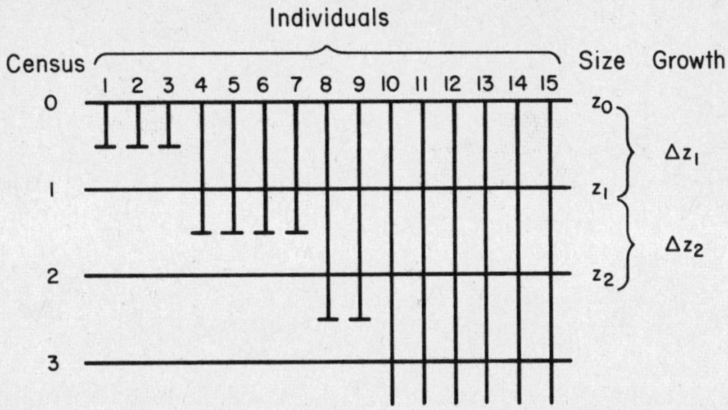

Fig. 1. A schematic for a longitudinal analysis of size and growth data on 15 hypothetical individuals. *Cross-hatches* denote times of death. Measurements are made at censuses 0, 1, and 2, allowing the computation of three size and two growth measures. Measurements are not necessarily taken at census 3, but this census is required to evaluate survivorship of the individuals that were alive at census 2. A later census could be taken to evaluate total lifetime fitness of individuals that survived to census 3

expressing all five size and growth characters, and a later survey may be taken to evaluate reproductive performance. Since three individuals died prior to the second census, size at birth is the only character available for the initial cohort of 15 individuals. The reduced cohort of 12 individuals alive at census 1 can be characterized by z_0, z_1, and Δz_1. The remaining 8 individuals at census 2 (6 of which have nonzero fitness) are fully characterized for traits z_0, z_1, z_2, Δz_1, and Δz_2. The individuals alive at the ith census are assigned a conditional relative fitness measure w(i) of 0 or 1/p(i), where p(i) is the fraction of the cohort that survives to i+1, depending on whether they died or survived in the interval (i, i+1).

For each of the three censuses, a conditional phenotypic covariance matrix **P**(i) can be constructed for the subset of individuals alive at that census. The diagonal elements of this matrix are variances, while the off-diagonal elements are covariances between characters. Due to the expression of new characters at each census, the dimensionality of this matrix will increase with census number. For example, **P**(0) consists of a single element, the variance of z_0 for all of the members of the initial cohort. **P**(1) contains additional variances and covariances. Since these are computed from the subset of individuals that survived to census 1, the variance of z_0 in **P**(1) is likely to differ from that in **P**(0) as well as that in **P**(2) and later matrices.

In reality, the complete set of size and growth characters cannot be analyzed simultaneously. Because the growth increments are computed from sequential sizes, inclusion of all of the available size and growth characters in **P**(i) would cause this matrix to be singular. This is a problem since the analysis of multivariate selection requires that the conditional covariance matrices be invertible. The choice of which characters to include will depend somewhat on the objectives of the analysis, but the elimination of the singularity problem requires that none of the characters employed in the analysis be linear functions of any of the other characters included. This means that with n censuses, no more than n size and/or growth characters can be employed in an analysis. For example, if size were measured on two occasions, a selection analysis could be performed on z_0 and z_1 or on z_0 and Δz_1 or on z_1 and Δz_1, but not on z_0, z_1, and Δz_1 since $\Delta z_1 = z_1 - z_0$. When measurements are taken on three occasions, the maximum number of characters that can be analyzed at any one time is three, but there are seven combinations of characters that can be analyzed: (z_0, z_1, z_2), (z_0, z_1, Δz_2), (z_0, z_2, Δz_1), (z_0, z_2, Δz_2), (z_0, Δz_1, Δz_2), (z_1, Δz_1, Δz_2), and (z_2, Δz_1, Δz_2).

For each census, it is also possible to calculate a conditional directional selection differential **S**(i). This differential is conditional in the sense that it is based only on those individuals alive at census i. Each element of the vector **S**(i) represents a different character and is the difference in mean phenotype (at time i) between individuals that survived to census i+1 and those that were present at census i. For example, the conditional directional selection differential for birth weight at census 1, is the difference in birth weight of individuals alive at census 2 and those alive at census 1. As in the case of **P**(i), the dimensionality of **S**(i) will usually increase with census number, as one has the opportunity to include additional characters.

The conditional directional selection gradient for each interval of time is the vector $\boldsymbol{\beta}(i) = \mathbf{P}^{-1}(i) \cdot \mathbf{S}(i)$. Since the elements of **S**(i) are equivalent to the covariance of relative fitness and phenotypic value (Robertson 1966), this formula is equivalent to the standard least-squares solution to a multiple regression, in this case of w(i) on the phenotypic traits measured at census i. Each element of $\boldsymbol{\beta}(i)$ is a partial regression coefficient, providing a measure of the slope of the regression of relative fitness on the respective character when all other characters are held constant. Provided that the conditional phenotype distribution is multivariate normal at census i, the elements of $\boldsymbol{\beta}(i)$ can be interpreted as the average gradient of the relative fitness surface weighted by the phenotype distribution (Lande and Arnold 1983). The preservation of multivariate normality throughout the period of study requires that the initial phenotype distribution be multivariate normal and that all subsequent selection functions be Gaussian in form, a rather unlikely situation. It should be noted, however, that the actual analysis of data is not contingent upon multivariate normality.

By computing all of the elements of $\boldsymbol{\beta}(i)$ at each census, we allow for the possibility that age-specific size and growth traits, although expressed only once, may influence fitness at all future ages. This will be important if unobserved characters that are acted upon by natural selection are correlated with earlier size and or growth rate. It is conceivable that the conditional selection gradients computed for the same character at different censuses may vary substantially in magnitude and/or direction.

A conditional stabilizing selection differential **C**(i) can also be calculated for each census. This is a square matrix, each element being the covariance of relative fitness and the cross-product of two traits measured as deviations from the mean phenotype at census i. For example, the diagonal element involving size at birth for individuals alive at census 1 is $\mathrm{Cov}[w(1),[z_0 - \bar{z}_0(1)]^2]$, where $\bar{z}_0(1)$ is the mean size at birth for the subset of individuals alive at census 1. For the population in Fig. 1, 12 individuals would be involved in this computation, four (individuals 4–7) with $w(1) = 0$ and eight (individuals 8–15) with $w(1) = 1/(8/12) = 1.5$. For the same census, the off-diagonal element involving size at birth and growth during the first increment is $\mathrm{Cov}[w(1), \{[z_0 - \bar{z}_0(1)][\Delta z_1 - \overline{\Delta z_1(1)}]\}]$. Each element of **C**(i) is an estimate of the change in the variance (diagonal element) or covariance (off-diagonal element) of characters resulting from selection. Using **C**(i) we can now compute the conditional stabilizing selection gradient, $\gamma(i) = \mathbf{P}^{-1}(i)\mathbf{C}(i) \times \mathbf{P}^{-1}(i)$ [cf. Lande and Arnold 1983, Eq. (14a)].

The elements of $\gamma(i)$ describe the average curvature of the conditional fitness surface weighted by the phenotype distribution. Positive and negative diagonal elements in $\gamma(i)$ indicate disruptive and stabilizing selection on the respective characters, whereas the signs of the off-diagonal elements indicate whether pairs of characters are being selected to become positively or negatively correlated. A detailed discussion of the relationship between these coefficients and the topography of the fitness surface is provided in Phillips and Arnold (in prep.).

3 Reconstruction of the Phenotypic Covariance Matrix

While the conditional selection gradients introduced in the previous section describe specific episodes of selection operating on a cohort, it will usually be of interest to quantify the total forces of selection over the entire period of study. This requires an estimate of the phenotypic covariance matrix for all of the characters prior to selection. Only at the final census do we have an estimate of the complete covariance matrix, and this will be a biased estimate of the matrix that existed before selection. Nevertheless, in conjunction with the conditional selection gradients, the final phenotypic covariance matrix can be used to derive an estimate of the initial matrix.

Let us now represent the full covariance matrix for individuals alive at census i as $\mathbf{P}_i$, with $\mathbf{P}_n = \mathbf{P}(n)$. (From here on, conditional matrices will be denoted with parentheses while complete matrices will be subscripted.) Each of these matrices has dimensionality $(n+1)\times(n+1)$. However, for the ith matrix, only $(i+1)\times(i+1)$ of the elements are known; these are the elements that were included in **P**(i). At each census, the complete directional selection gradient, $\boldsymbol{\beta}_i$, is simply the vector $\boldsymbol{\beta}(i)$ with zeroes added to yield the desired dimensionality of $(i+1)\times 1$. Such treatment entails the assumption that selection does not operate on growth or size characters prior to their expression. By similar reasoning, the complete stabilizing selection gradient at census i, $\mathbf{C}_i$, is simply **C**(i) with zeroes filled in for the missing elements.

Estimates of the missing elements of **P**(i) are obtained by noting that the within-generation dynamics of the elements of the phenotypic covariance matrix are defined completely by the directional and stabilizing selection gradients [Lande and Arnold 1983, Eq. (13)],

$$\mathbf{P}_i = \mathbf{P}_{i-1} + \mathbf{P}_{i-1}\gamma_{i-1}\mathbf{P}_{i-1} - \mathbf{P}_{i-1}\boldsymbol{\beta}_{i-1}[\mathbf{P}_{i-1}\boldsymbol{\beta}_{i-1}]^T , \quad (2)$$

where T denotes transpose. Since $\mathbf{P}_i$ is a symmetrical matrix, Eq. (2) amounts to a set of $(n+1)(n+2)/2$ equations with the unknowns being the missing elements of $\mathbf{P}_{i-1}$. Starting with $i = n$, the first step is to solve for the $n+1$ missing elements of $\mathbf{P}_{n-1}$; these are the elements that occupy the last row (and column) of $\mathbf{P}_{n-1}$. The elements of $\mathbf{P}_n$ that are also observed directly in $\mathbf{P}_{n-1}$ are defined solely in terms of the latter and are of no use in this solution. Thus, filling out the missing elements of $\mathbf{P}_{n-1}$ involves the solution of n+1 equations in n+1 unknowns.

If it is necessary to back-track a further step to $\mathbf{P}_{n-2}$, then the newly completed matrix $\mathbf{P}_{n-1}$ is substituted for $\mathbf{P}_i$ in Eq. (2), and the entire procedure is repeated with $\mathbf{P}_{n-2}$, β_{n-2}, and γ_{n-2} on the right. This analysis would involve the solution of 2n+1 equations in 2n+1 unknowns (the last two rows (and columns) of $\mathbf{P}_{n-2}$). By this means, information on the conditional selection gradients allows a progressive extrapolation of $\mathbf{P}_n$ back to $\mathbf{P}_0$. At each (xth) step in the process there are $\sum_{i=1}^{x} (n+2-i)$ equations to solve in the same number of unknowns. This procedure involves no assumptions regarding the form of the phenotype distribution or of the selection function.

4 Total Selection on Growth and Size Caused by Selective Mortality

Having reconstructed the phenotypic covariance matrix throughout the period of study, the stage-specific conditional selection gradients can be assembled into composite measures describing the total forces of selection operating on the cohort. It was necessary at the outset to break the analysis down into intervals of selection to account for the fact that even those individuals that die before expressing a character are relevant to the evolution of that character if it is correlated with size and growth components that have been expressed and selected upon. The total changes in the means and variances of size and growth attributes caused by selection can be partitioned into components describing the effects of direct selection on the characters and those resulting indirectly from selection on phenotypically correlated characters. Letting l(i) be the proportion of the initial cohort that survives to the ith census, the total directional selection differential for the jth character is, using Eq. (6c) from Lande and Arnold (1983),

$$S_j = \sum_{i=0}^{n-1} l(i)[\beta_j(i)\sigma_{jj}(i) + \sum_{\substack{k=1 \\ k\neq j}}^{n+1} \beta_k(i)\sigma_{jk}(i)] \ , \tag{3}$$

where $\beta_j(i)$ is the conditional directional selection gradient on the jth trait over the period (i, i+1) and $\sigma_{jk}(i)$ is the jk^{th} element of $\mathbf{P}_i$. The first set of terms on the right is the sum of direct effects of selection acting on trait j over the various census periods, while the second set of terms is the sum of indirect effects of selection acting on correlated traits. Letting $\mathbf{S}$ be the vector of the estimates of S_j, the total directional selection gradient caused by mortality is $\boldsymbol{\beta} = \mathbf{P}_0^{-1}\mathbf{S}$.

Similarly, the total correlational selection differential for the jth and k^{th} characters is, using equation 14a from Lande and Arnold (1983),

$$C_{jk} = \sum_{i=0}^{n-1} l(i)\left[\sum_{k=1}^{n+1}\sum_{l=1}^{n+1}\sum_{m=1}^{n+1} \sigma_{jk}(i)\,\gamma_{kl}(i)\,\sigma_{jm}(i)\right] \ , \tag{4}$$

where $\gamma_{kl}(i)$ is the conditional stabilizing selection gradient for the kth trait (k = l) or the conditional correlational selection gradient for traits k and l (k ≠ l).

One of the terms in this expression represents the change in covariance of traits j and k due to selection acting directly on the trait product $[z_j - \bar{z}_j(i)][z_k - \bar{z}_k(i)]$, while the remaining sum of terms represents indirect effects on the covariance of traits j and k due to stabilizing and correlational selection on correlated traits. When $j = k$, C_{jk} is the stabilizing selection differential for the jth trait. Letting $\mathbf{C}$ be the matrix of estimates of C_{jk}, the total stabilizing selection gradient due to mortality selection is $\gamma = \mathbf{P}_0^{-1}\mathbf{C}\mathbf{P}_0^{-1}$, with stabilizing selection gradients on the main diagonal and correlational selection gradients elsewhere.

5 Selection on Size Versus Growth Rate

The problem of whether natural selection operates on size per se and/or growth rate is a fundamental issue that has attracted little attention from field biologists. In the previous sections, we have developed a general method that can be applied to the entire lifespan of a cohort. We now point out a useful simplification for evaluating the selective pressures operating on average size and growth rate within any interval.

Suppose that a cohort of individuals is measured on two occasions (censuses 0 and 1) and a third survey is taken to identify the survivors among those individuals that lived to census 1. There are two ways to parameterize the data. The procedures outlined in the preceding sections can be applied directly to the size characters z_0 and z_1 to obtain estimates of the elements of the directional selection gradient, β_0 and β_1, as well as of the elements of the stabilizing selection gradient, γ_{00}, γ_{01}, and γ_{11}. Alternatively, the data can be transformed to yield $z_s = (z_0 + z_1)/2$ as a measure of size and $z_g = z_1 - z_0$ as a measure of growth rate. These transformed variables can be used to obtain the selection coefficients β_s, β_g, γ_{ss}, γ_{sg}, and γ_{gg}, where the subscripts s and g denote size and growth rate respectively.

Recalling that the selection gradients are equivalent to the partial regression coefficients of relative fitness (w) on phenotype (Lande and Arnold 1983), we have the alternative expressions

$$E(w) = \alpha + \beta_0 z_0 + \beta_1 z_1 + \tfrac{1}{2}\gamma_{00} z_0^2 + \tfrac{1}{2}\gamma_{11} z_1^2 + \gamma_{01} z_0 z_1 \ , \tag{5a}$$

$$E(w) = \alpha + \beta_s z_s + \beta_g z_g + \tfrac{1}{2}\gamma_{ss} z_s^2 + \tfrac{1}{2}\gamma_{gg} z_g^2 + \gamma_{sg} z_s z_g \ , \tag{5b}$$

where E denotes expectation. The intercept α is obtained by substituting 1 for E(w) and the means for the other parameters, e.g., $\bar{z}_0$ for z_0, and $\overline{z_0 z_1}$ for $z_0 z_1$. These formulae yield the relationships

$$\begin{aligned}
\beta_s &= \beta_0 + \beta_1 \ , \\
\beta_g &= \tfrac{1}{2}(\beta_1 - \beta_0) \ , \\
\gamma_{ss} &= \tfrac{1}{2}\gamma_{00} + \gamma_{01} + \tfrac{1}{2}\gamma_{11} \ , \\
\gamma_{sg} &= \tfrac{1}{2}(\gamma_{11} - \gamma_{00}) \ , \\
\gamma_{gg} &= \tfrac{1}{2}\gamma_{00} - \gamma_{01} + \tfrac{1}{2}\gamma_{11} \ .
\end{aligned}$$

Thus, the force of directional selection on average size is simply the sum of the forces of directional selection acting on size at the two ages. The force of directional selection on growth rate is one half the difference between the force of directional selection at the second and first ages. Positive selection for growth rate is equivalent to greater directional selection for size at the second age. In other words, there is no directional selection on the rate of growth when the directional selection gradients for size at the two ages are equal. Similar logic can be used to explain the existence of stabilizing, disruptive, or correlational selection on mean size and growth rate.

By use of Eq. (5a) [or equivalently (5b)], the selection gradients can be used to portray a fitness surface that describes the relationship between relative fitness, size, and growth rate. Starting with a bivariate surface in z_0 and z_1, the axes for mean size and growth rate are obtained by a 45° rotation of the original axes (Fig. 2). The topographical features of the fitness surface can be described by a series of contours connecting character combinations with equal fitness (Fig. 3).

We first consider some simple situations in which there is no stabilizing, disruptive, or correlational selection ($\gamma_{00} = \gamma_{01} = \gamma_{11} = 0$). Under these circumstances the selection surface is a plane whose orientation depends on the magnitudes of β_0 and β_1 (or equivalently β_s and β_g). Figure 3a shows selection for increased size at age 0 but no selection on size at age 1. When viewed on the z_s, z_g coordinate system, it is seen that such a fitness surface corresponds to selection for an increase in average size but a decrease in the growth rate in the interval (0, 1). Figure 3b illustrates the fitness surface when there is selection for increased size at the second age but no selection on size at the first age. This corresponds to selection for increased average size and higher growth rate. In Fig. 3c,

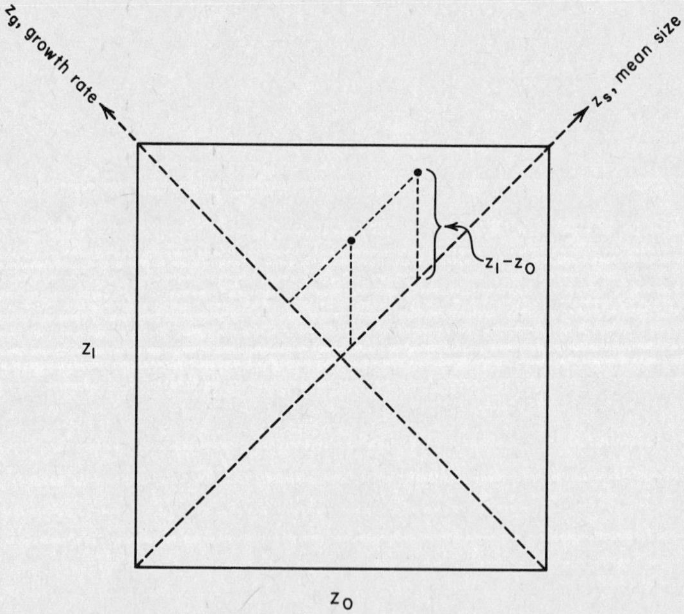

Fig. 2. Alternative coordinate systems for age-specific sizes (*solid lines*) and mean size and growth rate (*dashed lines*) The two points refer to individuals with different age-specific sizes (z_0, z_1) but the same growth rate ($z_1 - z_0$)

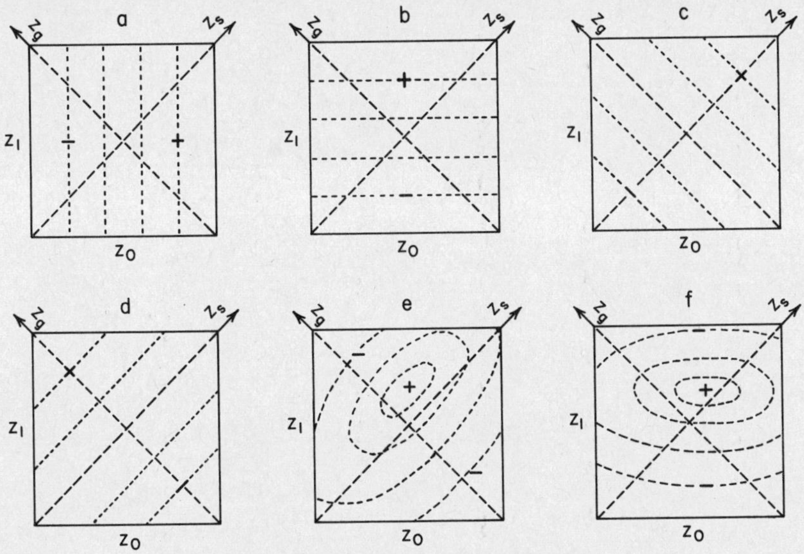

Fig. 3a–f. Selection surfaces (*dotted lines*) on the age-specific size (*solid lines*) and mean size-growth rate (*dashed lines*) coordinate systems. The gradient of fitness is uphill from contours marked with − to those marked with +. **a** $\beta_0 > 0$, $\beta_1 = 0$. **b** $\beta_0 = 0$, $\beta_1 > 0$. **c** $\beta_0 = \beta_1 > 0$. **d** $\beta_0 = -\beta_1 < 0$. **e** Selection towards intermediate optimum sizes at both ages; the axes of the selection surface are aligned with the axes of the mean size-growth rate coordinate system. **f** Selection towards intermediate optimum sizes at both ages; the axes of the selection surface are aligned with the axes of the age-specific size coordinate system

selection favors increased size at both ages. However, because selection for size is equally strong at both ages, there is no selection on the rate of growth. Figure 3d illustrates the opposite case, in which selection favoring increased size at age 1 is balanced by an equal intensity of selection for decreased size at age 0. This is equivalent to selection for high growth rate with no selection on mean size.

When the stabilizing and/or correlational selection gradients are nonzero, the fitness surface will contain a peak, a basin, a saddle, or a ridge (Phillips and Arnold in prep.). We will only consider two simple situations in which there is selection for an intermediate optimum size at both ages. In the first case (Fig. 3e), there is no correlational selection for growth and size ($\gamma_{sg} = 0$), but there is stronger stabilizing selection on growth rate than on mean size ($\gamma_{gg} < \gamma_{ss} < 0$). This disparity in the strengths of stabilizing selection operating on z_s and z_g translates into positive correlational selection between the sizes at both ages. In the second case (Fig. 3f), there is no correlational selection between size at the two ages, but stabilizing selection on size at the second age is stronger than that at the first age. This is equivalent to negative correlational selection between average size and growth rate. In other words, selection promotes a negative correlation between average size and growth rate.

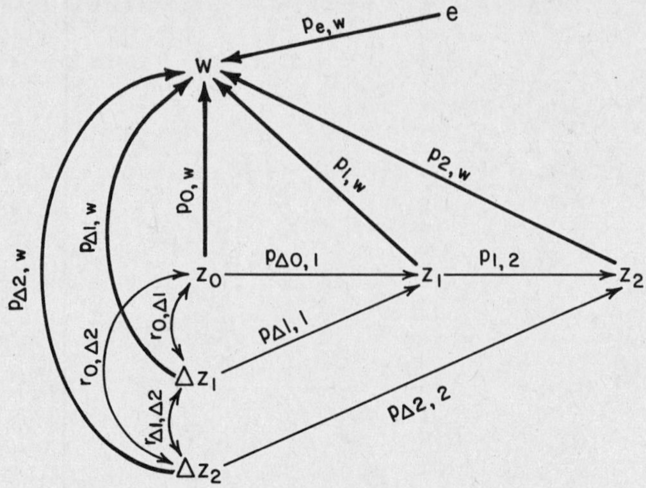

Fig. 4. A path diagram showing the effects of directional selection on relative fitness, w. The figure shows only the size, z_i, and growth variables, Δz_i, that could be computed by taking measurements at three ages (including at birth), as in Fig. 1. Correlations between growth variables are indicated with *narrow, double-headed arrows.* The path from e represents residual (unexplained) effects on fitness. Definitions of the remaining path coefficients are given in the text

6 Path Analysis of the Variance in Fitness Attributable to Size and Growth

The techniques outlined above are readily merged with the path analytical techniques described in Lynch (this Vol.) to evaluate the fractions of the variance in fitness that are associated with direct selection on size and growth components and those that are associated with correlations between characters. For clarity, just the directional effects of selection are summarized in Fig. 4. The squared path coefficients on the arrows leading from the size and growth characters to relative fitness (w) define the fraction of the variance in fitness caused by the direct effects of selection on the character. For the jth character, the path coefficient to w is simply the total directional selection gradient on the character multiplied by the ratio of phenotypic standard deviations of the character (obtained from $\mathbf{P}_0$) and fitness. For example, $p_{0w} = \beta_{z_0}[\sigma_0^2(z_0)/\sigma_0^2(w)]^{1/2}$ where $\sigma_0^2(z_0)$ is the estimated variance of size at birth among the initial members of the cohort and $\sigma_0^2(w)$ is the variance of relative fitness. The fraction of the variance in fitness not explained by the model is $p^2_{e,w}$.

The portion of the path diagram denoted by narrow lines defines the correlational structure of the growth and size attributes. The quantities on the double-headed arrows are ordinary phenotypic correlations obtained from the elements of $\mathbf{P}_0$. The path coefficients are simply the ratios of the standard deviations of the variables at the blunt and pointed tips of the arrows. The contribution of any indirect path to the variance in fitness is twice the product of the coefficients along the path. For example, the proportion of variance in fitness attributable to joint selection on size at birth (z_0) and growth in the first interval (Δz_1) is $2p_{0,w}r_{0,\Delta 1}p_{\Delta 1,w}$. The fractional contribution attributable to joint selection on

Δz_1 and z_2 is obtained by tracing down two paths: $2p_{\Delta 1,w}p_{1,2}p_{2,w}(p_{\Delta 1,1}+r_{0,\Delta 1}p_{\Delta 0,1})$.

The complete path diagram is given in Fig. 4 only for heuristic purposes. As noted above, a complete selection analysis of all of the size and growth components is not possible, so all of the paths cannot be computed. However, analyses involving different combinations of characters can be evaluated for their relative abilities to explain the variance in fitness by comparing the squared multiple regression coefficients, which in the context of Fig. 4 are equivalent to $1-p_{e,w}^2$. If, as in the previous section, the study involves only estimates of mean size and growth during one interval, a complete analysis is possible. The path diagram contains only two variables, z_s and z_g, connected by a correlation.

7 Discussion

The primary motivation of this paper has been to develop an approach for estimating selection on characters that undergo developmental change and/or are missing from selected members of the population. Although we have focused on size and growth, our technique for the reconstruction of the phenotypic covariance matrix provides a general solution to the problem of missing data and can be applied to any phenotypic traits that change with age. For example, if the dimensions of different body parts are measured at each age, selection on shape and shape change can be analyzed. Maturational changes in behavior such as feeding rate, movement pattern, or territorial defense can also be analyzed with the present approach. With plants, the technique can be used to study the form of selection operating on the mean time of flowering versus the duration of flowering.

Thus far our discussion has been limited to viability selection. In certain circumstances, the technique can be extended to the analysis of fecundity selection. For example, in semelparous organisms such as annual plants, in which a single reproductive bout follows viability selection, the phenotypic covariance matrix for individuals attaining sexual maturity can be estimated directly. Using that matrix, the conditional selection gradients corresponding to fecundity selection could be estimated as described by Lande and Arnold (1983).

We also emphasize that our focus has been on the estimation of parameters. Of equal importance is the need for testing for the significance of sample statistics. Substantial sampling problems may arise with the present approach as a consequence of estimating one statistic from many others. For example, the estimation of $\mathbf{P}_0$ will inevitably compound sampling errors, since $\mathbf{P}_0$ is computed from estimates of variances and covariances at several censuses. This problem may be minimized by having relatively few censuses (ages) and large samples at each census, but it should not be taken lightly. A resampling procedure (Efron 1982; Mitchell-Olds and Shaw 1987) may aid in evaluating the confidence that can be placed in the final estimates of $\boldsymbol{\beta}$ and $\boldsymbol{\gamma}$.

There are other methods for dealing with incomplete data sets when the missing data are a consequence of selection (Curnow 1961; Thompson 1973; Rubin 1976; Little and Rubin 1987). As maximum likelihood procedures, these methods have some restrictive assumptions, such as multivariate normality, which may not adequately describe the growth process. Our procedure is more general in the sense that it makes no assumption regarding the phenotype distribution. However, if the normality assumption is justifiable, the maximum likelihood approach should yield a more reliable estimate of $\mathbf{P}_0$ than our technique, since the former utilizes all of the data simultaneously.

High correlations between traits can sometimes frustrate the estimation of selection gradients by causing the phenotypic covariance matrix to be singular. For example, size at one census and growth during the next time interval may be highly correlated if growth is a linear function size. In such circumstances, it may not be possible to measure simultaneously selection on age-specific size and growth rate in adjacent intervals. This problem can usually be solved by deleting some highly correlated traits from the analysis (Lande and Arnold 1983; Mitchell-Olds and Shaw 1987).

In closing, we emphasize the obvious. The present results are directly applicable to laboratory or field populations in which individuals can be unambiguously classified as alive or dead at each census. This is not a problem for most plants, but in many natural animal populations, live individuals may elude the census-taker. If the phenotypes of dispersing and/or missing individuals are not random with respect to the population prior to selection, selection theory should not be applied.

Acknowledgments. We thank H. Caswell, M. Kirkpatrick, R. Lande, P. Phillips, and K. Spitze for helpful comments. Support was provided by NSF grants BSR 86-00487 to ML and BSR 85-06766 to SJA and NIH grant 1 RO1-GM35492-01 to SJA.

8 References

Atchley WR (1984) Ontogeny, timing of development, and genetic variance-covariance structure. Am Nat 123:519–540

Brooks JL, Dodson SI (1965) Predation, body size, and the composition of the plankton. Science 150:28–35

Curnow RN (1961) The estimation of repeatability and heritability from records subject to culling. Biometrics 17:553–566

Dirzo R, Sarukhán J (1984) Perspectives on plant population biology. Sinauer, Sunderland, MA

Efron B (1982) The jackknife, the bootstrap, and other resampling plans. Soc Ind Appl Math, Philadelphia, PA

Harper JL (1977) Population biology of plants. Academic Press, London New York

Kerfoot WC (ed) (1980) Ecology and evolution of zooplankton communities. Univ N Engl Press, Hanover, NH

Lande R, Arnold SJ (1983) The measurement of selection on correlated characters. Evolution 37:1210–1226

Little RJA, Rubin DB (1987) Statistical analysis with missing data. John Wiley & Sons, New York

Lynch M (1980) The evolution of cladoceran life histories. Q Rev Biol 55:23–42

Mitchell-Olds T, Shaw RG (1987) Regression analysis of natural selection: statistical inference and biological interpretation. Evolution 41:1149–1161

Monteiro LS, Falconer DS (1966) Compensatory growth and sexual maturity in mice. Anim Prod 8:179–192

Pearson K (1903) Mathematical contributions to the theory of evolution. XI On the influence of natural selection on the variability and correlation of organs. Philos Trans R Soc London Ser A 200:1–66

Riska B, Atchley WE, Rutledge JJ (1984) A genetic analysis of targeted growth in mice. Genetics 107:79–101

Robertson A (1966) A mathematical model of the culling process in dairy cattle. Anim Prod 8:65–108

Rubin DB (1976) Inference and missing data. Biometrika 63:581–592

Thompson R (1973) The estimation of variance and covariance components when records are subject to culling. Biometrics 29:527–550

Werner EE (1986) Amphibian metamorphosis: growth rate, predation risk, and the optimal size at transformation. Am Nat 128:319–341

Werner EE, Gilliam JF (1984) The ontogenetic niche and species interactions in size-structured populations. Annu Rev Ecol Syst 15:393–425

Wilbur HM (1980) Complex life cycles. Annu Rev Ecol Syst 11:67–93

Size, Scaling, and the Evolution of Complex Life Cycles

E. E. WERNER

1 Introduction

J.T. Bonner (1965) noted that the ultimate description of an organism is that of the life cycle. An organism is not the static representation of the adult we associate with taxonomic characterizations and most ecological theory, but the dynamic unfolding of the genome over ontogeny, and the consequent succession of life history stages or forms. The life cycle, of course, is also the fundamental unit of demographic analyses, and therefore a focus for considerations of ecological processes and their manifestation in evolutionary change. As Istock (1984) put it, every inference we make about the evolutionary process has some equivalent rendering within this demographic framework. Because the life cycle assumes a central position in the structure of biology, it is useful to order patterns in life cycle organization, and to ask what processes have shaped these patterns.

Animal phyla exhibit an enormous variety in life cycle organization (e.g., Pearse et al. 1987). This essay concerns the evolution of complex life cycles, which are a major feature of this variety, and develops some ideas about the role of body size and ontogeny in shaping life cycle structure. Wilbur (1980) defined a complex life cycle as a life history that includes an abrupt ontogenetic change in an individual's morphology, physiology, and behavior, usually associated with a change in habitat, i.e., a metamorphosis. It can be argued that metamorphosis is inevitably associated with a shift in ecology, and that the salient point is how this shift in ecology affects demographic parameters indexing life cycle performance (Werner 1986a). Because ontogenetic shifts in niche not associated with a metamorphosis will affect demographic parameters and present parallel conceptual issues, I also consider these to be complex life cycles (Werner and Gilliam 1984). Complex life cycles epitomize the problems we confront in considering the ecology and evolution of size-structured populations; not only do individuals in a population differ in size and age, but they also differ radically in form, function, and ecological life style.

Remarkably little theory has been advanced on the evolution of complex life cycles. The last 20 years or so have witnessed a burgeoning interest in life history theory, and a large literature exists addressing questions such as when should a species first reproduce, should it reproduce more than once, and how should it allocate resources to growth and reproduction. Despite these advances, comprehensive reviews do not even mention complex life cycles (e.g., Stearns 1976, 1977)! This is a startling omission, considering that complex life cycles are the

Department of Biology, University of Michigan, Ann Arbor, Mi 48109, USA

Size-Structured Populations
B. Ebenman and L. Persson (Eds.)

dominant life style among animals (see below). Surely metamorphosis is as dramatic an event in the life cycle as attaining sexual maturity, or adopting a profound ontogenetic shift in ecology as significant as adopting iteroparity. Complex life cycles formed by alternation of generations in plants present a parallel problem, and Willson (1981) has similarly observed "so far as I can determine virtually nothing has been written on the ecology (aside from genetics) of complex life cycles in plants". The few exceptions to this trend are papers by Istock (1967, and replies to this paper, see Wilbur 1980), Wilbur and Collins (1973), Wassersug (1975), Wilbur (1980), Lubchenco and Curbit (1980), Caswell (1982), and Werner (1986a).

My thesis is that the exigencies of scaling which attend increase in body size over a species' ontogeny often necessitate a shift in ecological niche, and that this necessity may be a major impetus to the evolution of complex life cycles. The ideas presented here are of a preliminary and speculative nature; I hope they will stimulate others to examine the problem and to challenge the perspective that I develop.

2 The Problem

Figure 1 presents the general patterns of development expressed in animal life cycles and an example of variation in one taxon. Following union of egg and sperm, the zygote undergoes a period of embryogenesis terminated by hatching or birth. In the majority of animals, hatching or birth produces a larval form (indirect development) which metamorphoses to a juvenile that is similar to the adult and gradually achieves adult size. The larva and adult typically have quite different phenotypes and function in very different ecological settings or roles. Alternatively, hatching or birth produces a juvenile form directly (direct development). In either case larvae or juveniles may be ecologically independent or the parent may feed or provision them. In addition to these two basic modes of development, asexual reproduction may occur at any of these stages, i.e., when an embryo, larva, juvenile, or adult, further compounding possible life cycle variety.

How pervasive are these modes of development in animal life cycles? Pearse et al. (1987) list some 33 phyla in the kingdom Animalia. In 25 of these phyla metamorphosis is a major life cycle feature, or at least important classes undergo a metamorphosis (Table 1). Of the seven phyla that appear not to exhibit metamorphosis, only the Nematoda is speciose. Due to the dominance of insects and the marine phyla, I have estimated that perhaps 80% of animal species undergo a metamorphosis during the life cycle! Additionally, a large proportion of direct developing species exhibit marked shifts in niche during ontogeny and therefore effectively have compley life histories (Werner and Gilliam 1984). Clearly, complex life cycles are the norm among animals, and ontogenetic shifts in ecology and attendant changes in body plan represent a pervasive and characteristic element of life history structure.

Why are complex life cycles ubiquitous across taxa that experience such widely different ecological circumstances as, e.g., marine invertebrates, terrestrial insects, and amphibians? What are the common elements that transcend the unique biol-

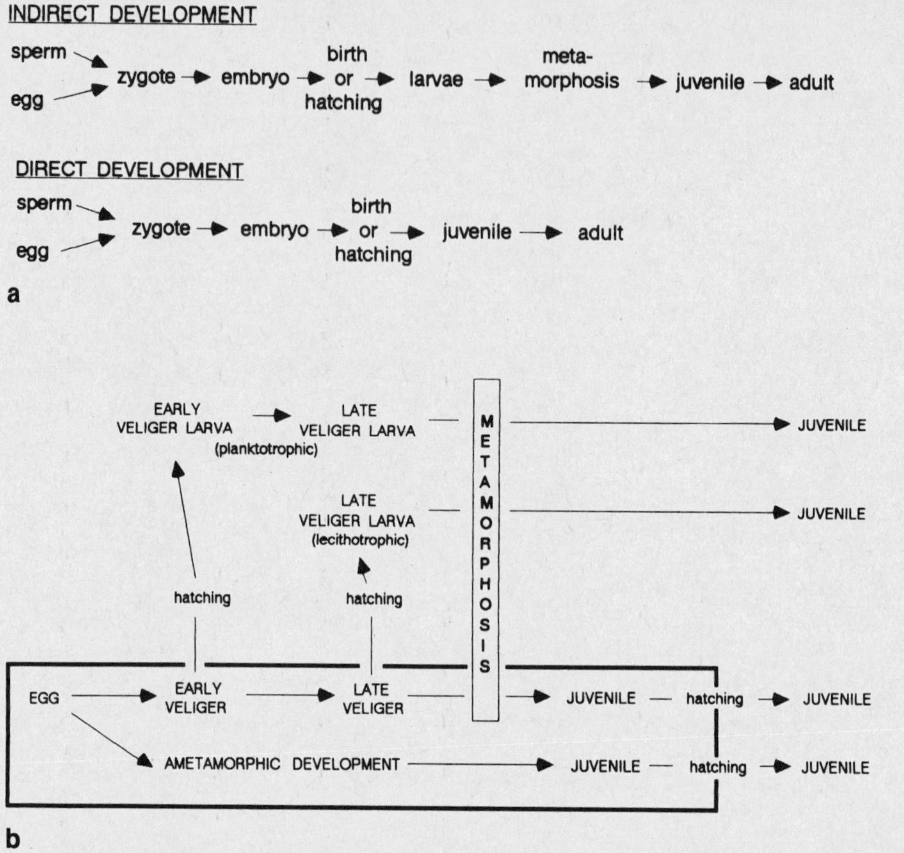

Fig. 1. a Schematic of the typical patterns in the development of animals. **b** Example of the variation in patterns of early development in a group of molluscs (Opisthobranchia). The *heavy-lined rectangle* represents development occurring within the egg capsule. (After Bonar 1978)

ogy of these taxa and lead to complex life cycles? Obviously, the increase in size that enables attainment of adult mass and reproductive condition is a fundamental process of ontogeny or life cycle development. Whatever the reasons for adoption of sexual reproduction by most organisms (Bell 1982; Williams 1975), this mode of replication (through oogenesis) reduces the individual to a single cell at some point in the life cycle. However, there is clearly a conflicting tension or advantage to achieve a much larger size to affect replication. This results, for instance, in mammals – which range over about 7–8 orders of magnitude in size but whose eggs are about 1/10 mm in diameter (Brody 1945) – having to increase from several to 10–11 orders of magnitude in size to reach adult condition. Similarly, many species of fish must increase 5 orders of magnitude in mass from ovum to adult. These extraordinary ontogenetic increases in size must impose constraints on organization of the life cycle. Below I ask what the consequences of this size increase are in terms of ecological life style, and whether there are

Table 1. Presence of metamorphosis in animal phyla, and examples of larval forms represented in these phyla

Phylum	Metamorphosis	Example of Larva
Porifera	Yes	Amphiblastula
Placozoa	?	
Orthonectida	Yes	Bilobed
Dicyemida	Yes	Infusoriform
Cnidaria	Yes	Planula
Ctenophora	Yes	Cydippid
Platyhelminthes	Yes	Muller's, miricidium, cercariae, coracidium, procercoid
Gnathostomulida	No	
Nemertea	Yes	Pilidium
Nematoda	No	
Nematomorpha	Yes	Gordioid
Acanthocephala	Yes	Acanthor
Rotifera	Yes	
Gastrotricha	No	
Kinorhyncha	No	
Loricifera	Yes	Higgins
Tardigrada	No	
Priapula	Yes	Lorica
Mollusca	Yes	Trochophore, veliger
Kamptozoa	Yes	
Pogonophora	Yes	
Sipuncula	Yes	Trochophore type
Echiura	Yes	Trochophore type
Annelida	Yes	Trochophore
Onychophora	No	
Arthropoda	Yes	Nauplius, zoea, cypris, trilobite, caterpillar, nymph
Chaetognatha	No	
Phoronida	Yes	Actinotrocha
Brachiopoda	Yes	Articulate larva
Bryozoa	Yes	Cyphonautes, coronate larva
Echinodermata	Yes	Dipleurula, pluteus, bipinnaria, auricularia
Hemichordata	Yes	Tornaria
Chordata	Yes	Tadpole, amphioxus, ammocoete, leptocephalus

limits to the size range over which a particular body design or life style is viable. I then explore the implications of such limits to structure of the life cycle. In a subsequent section I outline a conceptual approach to determining these limits and predicting ontogenetic shifts in ecology.

3 Size, Scaling, and the Limits on Life Style

It has long been appreciated that changes in size generate scaling problems arising from the interplay of an organism's geometry and physical, chemical, and biological processes. These ideas have a venerable history, going back to the seminal works of D'Arcy Thompson (1961, first published in 1917) and Huxley (1932), and in recent years have received renewed attention (Peters 1983; Calder 1984; and

Table 2. Crucial physical parameters and structures for small versus large organisms. (After Horn et al. 1982)

Function	Small organism	Large organism
Swimming or flying	Viscosity	Inertia and streamlining
Locomotion and support	Surface tension, molecular and electrostatic cohesion	Gravity
Structure (nonaquatic species)	Exoskeleton	Endoskeleton with tensile elements
Physiological transport	Diffusion	Convection, stirring, and circulation

Schmidt-Nielsen 1984). Scaling relations provide critical insight on the manner in which size interacts with other factors to influence the ecological performance of an animal. For example, physical parameters crucial to performance for very small and large organisms can be quite different (Table 2). At low Reynolds numbers an animal swims with friction as the propulsive mode and at high Reynolds numbers uses the inertia of water to propel itself. For organisms where surface area controls gas and nutrient exchange, scaling places severe constraints on the ultimate size of a organism of a given structural configuration (Gould 1966). By the law of geometric similarity isometric growth (maintaining the same geometry with increase in size) results in surface increasing as the square of linear dimensions and volume as the cube. Consequently, most organisms grow allometrically (i.e., depart from the geometric similarity principle) to meet structural needs, or exhibit convolution of surfaces, flattening or elongation of the body, or the incorporation of non-respiring material into the body mass to maintain a positive allometry of surface to volume.

Scaling properties will generally impose a hierarchical series of constraints on organisms, limiting the size range over which a particular morphology or life style can be exploited. First, physical or mechanical constraints set feasibility limits on size and morphology, and then a series of more restrictive limits are imposed by ecological factors that limit life style. For example, it appears that given the properties of vertebrate muscle, methods of oxygen supply to the muscles, and the scaling of metabolic power with mass, birds lighter than 2 to 3 g and heavier than about 12 to 13 kg cannot fly (Schmidt-Nielsen 1984). Within these limits, however, further ecological constraints are levied on life style as each specialized form tends to exhibit a narrower range of body sizes. Whereas birds capable of substained flight span about a 6250-fold range in mass, diving birds span only a 2000-fold range, and hummingbirds only an 11-fold range (Calder 1984). Most avian orders span a 100-fold range or less. Thus it is clear that definite limits exist on the size range over which a particular life style can be exploited, though these limits vary greatly and depend on ecological considerations.

Hence, if an organism is to span a large size range during ontogeny, it often will be advantageous or necessary to alter body design or life style in order to maintain performance. There are obviously strong limits on the extent to which allometric growth can compensate for scaling constraints, and metamorphosis or ecological shifts in the life cycle may be mechanisms to obviate or cope with these limits. If during ontogeny an organism spanned size ranges where, for example, Reynolds number changed markedly, radical changes in selection pressures on form at different sizes would result (the reason for large shape changes between the nauplius and adult in copepods?, see Strickler 1975). An example of how scaling of mechanical and morphological features interacts with a species ecology to condition ontogenetic shifts in resource use is provided by Wainwright (1987, unpubl.). He showed that three Caribbean wrasses exhibited marked ontogenetic shifts in diet to hard-bodied prey (primarily molluscs) at sizes ranging from 70 to 120 mm. Crushing strength of the pharyngeal jaw limited the ability to feed on hard-bodied prey, and ontogenetic profiles indicated that the shift in all species occurred at the same crushing strength of about 3 to 5 newtons. Though there is clearly some range in size over which different species achieve this ability, it cannot be achieved at a small size. That is, a complex life cycle is required to gain access to this resource because other resources have to be exploited when small. Studies integrating functional morphology and ecology are critical to elucidate the relation of scaling to such life history switches.

Thus the fundamental questions of scaling as posed by Schmidt-Nielsen (1975, 1977) are quite apropos to our concerns. What are the size-limits for a certain type of organism? For a given size, how must an organism be designed? For an organism of a given structure, what are the consequences of a change in size? Unfortunately, these questions are rarely applied to the scaling problems most organisms face during ontogeny, and there is little recognition of their implications to life cycle structure. The vast majority of allometric relations available are interspecific comparisons at a comparable stage, or static allometry (and mostly for birds and mammals which do not have complex life cycles – see below). Ontogenetic or intraspecific allometry is virtually ignored. Further, it is not clear that interspecific relations are an accurate portrayal of ontogenetic relations in most taxa (Heusner 1982a, b; and Hughes 1977). Given the marked scaling effects body size exhibits in interspecific analyses, and the fact that most animals span an extensive size range during ontogeny, these are unfortunate lacunae.

If scaling presents a pervasive problem during ontogeny and life styles are constrained to definite size limits, we should see patterns suggesting ways in which various taxa cope with the problem. The broad scope of life history strategies has rarely been examined from this perspective. Cohen (1985), however, listed some of the tactics animals apparently employ in various combinations to obviate ontogenetic scaling problems. For example, animals may avoid scaling problems by (1) specializing as a very small adult, (2) multiplying the basic trophic apparatus as a unit to grow larger rather than increasing the size of each, (3) extensively provisioning the egg and/or juvenile with transformable mass (yolk, maternal fluids, bodies of prey) so that independent existence is delayed until the young can take up the parental life style, or (4) adopting a complex life cycle or a succession of life styles or forms commensurate with the increase in size (a sort of sequential polymorphism).

Tactic (1), remaining small as an adult, is exemplified by animals such as rotifers, nematodes, and protozoans. Cohen speculates that this tactic is effective if the increase in size is about an order of magnitude or less over the life cycle. Tactic (2), multiplying a basic trophic unit, enables an animal to maintain the essential scaling relations of this unit with its resources and still grow larger as a whole. Coral polyps and tapeworm segments would perhaps be examples of this tactic. Tactic (3), provisioning the offspring, is evident in a wide range of taxa but preeminent in mammals and birds. Mass at weaning in mammals ranges from about 3 to 60% of adult mass (Peters 1983; Calder 1984; Blueweiss et al. 1978). Birds hatch at about 1 to 14% of adult mass but can fledge at adult size or even larger in altricial species (Calder 1984; Blueweiss et al. 1978). Thus mammals typically increase only by a factor of 2 to 33 when independent of the parent and birds by a factor of 1 to less than 100. A number of insects also use this tactic, e.g., many hymenoptera and certain flies either feed or provision their larvae. Tactic (4), adopting a series of life styles and body forms seriatim, is the tactic adopted by most animals. Contrast with birds and mammals a fish of roughly median size (about 100 g) that emerges from an egg whose mass is 0.002% of adult mass (Peters 1983). This fish must increase by about a factor of 50000 while ecologically independent. A 50-kg fish typically must increase by a factor of over 200000. A 10-mg crustacean must increase by a factor of nearly 6000 to achieve adult size (Peters 1983). Organisms such as these face significant scaling challenges during ontogeny, and these scaling considerations are most likely the reason complex life cycles have been so widely adopted.

The following section outlines a preliminary approach that integrates the effects of physical and biological scaling factors on the limits to life style, and that permits predictions of the consequences to life cycle structure. I first examine how ecological relations scale with size for organisms that do not experience major modification of body plan. In this case the dictates of efficient performance often necessitate ontogenetic shifts in food or habitat. The logic underlying these shifts then provides insight into the factors that may contribute to the evolution of more spectacular shifts involving metamorphoses.

4 Size, Ontogeny, and Niche Shifts

The significance of ecological shifts during ontogeny lies in changes in size-specific growth, survivorship, and fecundity associated with these shifts. In order to assess the consequences of such shifts (and their timing) to life cycle performance, we need an approach that provides theoretical and/or empirical relationships predicting how vital rates scale with ontogenetic increase in size. Many of these shifts occur during pre-reproductive phases when we are concerned principally with size-specific growth and mortality rates. Consequently, I first examine how the potential, or scope, for growth scales with body size for an organism, and how this information can be used to predict ontogenetic shifts in habitat or resources. I subsequently show how size-specific mortality rates modify the predictions based on growth rates. Together size-specific growth and mortality curves begin to quantitatively delimit the bounds on life style for a given design, and to provide the foundation for a theory of ontogenetic shifts.

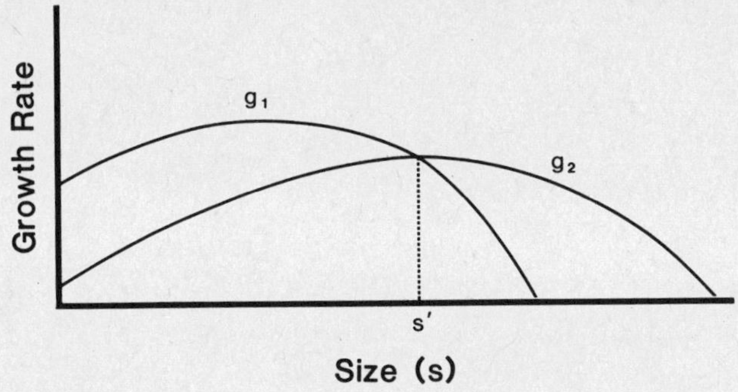

Fig. 2. Absolute growth rates (g) for a hypothetical species as a function of body size(s) in two habitats, or on two resources. s' denotes size at which a habitat shift maximizes growth rate across the life history

I start by hypothesizing that ontogenetic shifts are directed to maximizing growth rates across the life history. Because higher growth rates reduce time to reproductive size, and often increase survivorship and size at reproduction, it has been argued that there will be strong selection to maximize growth rate (Wilbur 1980; Werner and Gilliam 1984). A mechanistic foundation for predicting growth rates in different environments or on different resources is obtained by generating the foraging gain and cost curves as a function of size. The difference between these curves, i.e., the surplus energy curve, provides a quantitative measure of the scope for growth or the energy available for investment in reproduction or allied fitness-related activities at each size. If the size-specific growth rate curves are generated for different habitats or resources, shifts should occur at their intersection (Fig. 2). A shift at size s′ maximizes growth rate and minimizes the time it takes to reach each size (including reproductive size). (In some cases the range of foods or habitat types utilized changes continuously with size and no shift is predicted – see Werner and Gilliam 1984).

The manner in which the rate of energy metabolism, E_m, or the cost of operation, scales with size has been studied in some detail (reviewed in Calder 1984 and Schmidt-Nielsen 1984). Metabolism scales with body mass by the allometric equation:

$$E_m = kM^d ,$$

where M is mass, d is a constant that averages 0.75, and k is a species-specific constant that depends on factors such as temperature and activity. These relations, however, have been developed primarily across species and the values obtained may not accurately reflect intraspecific values across ontogeny. Preliminary work indicates that d may range between 0.5 and 1.0 intraspecifically (Heusner 1982a, b; Hughes 1977; Sebens 1987).

How energy gain curves scale with size is less obvious. It is unlikely general relations will be discovered here given the diversity of feeding modes in animals and subtle ways in which behavior and morphology affect foraging relationships. Nonetheless, in certain taxa energy gain has been found to scale by the allometric equation as well, e.g., in a number of filter-feeding organisms (zooplankton, molluscs, anemones, and corals), though with quite variable exponents (Werner and Gilliam 1984; Sebens 1982, 1987). Surplus energy, E_s, then is:

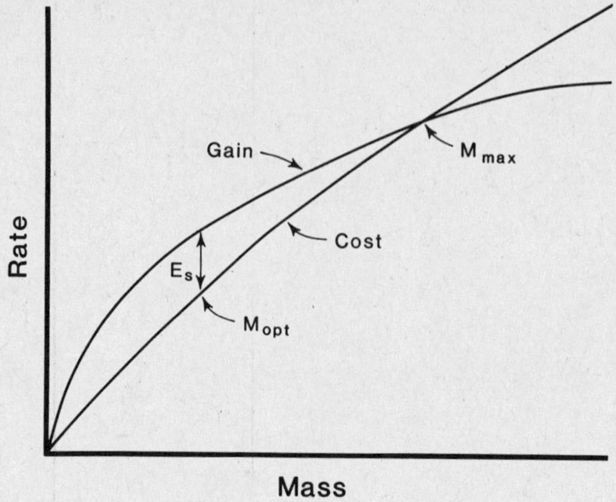

Fig. 3. Schematic of the metabolic cost and foraging gain curves as a function of body mass. M_{max} is the maximum size attainable in a given habitat (on a given set of resources), E_s is the surplus energy available for growth and reproduction, and M_{opt} is the size at which surplus energy is maximized. (See Werner and Gilliam 1984 for details)

$$E_s = hM^a - kM^d \ ,$$

where h and a are fitted constants that will be habitat- and species-specific. Gain and cost curves generally take the form shown in Fig. 3; conversion efficiencies are employed to obtain growth rates from surplus energy and will usually result in curves such as in Fig. 2. These relations have been the foundation for a number of growth models, e.g., those of Putter (1920) and von Bertalanffy (1938). The maximum size for an animal of given design on a specific resource (where gains and costs balance), is $M_{max} = (k/h)^{1/(a-d)}$; the optimal body size, M_{opt}, defined as the size providing maximum surplus energy, is $(dk/ah)^{1/(a-d)}$ (Sebens 1987).

Gain curves for particulate feeding organisms have been measured less often. Habitat shifts by fish have been predicted by estimating the surplus energy curve for different habitat types using foraging theory (Mittelbach 1980, 1981, and Werner et al. 1983 b). Other investigators have taken a similar approach to ask different questions (e.g., Schoener 1969; Ware 1978; Sebens 1982) or have measured part of the necessary size-specific functions (e.g., reviewed in Hassell 1978 and Werner and Gilliam 1984). Scaling or energy intake may be influenced by a host of mechanical, physiological, and ecological considerations and need not take the form of a power function, but will generally result in a humped-shaped surplus energy curve. Whether employed for filter or particulate feeding, prey capture devices, like any mechanical tool, are bound to have certain size ranges over which they operate efficiently and performance will decline outside that range.

Given the cost and gain curves for a series of discrete habitat or resource types, we can predict the sizes at which a switch in habitat or resource would be advantageous, or in fact necessary to grow to a larger size (Fig. 2). The important point is that these curves integrate the net effect of mechanical, physiological, and ecological constraints on performance and therefore provide a precise and quantitative way to define ontogenetic size limits of a given body design or life style for a particular ecological context.

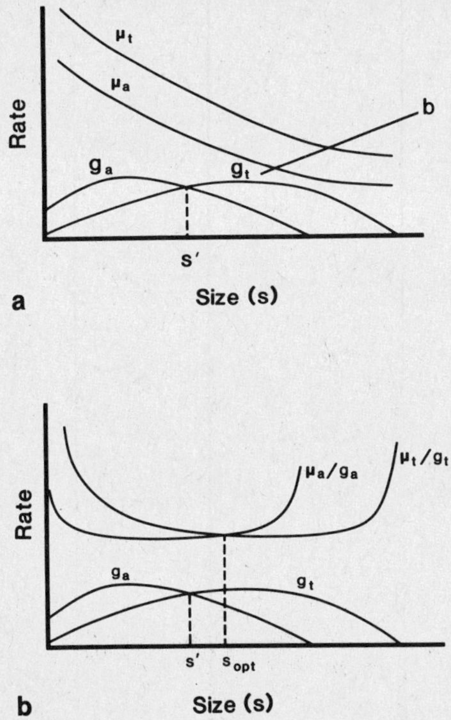

Fig. 4. **a** Hypothetical size-specific vital rates for an organism in two habitats (a, t); g_a and g_t are growth rates (g per day) in the two habitats; u_a and u_t are mortality rates (probability of death per day); b is birth rate (eggs per day); s' is the switch size that maximizes growth rate across the life history. **b** Ratios of mortality to growth rates for the curves portrayed in **a**. The optimal size to switch habitats (s_{opt}), i.e., the switch size that minimizes u/g at each size, is delayed to a larger size than the one that maximizes growth rates (s') because of the trade-off between growth rate and risk of mortality

Mortality rates, however, also may scale with size (e.g., risk from predators, Werner and Gilliam 1984) and in many cases will differ among habitats because of differences in predators or habitat structure, or when feeding on foods that require different foraging tactics. How the scaling of mortality rates alters the above predictions of optimal size to switch habitats is a difficult problem because growth and mortality are in different units and therefore have to be related through common units of fitness. We are beginning to see development of theory that attacks this question (Gilliam 1982; Clark and Levy 1988). Consider Fig. 4, where hypothetical growth, mortality, and birth rates are presented as a function of size in two habitats. The figure is drawn such that a tradeoff between growth and mortality rates exists (though the curves could take any form), i.e, by switching habitats at s′ where growth is maximized, the animal raises its mortality rate from u_a to u_t. We would expect that under these circumstances fitness is maximized by delaying the switch to a larger size; if mortality rates were lower in the second habitat, then the switch should occur at a size smaller than s′ (similarly with differences in growth rates). The question is what is the optimal size to switch habitats?

For time-invariant rates, Gilliam (1982) showed that to maximize fitness (r), a pre-reproductive individual should minimize the ratio, $(u+r)/g$, at each size (the solution is more complex for adults – see Gilliam 1982). If the population is stable (r = 0), the rule further simplifies to minimize u/g at each size. Figure 4 depicts the u/g curves and indicates how the predicted switch size differs from that based only on maximizing growth rates. Thus we now can predict the switch size that balances effects of size-specific growth and mortality rates so as to maximize fitness.

The logic developed thus far permits several tentative conclusions. First, ecological performance of an organism will strongly scale with size. As a consequence of this scaling, there are definite limits on the size range over which a particular body design will prove effective at a given ecological task. Because of these limits, an organism will often benefit from a shift in habitat or food at some point during ontogeny. Second, these limits can be quantitatively determined from analyses such as those presented above by decomposing performance into size-specific mechanistic descriptions of vital rates. How can we interpret the evolution of metamorphosis and subsequent modifications of the life cycle in terms of these arguments?

5 The Evolution, Maintenance, and Loss of Metamorphosis

If ontogenetic switches are exhibited by a species for the reasons outlined above, and the two habitats (or resources) are of sufficiently different character, there will be selection to modify body plan and behavior in the respective stages to perform more effectively in each. This selection can be visualized as follows. An organism that minimizes u/g at each size maximizes fitness; over the life cycle this is equivalent to minimizing the area under the u/g curves for the different habitats exploited. Consider the u/g curves in Fig. 4 and assume a habitat switch is made at the predicted size without alteration in body design. If we hypothetically allowed a reorganization of the morphology at the transition point (ignoring for the moment the costs of metamorphosis), the body plan could be uniquely tailored to the ecological environment of each stage. Because the original body design is unlikely to be the best at exploiting both habitats, this alteration would lower the u/g curves in both stages reducing area under the curves. The more radical the reorganization, the more completely the organism could be refashioned to perform better (lower u/g) in each habitat.

This argument assumes that the gradual changes required to arrive at metamorphosis would be selected for (and, of course, that constraints on genetic variation and canalization of developmental processes would not limit the changes possible). That is, any changes would have to be of immediate value and not just significant in terms of the net result of the developing trend. Any changes, however small, reducing area under the u/g curves should be selected for. Initially, these changes could be accomplished without being greatly disruptive, simply through behavioral alterations or changes in allometric growth that enable an organism to adapt to succeeding habitats. We have numerous examples of the tinkering with established canalized development to produce markedly different forms during the ontogeny that clearly are adaptive, e.g., through alterations in the timing of developmental events a melange of juvenile and adult traits can occur together at different times (Gould 1977; Raff and Kaufman 1983). Over time, more and more substantial changes could lead to a metamorphosis. Wald (1981) has made the point that once we comprehend the elements of dramatic metamorphoses, we begin to see the more subtle and partial manifestations of these elements everywhere, and that they form a continuous gradient. Thus there seems to be no difficulty in imagining the gradual evolution of a metamorphosis. The tendency to make dramatic alterations, of course, would be countered by a number of costs that escalate as reorganization during the ontogeny becomes more extreme. It can take a substantial amount of energy to metamorphose (Pandian and Marian 1985), there are significant risks to metamorphosis from predators (Wassersug and Sperry 1977) and disease, and developmental mistakes can be a hazard (Raff and Kaufman 1983).

We see abundant evidence of the continuous gradient from simple habitat shifts to metamorphosis in many groups. Many species of fish switch from feeding on plankton to benthos as they increase in size. Often this switch is made with little change in morphology simply by changing habitats (Werner 1986b). In other cases, species exhibit modest morphological alterations when the switch is made. Postlarval spot (*Leiostomus xanthurus*) have a terminal mouth and feed on zooplankton, but when 22–32 mm in length, the body deepens and the mouth becomes subterminal as the spot switches to feeding on benthos (Chestnut 1983). At the extreme, flatfishes (Pleuronectiforms) exhibit a substantial metamorphosis when the normal, pelagic, cod-like larva exhibits a 90° rotation of the neurocranium, carrying one eye with it to form an asymmetrical flatfish that settles to the bottom and takes up a benthic life style (Policansky 1983). Insects provide parallel examples in ametabolous taxa that simply increase in size at each molt, through hemi-, holo-, and even hypermetabolous groups that undergo more and more radical metamorphoses.

Unless traits can be dissociated as changes are made, however, critical tradeoffs often exist that reduce performance in one stage in proportion to advances made in the other. For instance, planktivorous and piscivorous fish require very different morphological features to be effective (Werner 1977). Yet most piscivorous species must pass through an early life history stage where they feed on plankton. Thus selection on morphology will be conflicting because of the tradeoffs; modifications enhancing abilities as a planktivore early in life will generally diminish abilities as a piscivore later in life (u/g will be lowered in the planktivorous stage but raised in the piscivorous stage). Whether a particular modification will be selected for or not depends on the change in area under the u/g curves, not just the improvement in each habitat. One advantage of metamorphosis may be that it allows isolation of different parts of the life cycle, permitting relatively independent responses to the different selective pressures confronted in each stage, i.e., the covariance structure of the genome can be broken up during development avoiding the sorts of tradeoffs envisioned above (Freeman 1982; Horn et al. 1982; see Kirkpatrick this Vol.). There is substantial evidence that in some animals the same structure is separately encoded in different stages. For example, Willis et al. (1981) found that cuticular proteins were distinct in larval, pupal, and adult stages of certain insects. This independence may be facilitated in holometabolous insects, where adult instructions are sequestered away in imaginal discs during the larval phase and the larval body is largely histolized and serves as "yolk" for development of the adult.

How can we visualize selection for subsequent changes in life cycle structure through paedomorphosis or direct development? Life cycle structure in most taxa has been very dynamic over evolutionary time as new stages evolve and previous stages are lost (e.g., Fig. 1b). The larval stage is often lost via the evolution of direct development and in some cases it is argued that a secondary larva (and therefore a secondary metamorphosis) has evolved subsequent to loss of the original larval stage (e.g., the tadpole larva of ascidians and amphibians, see Jagersten 1972; Matsuda 1987). Similarly, the adult stage has been lost in a number of taxa via paedomorphosis (either neoteny or progenesis, Gould 1977). Both direct development and paedomorphosis have arisen repeatedly and independently in a large number of taxa as divergent as molluscs and amphibians. Direct development has evolved at least 12 independent times in the anurans alone (Duellman and Trueb 1986). Additional variants on this theme include the evolution of several larval stages (e.g., some Cnidaria, Mollusca, and Echinodermata) and the addition of yet another distinct form in the life cycle through hypermorphosis (e.g., the cyprid stage of barnacles, or those exhibited in many Hemiptera, Neuroptera, and Hymenoptera – see Highnam 1981).

Once metamorphosis is in place in a life cycle, there will be continued pressure to lower u/g in each habitat. It is unlikely that modification will proceed uniformly in each stage, but scaling considerations and the sorts of tradeoffs outlined above will usually prevent elimination of one stage. If, however, an innovation were to occur, say in the larval stage, that vastly improved the larva's ability to grow (or escape mortality agents), this would lower the u/g curve much more rapidly in the larval than in the adult phase and select for a protracted part of the life cycle in that phase (Fig. 4). If the ability to grow (or lower u/g) in the larval habitat greatly outstripped that in the adult stage, so that u/g was everywhere lower, or at some size other selective pressures such as on size at first reproduction interceded, there would be selection to reproduce in the larval habitat, i.e., for paedomorphic individuals. (This argument in a sense is the genesis of "Istock's paradox" that complex life cycles are unstable; i.e., because evolution is unlikely to proceed at equal rates in the two phases, there would be selection to eliminate a bottleneck stage.) If, however, there were additional constraints such as the necessity to disperse from the larval habitat, metamorphosis and the adult stage would not be abandoned. The lower u/g in the larval stage would select for spending all sizes up to reproductive size in this stage, and the adult stage would be relegated to a dispersal and reproductive stage. At this point, the adult stage would no longer experience selection for ability to grow, and consequently would be released to respond solely to selection for effective dispersal and reproduction. Numerous examples of this may be seen in aquatic insects and this scenario may explain some of the specific adaptations for reproduction and dispersal that we see in the adult stages of many insects.

Alternatively, if u/g was greatly reduced in the juvenile-adult stage (or raised in the larval stage), the above argument would be reversed and direct development would be selected for. Both of the above scenarios result in either reduction of the life cycle to a single stage or retention of both stages, but with more complete separation of function in each. If the life cycle were reduced to a single stage, a second stage may again evolve as new ecological conditions are encountered (e.g., a secondary larva). Because of these subsequent modifications, we inevitably will find it difficult to infer the selective regimes that have led to many such complex life cycles from present forms. Distinguishing or differentiating reasons for the origin of a feature from current utility, of course, is an ever-present problem in evolutionary biology.

6 Several Examples

We have used the theory developed thus far to interpret variation in life history structure both in species that simply change allometrically with size (fish) and species that undergo a metamorphosis (amphibians). In fish we have abundant evidence of size-dependent, ontogenetic shifts in resources or habitat (Werner and Gilliam 1984; Werner 1986b; Persson this Vol.) and these shifts are often accompanied by distinct spurts of growth (e.g., Palaheimo and Dickie 1966; Osenberg et al. 1987). Particular resource types (or characteristic sizes) set definite limits on maximum body size which can be alleviated if fish switch resources or are transplanted to different habitats (e.g., Martin 1966). Further, it has been shown that size-specific predation risk also influences shifts in habitat (Cerri and Fraser 1983; Werner et al. 1983a; Power et al. 1985).

For example, we sampled five small lakes and showed that bluegill sunfish (*Lepomis macrochirus*) switched from littoral zone vegetation to the pelagic zone open water at sizes ranging from 51 to 83 mm standard length (Fig. 5). At the extremes this represents a tenure of 4 years (Warner Lake) vs. 2 years (Lawrence Lake) in the littoral zone and over a 4.4-fold difference in mass when the shift occurred. In midsummer the open water is the more profitable habitat for all size classes of bluegill; we have shown this by using foraging theory to predict size- and habitat-specific net energy gain, by estimating growth rates in each habitat, and by caging small fish in the open water (Mittelbach 1981; Werner and Hall 1988). However, we also have demonstrated that risk of predation by largemouth bass (*Micropterus salmoides*) was 40 to 80 times greater for small bluegills in open water, but declined as the bluegill became larger, and that switch sizes in the natural lakes were directly correlated with abundance of bass (Werner and Hall 1988). Further, we have direct experimental evidence for the ability of small bluegill to switch to open water when predicted on the basis of foraging rate in the absence of predators, and to respond to presence of the predator by moving into littoral vegetation (Werner et al. 1983a, b; Ehlinger 1986; Werner and Hall 1988). We do not as yet have a direct test of whether the switch sizes in Fig. 5 are those that minimize u/g. Recent experiments of Gilliam and Fraser (1987 and this Vol.) indicate that fish are capable of balancing these two factors and both are probably involved in shifts by bluegill (Werner and Hall 1988). The approach outlined above promises to be a powerful tool in elucidating such shifts in organisms where costs and gains scale with size and there is no discontinuity around a metamorphosis.

I have extended this approach to provide a framework for examining variation in life history structure in amphibians (Werner 1986a). Unlike most insects, amphibians usually grow a great deal after metamorphosis. North American frogs,

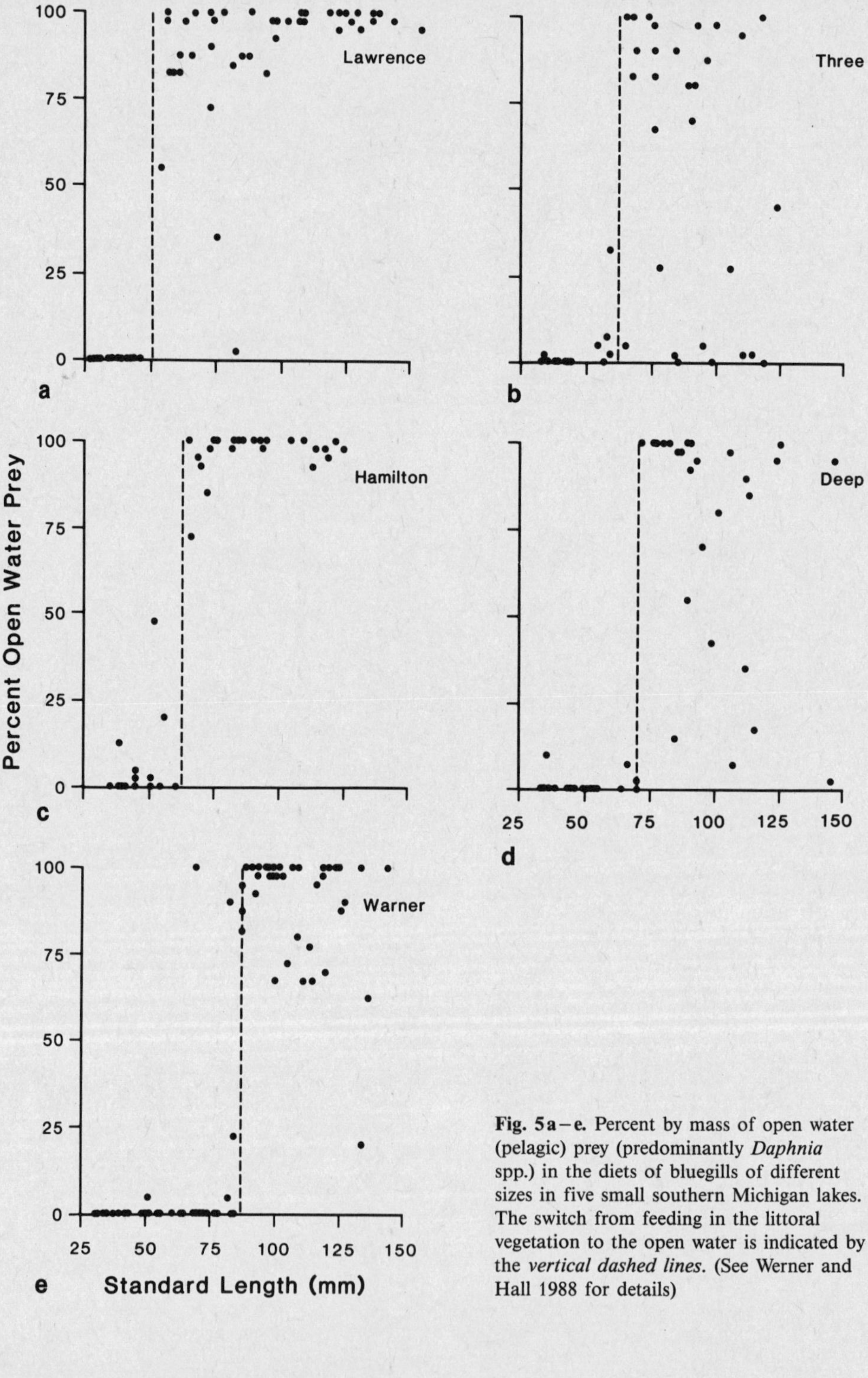

Fig. 5a–e. Percent by mass of open water (pelagic) prey (predominantly *Daphnia* spp.) in the diets of bluegills of different sizes in five small southern Michigan lakes. The switch from feeding in the littoral vegetation to the open water is indicated by the *vertical dashed lines*. (See Werner and Hall 1988 for details)

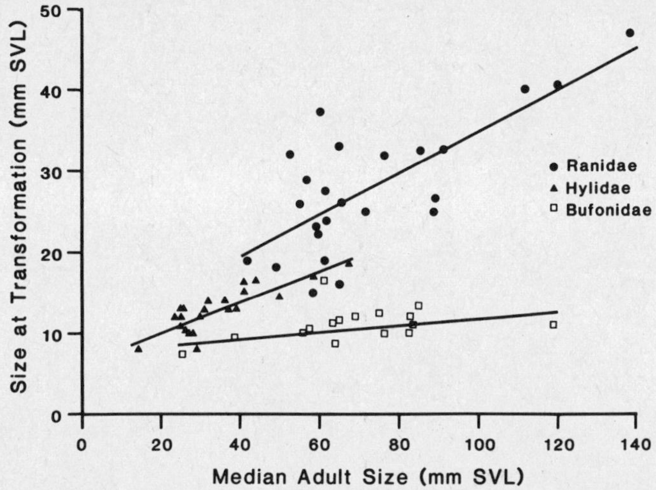

Fig. 6. Size at transformation versus median adult size for three families of North American frogs. The regression for the bufonids differs significantly from those of the ranids and the hylids; the slopes of the regressions for the ranids and hylids are not different, but the two regressions differ in elevation. (See Werner 1986a for details)

for instance, typically accomplish 90 to 99% of their growth after transformation (Werner 1986a). Therefore it seems plausible that timing of metamorphosis is influenced by the u/g tradeoff; i.e., selection on size at metamorphosis would be influenced by the relative fitness consequences of size-specific growth and mortality rates in both aquatic and terrestrial habitats.

Figure 6 portrays the variation in size at metamorphosis in some of the North American frogs. One of the more striking patterns is that bufonids transform at much smaller sizes than similar-sized ranids and hylids. The model suggests that selection for a reduced size at metamorphosis in bufonids would occur if u/g was relatively high in the larval habitat and/or low in the adult habitat as compared to the two other groups. Especially in the terrestrial phase, bufonids possess a variety of toxins that are effective against birds and most mammals. These toxins may lower u (and therefore u/g) in the terrestrial habitat relative to most ranids and hylids, and select for a smaller size at metamorphosis. Bufonids could then emerge at a smaller size and take advantage of potentially higher growth rates on land (Werner 1986a). (Here again additional constraints are suggested as bufonids all transform around 10 mm in length regardless of habitat or life style – see discussion in Werner 1986a). There are few data on growth or mortality rates in both stages of an amphibian species that permit more quantitative use of the theory. I was, however, able to find data for two large ranids, the bullfrog (*Rana catesbeiana*) and the green frog (*R. clamitans*), to compare growth rates. These comparisons, though crude, suggested that size-specific growth was higher in terrestrial juveniles than in aquatic larvae, and I hypothesized that the u/g relations for these two species looked like those portrayed in Fig. 7. The importance of this observation, if general, is that we would have to rethink some of the current ideas on the adaptive significance of the tadpole stage. It has been suggested that the

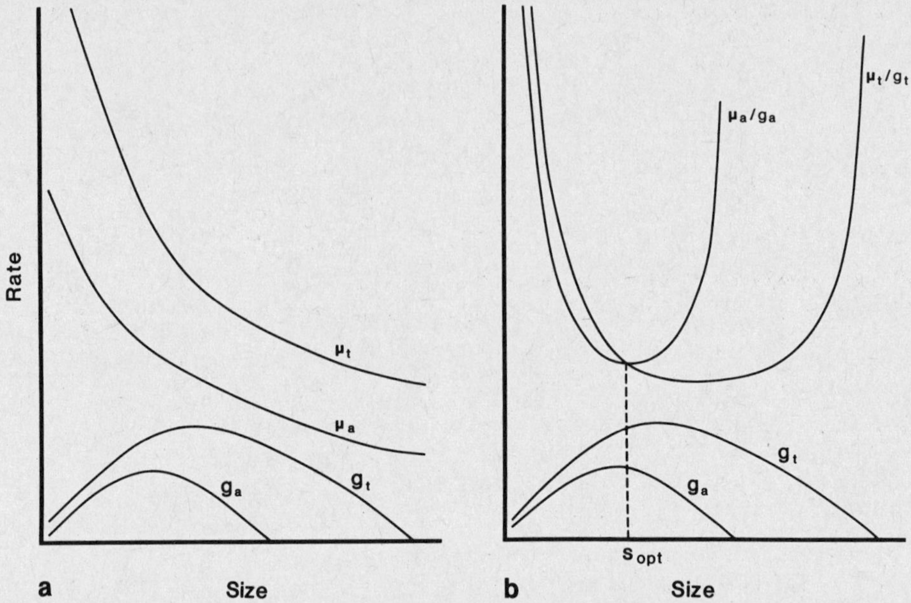

Fig. 7. **a** Hypothetical size-specific growth (g) and mortality (u) rates for the bullfrog or greenfrog in the aquatic (a) and terrestrial (t) habitats. **b** Ratios of mortality to growth rates for the curves in **a** and the predicted size at transformation (S_{opt}). (See Werner 1986a for details)

tadpole stage is maintained in the anuran life history because of the large growth opportunity afforded by pulses of primary productivity in temporary ponds (e.g., Wassersug 1975). The above analyses suggest that size-specific mortality rates may explain the maintenance of the tadpole stage in these species. Arguments for the evolution of direct development and paedomorphosis in amphibians can be made that parallel those presented in the previous section (Werner 1986a). Though we again cannot provide direct tests of the theory, it has provided a useful framework for interpreting the diversity of amphibian life cycles and elaborating a series of testable hypotheses. Studies of the ecology of ontogenetic shifts in forms such as fish where shifts are under behavioral control and facultative, and in certain salamanders and insects that exhibit facultative paedomorphosis, could provide direct tests of the theory and insight into factors responsible for differences among species in life cycle structure.

7 Some Concluding Remarks

In recent years the relation between ontogeny and evolutionary phenomena has received renewed attention (e.g., Gould 1977; Bonner 1982; Raff and Kaufman 1983). For evolution to occur as expressed in morphological change, this change must occur during ontogeny. While it is clear that patterns of development are in general more conservative than adult structures and some are inevitable consequences of physical constraints, developmental architecture, or ancestral legacies, it is also clear that some are adaptive (Horn et al. 1982). In fact, in some groups the larval form has radiated extensively while adult structure appears relatively conservative, e.g., some frogs exhibit extensive modification of the tadpole stage to utilize habitats from bromeliads to mountain brooks and foods from planktonic algae to other tadpoles (Wassersug 1980). Yet many of these variants generally lead to what Horn et al. (1982) termed "reagent-grade" frogs. In pyralid moths the larva has radiated to use stems of terrestrial plants, dried vegetable matter and stored grains, roots of plants, wax of beehives, and eggs and young of scale insects. Others feed on aquatic plants and breath through gills, and have become parasites of mammals (Borror and DeLong 1971).

The importance of this apparently independent modification of larval and adult forms is the raw material it provides for novel mingling of adaptations (or nonadaptive "quirks") that develop in the respective stages of complex life histories (Gould 1977). For example, at some point, nearly all larvae form adult action systems, the veliger is a modified trochophore in which adult action systems have appeared, and therefore adultation may provide an important mechanism for introducing novel features into the larval niche that initially evolved in another context in the adult stage (Freeman 1982). If evolution can proceed relatively independently in the separate stages of a complex life cycle and subsequently the products of this evolution moved into new adaptive contexts in other stages through heterochrony or adultation, life cycles with metamorphosis may have very different evolutionary potentials and constraints than those with more gradual transitions. We will be in a position to more fully comprehend these processes if we understand the ecological forces that have shaped, and presently maintain, patterns in the life cycles of organisms.

At present there is little foundation on which to systematically build such an understanding. The initial attempt at a formal theory of complex life cycles concluded that they were unstable (Istock 1967). Some taxa with complex life cycles, however, are quite ancient (e.g., odonates), and in many taxa (e.g., frogs, trematodes, and insects) more evolutionarily advanced groups exhibit increasingly more divergent larval and adult forms with a more rapid metamorphic transition (Wassersug and Hoff 1982; Toft 1986). The extensive radiations that accompany complex life cycles in these groups does not suggest the brink of extinction. As noted above, a number of tradeoffs arise from scaling considerations that could stabilize a complex life cycle.

Wilbur (1980) has proposed that complex life cycles evolved as an adaptation to exploit transient opportunities for growth or dispersal. Willson (1981) also divided explanations into "dispersal and resource-acquisition" models. Resource acquisition is obviously an integral function of the life cycle and the timing of resource availability will be a critical consideration. As I argued above, however, size is generally a dominant parameter determining ability to acquire various resources, and therefore how abilities scale with size may have a broad influence on changes in form during the life cycle. Clearly a large number of taxa with complex life cycles exploit transient resources, and complex life cycles are a useful adaptation in this context, but it does not appear that transience alone provides a general explanation for the wide representation of such life cycles in the animal kingdom. For example, complex life cycles predominate in marine invertebrates, many of which inhabit very stable environments and can have larval periods of 6–7 months or longer (Chia and Rice 1978). Further, prevalence of a larval stage in many marine benthic taxa declines markedly (direct development increases) at higher latitudes (Thorson's rule, 1950) where planktonic production is often higher and very seasonal (Matsuda 1987; Steemann Nielsen 1975).

Adaptations for dispersal are ubiquitous in complex life cycles and dispersal is clearly advantageous, even in uniform and predictable habitats (Hamilton and May 1977). However, the specific stage where dispersal is affected differs in different taxa, e.g., larvae in most marine benthic organisms, recently metamorphosed juveniles in many amphibians, and adults in many insects. Thus dispersal appears to occur when expedient in the life cycle and so is not a consistent constraint on form of the life history. For example, a large number of marine taxa display remarkably similar feeding larvae regardless of whether the adults are planktonic or benthic, e.g., both planktonic and benthic opisthobranch gastropods have feeding veligers, planktonic and benthic shrimp have feeding zoeae, and copepods and barnacles (both benthic and planktonic forms) share a similar naupliar larval stage (Strathman 1985). Planktonic adults have ample opportunity for dispersal, so advantages for dispersal are not essential for persistence of an extended planktonic larval stage. These common larval forms, of course, could represent historical or phylogenetic constraints, but even within these groups the larval stage has been greatly modified, or lost to direct development, in closely related forms. Strathman (1985), reviewing the evidence for marine invertebrates, concluded that although dispersal can have profound evolutionary consequences, it does not appear that selection for dispersal maintains a feeding larval stage in the life histories of these taxa. He feels these stages are more likely maintained because of feeding and safety advantages of a planktonic life when small. That is, the larval forms may be maintained because they are effective morphological and physiological solutions to utilizing the plankton habitat when small providing a lower u/g than options available for exploiting the benthos when small. I hope the framework offered here will allow us to see more clearly how factors such as resource acquisition and dispersal can interact to shape the life cycle.

It appears to me that an understanding of the scaling problems confronting an organism during ontogeny promises to provide a broadly based hypothesis concerning the evolution and maintenance of complex life cycles. However, these ideas are largely conjecture at this point. We especially need detailed comparative surveys of the structure and ecology of complex life cycles across a number of taxa to assess these hypotheses. We also need a great deal more work on ontogenetic scaling, especially studies directed at the relation between size and ecological life style that would permit tests of the theory. Further, the theory itself is very incomplete; u and g clearly are not the only demographic parameters of concern. Minimally, the effects of size at first reproduction and of dispersal will have to be added to the model before a more complete theory can be elaborated. These latter factors also create tradeoffs that could maintain complex life cycles. Complex life cycles allow differential adaptation to foraging and growth, reproduction, predation, dispersal, and dormancy by spreading development of adaptations to different vital functions over the whole life span. The relative apportionment of these critical demographic functions to particular stages of the life cycle should be seen as a fundamental problem in life history theory. The advantage of the framework I have proposed is that it provides a tangible connection between mechanistic models of the consequences of ontogenetic scaling on ecological performance and demographic parameters, and therefore life cycle performance or fitness. Hopefully this approach will provide a general framework for elaborating testable hypotheses concerning life cycle structure.

Acknowledgments. I thank Lennart Persson and Bo Ebenman for the invitation to participate in this symposium, and for making the stay in Sweden such a pleasant one. I also thank the following colleagues for comments on an earlier version of this paper; Brad Anholt, Spencer Cortwright, Bo Ebenman, Jim Gilliam, Charles Kerfoot, Ake Norberg, Ronald Nussbaum, Lennart Persson, David Skelly, Henry Wilbur, and Kim Zerba. This study was supported by NSF grant BSR-8796176.

8 References

Bell G (1982) The masterpiece of nature: the evolution and genetics of sexuality. Univ Cal Press, Berkeley

Blueweiss L, Fox H, Kudzma U, Nakashima D, Peters R, Sams S (1978) Relationship between body size and some life history parameters. Oecologia (Berlin) 37:257–272

Bonar DB (1978) Morphogenesis at metamorphosis in opisthobranch molluscs. In: Chia F, Rice M (eds) Settlement and metamorphosis of marine invertebrate larvae. Elsevier, New York

Bonner JT (1965) Size and cycle. Univ Press, Princeton

Bonner JT (1982) Evolution and development. Springer, Berlin Heidelberg New York

Borror DJ, DeLong DM (1971) An introduction to the study of insects. Holt, Rinehart & Winston, New York

Brody S (1945) Bioenergetics and growth. Hafner, New York

Calder WA (1984) Size, function, and life history. Harvard Univ Press, Cambridge

Caswell H (1982) Stable structure and reproductive value for populations with complex life cycles. Ecology 63:1223–1231

Cerri RD, Fraser DF (1983) Predation risk in foraging minnows: balancing conflicting demands. Am Nat 121:552–561

Chestnut DE (1983) Feeding habits of juvenile spot *Leiostomus xanthurus* (Lacepede) in North Inlet Estuary, Georgetown, S.C. Ms Thesis, Univ S Car

Chia F-S, Rice ME (1978) Settlement and metamorphosis of marine invertebrate larvae. Elsevier, New York

Clark CW, Levy DA (1988) Diel vertical migrations by juvenile sockeye salmon and the antipredator window. Am Nat 131:271–290

Cohen J (1985) Metamorphosis: Introduction, usages, and evolution. In: Bulls M, Brownes M (eds) Metamorphosis. Clarendon, Oxford

Duellman WE, Trueb L (1986) Biology of amphibians. McGraw-Hill, New York

Ehlinger TJ (1986) Learning, sampling and the role of individual variability in the foraging behavior of bluegill sunfish. Ph D Thesis, Michigan State Univ

Freeman GL (1982) What does the comparative study of development tell us about evolution? In: Bonner JT (ed) Evolution and development. Springer, Berlin Heidelberg New York, pp 155–167

Gilliam JF (1982) Habitat use and competitive bottlenecks in size-structured fish populations. Ph D Thesis, Michigan State Univ

Gilliam JF, Fraser DF (1987) Habitat selection under predation hazard: Test of a model with foraging minnows. Ecology 68:1856–1862

Gould SJ (1966) Allometry and size in ontogeny and phylogeny. Biol Rev 41:587–640

Gould SJ (1977) Ontogeny and phylogeny. Harvard Univ Press, Cambridge

Hamilton WD, May RM (1977) Dispersal in stable habitats. Nature (London) 269:578–581

Hassell MP (1978) The dynamics of arthropod predator-prey systems. Univ Press, Princeton

Heusner AA (1982a) Energy metabolism and body size. I. Is the 0.75 mass exponent of Kleiber's equation a statistical artifact? Respirat Physiol 48:1–12

Heusner AA (1982b) Energy metabolism and body size. II. Dimensional analysis and energetic nonsimilarity. Respirat Physiol 48:13–25
Highnam KC (1981) A survey of invertebrate metamorphosis. In: Gilbert LI, Frieden E (eds) Metamorphosis. Plenum, New York, pp 43–73
Horn HS, Bonner JT, Dohle W, Katz MJ, Koehl MAR, Meinhardt H, Ragg RA, Reif W-E, Stearns SC, Strathmann R (1982) Adaptive aspects of development group report. In: Bonner JT (ed) Evolution and development. Springer, Berlin Heidelberg New York, pp 215–235
Hughes GM (1977) Dimensions and the respiration of lower vertebrates. In: Dedley TJ (ed) Scale effects in animal locomotion. Academic Press, New York London, pp 57–81
Huxley JS (1932) Problems of relative growth. Methuen, London
Istock CA (1967) The evolution of complex life cyle phenomena: An ecological perspective. Evolution 21:592–605
Istock CA (1984) Boundaries to life history variation and evolution. In: Price PW, Slobodchikoff CN, Gaud WS (eds) A new ecology: novel approaches to interactive systems. John Wiley & Sons, New York, pp 143–168
Jagersten G (1972) Evolution of the metazoan life cycle. Academic Press, New York London
Lubchenco J, Curbit K (1980) Heteromorphic life histories of certain marine algae as adaptations to variations in herbivory. Ecology 61:676–687
Martin NV (1966) The significance of food habits in the biology, exploitation, and management of Alqonquin Park, Ontario, lake trout. Trans Am Fish Soc 95:415–422
Matsuda R (1987) Animal evolution in changing environments with special reference to abnormal metamorphosis. John Wiley & Sons, New York
Mittelbach GG (1980) Foraging efficiency and size-class competition in the bluegill sunfish (*Lepomis macrochirus*). Ph D Thesis, Michigan State Univ
Mittelbach GG (1981) Foraging efficiency and body size: A study of optimal diet and habitat use by bluegills. Ecology 62:1370–1386
Osenberg CW, Werner EE, Mittelbach GG, Hall DJ (1987) Growth patterns in bluegill (*Lepomis macrochirus*) and pumpkinseed (*L. gibbosus*) sunfish: environmental variation and the importance of ontogenetic niche shifts. Can J Fish Aquat Sci 45:17–26
Paloheimo JE, Dickie LM (1966) Food and growth of fishes. II. Effects of food and temperature on the relation between metabolism and body weight. J Fish Res Board Can 23:869–908
Pandian TJ, Marian MP (1985) Time and energy cost of metamorphosis in the Indian bullfrog *Rana tigrina*. Copeia 1985:653–662
Pearse V, Pearse J, Buchsbaum M, Buchsbaum R (1987) Living invertebrates. Blackwell, Palo Alto Oxford
Peters RH (1983) The ecological implications of body size. Univ Press, Cambridge
Policansky D (1983) Size, age and demography of metamorphosis and sexual maturation in fishes. Am Zool 23:57–64
Power ME, Mathews WJ, Stewart AJ (1985) Grazing minnows, piscivorous bass, and stream algae: dynamics of a strong interaction. Ecology 66:1448–1456
Putter A (1920) Studien über physiologische Ähnlichkeit VI. Wachstumsähnlichkeiten, Pflüger's Arch Ges Physiol Mensch Tiere 180:298–340
Raff RA, Kaufman TC (1983) Embryos, genes, and evolution. Macmillan, New York
Schmidt-Nielsen K (1975) Scaling in biology: The consequences of size. J Exp Zool 194:287–308
Schmidt-Nielsen K (1977) Problems of scaling: Locomation and physiological corrolates. In: Pedley TJ (ed) Scale effects in animal locomotion. Academic Press, New York London, pp 1–21
Schmidt-Nielsen K (1984) Scaling: Why is animal size so important? Univ Press, Cambridge
Schoener TW (1969) Models of optimal size for solitary predators. Am Nat 103:277–313
Sebens KP (1982) The limits to indeterminate growth: An optimal size model applied to passive suspension feeders. Ecology 63:209–222
Sebens KP (1987) The ecology of indeterminate growth in animals. Annu Rev Ecol Syst 18:371–407
Stearns SC (1976) Life history tactics: A review of the ideas. Q Rev Biol 51:3–47
Stearns SC (1977) The evolution of life history traits. Annu Rev Ecol Syst 8:145–172
Steemann Nielsen E (1975) Marine photosynthesis with special emphasis on the ecological aspects. Elsevier, Amsterdam
Strathmann RR (1985) Feeding and nonfeeding larval development and life-history evolution in marine invertebrates. Annu Rev Ecol Syst 16:339–361

Strickler JR (1975) Swimming of planktonic *Cyclops* species (Copepoda, Crustacea): pattern, movements and their control. In: Wu TY-T, Brokaw CJ, Brennen C (eds) Swimming and flying in nature. Plenum, New York, pp 599–613

Thompson D'AW (1961) On growth and form. Univ Press, Cambridge

Thorson G (1950) Reproductive and larval biology of marine bottom invertebrates. Biol Rev 25:1–45

Toft CA (1986) Communities of species with parasitic life-styles. In: Diamond J, Case TJ (eds) Community ecology. Harper & Row, New York, pp 445–463

von Bertelanffy L (1938) A quantitative theory of organic growth. Human Biol 10:181–213

Wainwright PC (1987) Biomechanical limits to ecological performance: Mollusc-crushing by the Caribbean hogfish, *Lachnolaimus maximus* (Labridae). J Zool 213:283–297

Wald G (1981) Metamorphosis: An overview. In: Gilbert LI, Friedan E (eds) Metamorphosis. Plenum, New York, pp 1–39

Ware DM (1978) Bioenergetics of pelagic fish: Theoretical change in swimming speed and ration with body size. J Fish Res Board Can 35:220–228

Wassersug RJ (1975) The adaptive significance of the tadpole stage with comments on the maintenance of complex life cycles in anurans. Am Zool 15:405–417

Wassersug RJ (1980) Internal oral features of larvae from eight anuran families: functional, systematic, evolutionary, and ecological considerations. Univ Kansas Mus Zool, Misc Publ 68

Wassersug RJ, Hoff K (1982) Developmental changes in the orientation of the anural jaw suspension: A preliminary exploration into the evolution of anuran metamorphosis. Evol Biol 15:223–246

Wassersug RJ, Sperry DG (1977) The relationship of locomotion to differential predation on *Pseudacris triseriata* (Anura: Hylidae). Ecology 58:830–839

Werner EE (1977) Species packing and niche complementarity in three sunfishes. Am Nat 111:553–578

Werner EE (1986a) Amphibian metamorphosis: Growth rate, predation risk, and the optimal size to transform. Am Nat 128:319–341

Werner EE (1986b) Species interactions in freshwater fish communities. In: Diamond J, Case TJ (eds) Community ecology. Harper & Row, New York, pp 344–358

Werner EE, Gilliam JF (1984) The ontogenetic niche and species interactions in size-structured populations. Annu Rev Ecol Syst 15:393–425

Werner EE, Hall DJ (1988) Ontogenetic habitat shifts in the bluegill sunfish (*Lepomis macrochirus*): The foraging rate-predation risk tradeoff. Ecology (in press)

Werner EE, Gilliam JF, Hall DJ, Mittelbach GG (1983a) An experimental test of the effects of predation risk on habitat use in fish. Ecology 64:1540–1548

Werner EE, Mittelbach GG, Hall DJ, Gilliam JF (1983b) Experimental tests of optimal habitat use in fish: the role of relative habitat profitability. Ecology 64:1525–1539

Wilbur HM (1980) Complex life cycles. Annu Rev Ecol Syst 11:67–93

Wilbur HM, Collins JP (1973) Ecological aspects of amphibian metamorphosis. Science 182:1305–1314

Williams GC (1975) Sex and evolution. Univ Press, Princeton

Willis JH, Regier JC, Debrunner BA (1981) The metamorphosis of arthropodin. In: Bhastzaran G, Friedman S, Rodriguez JG (eds) Current topics in insect endocrinology and nutrition. Plenum, New York, pp 27–46

Willson MF (1981) On the evolution of complex life cycles in plants: A review and an ecological perspective. Ann Miss Bot Gard 68:275–300

Population Ecology and Demography in Size-Structured Populations

Approaching Size and Age in Matrix Population Models

H. CASWELL

1 Introduction

Matrix projection models are a simple and powerful way to analyze the life cycles of organisms whose demography is determined by size or developmental stage, rather than age. Here I address three problems: (1) choosing between age and size as state variables, (2) using size-classified models as approximations when both age and size are important, and (3) recovering age-specific information from size-classified models.

Matrix projection models are reviewed by Caswell (1986, 1988). The life cycle is described by a directed graph (Fig. 2, 3) called the life cycle graph (Caswell 1982). Nodes indicate developmental stages (e.g., age or size classes). Edges indicate transitions possible during the interval $[t, t+1]$; the coefficient on the edge from stage j to stage i gives the number of individuals in stage i at $t+1$ per individual in stage j at t. The projection matrix corresponding to this graph is $\mathbf{A}$, where a_{ij} is the coefficient on the edge from node j to node i. This matrix maps the stage distribution vector $\mathbf{n}$ from t to $t+1$.

$$\mathbf{n}(t+1) = \mathbf{A}\mathbf{n}(t) \quad . \tag{1}$$

Under most circumstances, continued application of Eq. (1) leads to exponential growth at a rate λ with a fixed stage distribution $\mathbf{w}$. λ is given by the dominant eigenvalue[1] of $\mathbf{A}$. The stable stage distribution and stage-specific reproductive value distribution are given by the corresponding right and left eigenvectors $\mathbf{w}$ and $\mathbf{v}$. The characteristic equation and the eigenvectors of $\mathbf{A}$ can be written down directly from the life cycle graph using z-transform methods (Hubbell and Werner 1979; Caswell 1982).

Perturbation analysis of the asymptotic growth rate λ plays a central role in evolutionary demography. The sensitivity of λ to changes in an element a_{ij} of the projection matrix is

$$\frac{\partial \lambda}{\partial a_{ij}} = \frac{v_i w_j}{(\mathbf{v}, \mathbf{w})} \tag{2}$$

(Caswell 1978) where v_i and w_j are the ith entry of $\mathbf{v}$ and the jth entry of $\mathbf{w}$, and $(\mathbf{v}, \mathbf{w})$ denotes the scalar product. Sensitivities to other parameters, which do not appear directly as matrix elements a_{ij}, are easily calculated. If some variable x affects the a_{ij}, then

Biology Department, Woods Hole Oceanographic Institution, Woods Hole, MA 02543, USA

[1] When it is necessary to distinguish the dominant from the other eigenvalues, it will be denoted λ_1.

Size Structured Populations
B. Ebenman and L. Persson (Eds.)

$$\frac{\partial \lambda}{\partial x} = \sum_{i,j} \frac{\partial \lambda}{\partial a_{ij}} \frac{\partial a_{ij}}{\partial x} . \tag{3}$$

The evolutionary analysis of life histories relies on the connection between demographic parameters and fitness. The genetic basis for this connection (the "secondary theorem of natural selection" (Robertson 1968)) has been developed by Lande (1982a, b). Consider a vector $\bar{\mathbf{z}}$ giving the mean of a set of phenotypic traits z_i. The rate of phenotypic change under natural selection is

$$\Delta \bar{\mathbf{z}} = \lambda^{-1} \mathbf{G} \nabla \lambda , \tag{4}$$

where $\mathbf{G}$ is the additive genetic covariance matrix and the "selection gradient" vector $\nabla \lambda$ is

$$\nabla \lambda = \left(\frac{\partial \lambda}{\partial z_1} \frac{\partial \lambda}{\partial z_2} \cdots \frac{\partial \lambda}{\partial z_k} \right)' , \tag{5}$$

which can be obtained directly from the projection matrix, by Eqs. (2) and (3) (see Chaps. by Kirkpatrick and Lynch and Arnold this Vol.). For applications in life history theory see Caswell (1983, 1984, 1985), Caswell et al. (1984).

This paper will focus on a simple size- or stage-classified model, which has been applied to a number of real populations, especially of trees (review in Caswell 1986); I will refer to it as the "standard size-classified model". The projection matrix (for an example with four size classes) is

$$\mathbf{A} = \begin{pmatrix} P_1 & F_2 & F_3 & F_4 \\ G_1 & P_2 & 0 & 0 \\ 0 & G_2 & P_3 & 0 \\ 0 & 0 & G_3 & P_4 \end{pmatrix} , \tag{6}$$

where F_i is the size-specific fertility, G_i the probability of surviving and growing into the next size class, and P_i the probability of surviving and remaining in the present size class. G_i and P_i can be written in terms of the size-specific survival probability σ_i and the size-specific growth probability γ_i:

$$G_i = \sigma_i \gamma_i \tag{7}$$

$$P_i = \sigma_i (1 - \gamma_i) . \tag{8}$$

2 Age and Size as State Variables

State variables in dynamic models must satisfy certain requirements (Zadeh 1969, Caswell et al. 1972, Metz and Diekmann 1986). First, the state and the environment at time t must determine (at least stochastically) the response of the system at time t. Second, the state at time t_1 and the environment over the interval $[t_1, t_2)$ must determine (at least stochastically) the state at t_2. Metz and Diekmann (1986) introduced the useful distinction between i-state variables (e.g., age, size, hunger), which describe the state of an individual, and p-state variables, which describe the state of the population. The p-state variable is a density function of individuals over the values of the i-state variable.

Age was the i-state variable of choice in classical demographic models, all of which used the age distribution as a p-state variable. However, the development rate of many species is so variable that age provides little or no information about the demographic status of an individual. For such organisms, the age distribution reveals little about the behavior of the population, and is an inadequate p-state variable. The question is, how can potential state variables be evaluated?

2.1 Evaluating State Variables by Loglinear Analysis

A successful i-state variable permits prediction of an individual's fate, or at least improves such prediction over that possible without the state information. This can be tested by following the fates of individuals. The corresponding procedure at the population level is more difficult, since it requires establishing populations with different p-states and seeing if the p-state determines the population's response to the environment. Metz and Diekmann (1986) show, however, that under most situations a distribution function over an adequate i-state variable is an adequate p-state variable. Thus data on the state transitions of individuals also provide a test of the adequacy of potential p-state variables.

When both the i-state variables and the possible fates of an individual are discrete, loglinear analysis (Bishop et al. 1975; Fienberg 1977; Fingleton 1984) provides a powerful statistical approach to this problem.

Individuals are classified according to their initial state, as measured by two or more criteria (e.g., age and size). Their demographic fate is recorded after some time interval. The fate categories may or may not be the same as the state categories (e.g., death often appears as a fate, but not usually as a state). Such an experiment yields a multi-way contingency table; the examples to follow produce tables classified by age (A), size (S), and fate (F). The statistical significance of interactions among factors is tested by examining the changes in the log likelihood ratio G^2 as terms are added to or deleted from a model. (Following standard practice, I consider only hierarchical models, which can be specified by writing down the highest order interactions necessary to imply all terms actually present, e.g., the model AS, F contains the terms A, S, F, and AS.)

The significance of an interaction is always measured relative to the model to which it is added. This requires the choice of an appropriate "null model" against which to test the interaction. In the present context, age and size are explanatory variables, and fate is a response variable. The appropriate null model is thus AS, F (Fingleton 1984), asserting that the structure of the table reflects the age-size distribution of the tested individuals and the frequency of different fates, but that fate is independent of age and size. Against this null hypothesis, we can evaluate the effects of adding terms representing the interaction of explanatory and response variables: AF (the effect of age on fate), SF (the effect of size on fate), and ASF (the effect of age-size combinations on fate).

2.2 Examples

Monocarpic Perennial Plants. Werner (1975) and Gross (1981) followed the fate (death, survival as a vegetative rosette, or flowering) of individuals of known size and age of *Dipsacus sylvestris* and *Verbascum thapsus*. Both species are short-lived monocarpic perennial plants.

The results are shown in Table 1. Using the null model AS, F, the effects of size (measured by the SF interaction) and age (measured by the AF interaction) are highly significant in both species. These tests, however, examine the effects of age and size relative to a model containing no other factors. More interesting is the question of whether age (size) provides any additional information, once size (age) has been included. This is tested by evaluating the SF interaction against the model AS, AF, and the AF interaction against the model AS, SF. For *Dipsacus*, the SF interaction is still highly significant after age is included, but the AF interaction is only slightly significant after size is included. For *Verbascum*, the SF interaction maintains its significance when age is included, but the AF interaction is no longer significant when size is included. In these species, age loses most or all of its importance once size is accounted for. Size is clearly the more important i-state variable in these species.

Reproductive Output in the Moose. A very different pattern appears in data on reproduction (nonreproductive, one calf, or twins) in the moose (*Alces alces*) in

Table 1. Results of loglinear analysis of age and size effects on fate of *Dipsacus sylvestris* and *Verbascum thapsus*

Model	*Dipsacus sylvestris*			*Verbascum thapsus*		
	df	G^2	P	df	G^2	P
AS, F	46	1253.19		22	648.8	
AS, SF	36	55.04		12	20.1	
SF	10	1198.15	0.0001	10	628.7	0.0001
AS, F	46	1253.19		22	648.8	
AS, AF	40	718.23		20	629.79	
AF	6	534.96	0.0001	2	19.01	0.0001
AS, AF	40	718.23		20	629.79	
AS, SF, AF	30	38.56		10	17.72	
SF	10	679.67	0.0001	10	612.07	0.0001
AS, SF	36	55.04		12	20.1	
AS, SF, AF	30	38.56		10	17.72	
AF	6	16.48	0.0114	2	2.38	0.3052
AS, SF, AF	30	38.56		10	17.72	
ASF	0	0		0	0	
ASF	30	38.56	0.1360	10	17.72	0.0598

Table 2. Results of loglinear analysis of the effects of age and size on reproductive fate of female moose (*Alces alces*)

Model	df	G^2	P
AS, F	28	247.21	
AS, SF	24	234.34	
SF	4	12.87	0.012
AS, F	28	247.21	
AS, AF	20	16.66	
AF	8	230.55	0.0001
AS, AF	20	16.66	
AS, SF, AF	16	12.05	
SF	4	4.61	0.3299
AS, SF	24	234.34	
AS, SF, AF	16	12.05	
AF	8	222.29	0.0001
AS, SF, AF	16	12.05	
ASF	0	0	
ASF	16	12.05	0.7404

Table 3. Loglinear analysis of age and size effects on colony growth and mortality of three genera of corals. (Data of Hughes and Connell 1987)

Model	*Acropora*			*Porites*			*Pocillopora*		
	df	G^2	P	df	G^2	P	df	G^2	P
AS, F	15	104.7		15	43.47		15	25.99	
AS, SF	9	34.03		9	10.82		9	8.72	
SF	6	70.67	0.0001	6	32.65	0.0001	6	17.27	0.0083
AS, F	15	104.7		15	43.47		15	25.99	
AS, AF	12	73.5		12	32.98		12	20.44	
AF	3	31.2	0.0001	3	10.49	0.0148	3	5.55	0.1358
AS, AF	12	73.5		12	32.98		12	20.44	
AS, SF, AF	6	14.86		6	1.3		6	2.31	
SF	6	58.64	0.0001	6	31.68	0.0001	6	18.13	0.0059
AS, SF	9	34.03		9	10.82		9	8.72	
AS, SF, AF	6	14.86		6	1.3		6	2.31	
AF	3	19.17	0.0003	3	9.52	0.0231	3	6.41	0.0934
AS, SF, AF	6	14.86		6	1.3		6	2.31	
ASF	0	0		0	0		0	0	
ASF	6	14.86	0.0214	6	1.3	0.9717	6	2.31	0.8893

Table 4. Loglinear analysis of age and size effects on colony mortality in three genera of corals. (Data of Hughes and Connell 1987)

Model	*Acropora*			*Porites*			*Pocillopora*		
	df	G^2	P	df	G^2	P	df	G^2	P
AS, F	5	41.69		5	20		5	18.54	
AS, SF	3	1.25		3	3.61		3	7.47	
SF	2	40.44	0.0001	2	16.39	0.0003	2	11.07	0.0039
AS, F	5	41.69		5	20		5	18.54	
AS, AF	4	40.98		4	18.76		4	12.94	
AF	1	0.71	0.3977	1	1.24	0.2664	1	5.6	0.018
AS, AF	4	40.98		4	18.76		4	12.94	
AS, SF, AF	2	0.37		2	1.15		2	0.63	
SF	2	40.61	0.0001	2	17.61	0.0001	2	12.31	0.0021
AS, SF	3	1.25		3	3.61		3	7.47	
AS, SF, AF	2	0.37		2	1.15		2	0.63	
AF	1	0.88	0.347	1	2.46	0.1165	1	6.84	0.0089
AS, SF, AF	2	0.37		2	1.15		2	0.63	
ASF	0	0		0	0		0	0	
ASF	2	0.37	0.8312	2	1.15	0.563	2	0.63	0.7301

Norway (Saether and Haagenrud 1983). They classified females into three weight classes and five age classes. Both the AF and SF interactions are significant when tested against the null model AS, F (Table 2), but the SF interaction loses its significance completely when tested against the model AS, AF. Age is thus much more important than body size in determining reproductive output in this species.

Colony Growth and Mortality in Reef Corals. Hughes and Connell (1987) report the effects of age and size on colony growth and survival of three genera of corals (*Acropora, Porites,* and *Pocillopora*) in Australia. Colonies were assigned to two age classes (<4.5 yr and >4.5 yr) and three size classes, and their growth and survival followed from 1970–1971. Colonies were assigned to one of four fate categories: negative growth, 0%–100% growth, >100% growth, and death.

For *Pocillopora* and *Porites,* size is clearly superior to age as a state variable (Table 3), while for *Acropora* both size and age effects are significant.

These results, of course, depend on the definition of "fate." Hughes and Connell (1987) also investigated the effects of size and age on mortality, without considering growth. The results of this loglinear analysis (Table 4) are quite different. Mortality of *Acropora* and *Porites,* is highly dependent on size, but almost totally unaffected by age, while both age and size have significant effects on the mortality of *Pocillopora.*

Shoot Demography in Rhododendron maximum. McGraw (1987) examined the demography of individual shoots (ramets) of this clonal shrub. Shoots were aged

Table 5. Loglinear analysis of the effects of shoot age and size on fate of ramets of *Rhododendron maximum*. (Data of McGraw 1987)

Model	df	G^2	P
AS, F	24	404.47	
AS, SF	18	270.59	
SF	6	133.88	0.0001
AS, F	24	404.47	
AS, AF	18	153.34	
AF	6	251.13	0.0001
AS, AF	18	153.34	
AS, SF, AF	12	23.11	
SF	6	130.23	0.0001
AS, SF	18	270.59	
AS, SF, AF	12	23.11	
AF	6	247.48	0.0001
AS, SF, AF	12	23.11	
ASF	0	0	
ASF	12	23.11	0.0268

by counting leaf scars, and measured by estimating leaf area. Individual shoots were assigned to one of four fates: death, branching, flowering, and survival without branching or flowering. Both age and size have highly significant effects on demographic fate (Table 5).

2.3 *Quantifying the Importance of State Variables*

This haphazard survey of discrete demographic data shows clearly that the importance of size and age as i-state variables varies among species. It is important to go beyond testing the significance of the interactions, and to quantify the relative importance of size and age, especially when both have significant effects. A powerful approach to this problem is provided by Goodman and Kruskal's (1954) τ, one of several measures of association for cross-classified data (Liebetrau 1983).

Consider a two-way table with age classes A_i as rows and fates F_j as columns. Select an individual at random and attempt to predict its fate. There will be a certain proportion of incorrect answers, depending on the marginal distribution of individuals among fates. Now choose an individual and attempt to predict its fate, given a knowledge of its age. The value of age as an i-state variable is measured by the proportionate decrease in the probability making an incorrect prediction. Goodman and Kruskal (1954) defined the index τ_{AF} to measure this reduction in incorrect predictions, by

$$\tau_{AF} = \frac{\sum_i \sum_j (m_{ij}^2/m_{i.}) - \sum_j (m_{.j}^2/m_{..})}{m_{..} - \sum_j (m_{.j}^2/m_{..})}, \tag{9}$$

where m_{ij} is the count in cell (i, j), $m_{i.}$ and $m_{.j}$ the marginal totals of row i and column j, and $m_{..}$ the total number of cases in the table. Perfect association ($\tau_{AF} = 1$) implies that knowledge of A totally determines the value of F; $\tau_{AF} = 0$ implies that the two variables are statistically independent. This index is not in general symmetrical; $\tau_{AF} \neq \tau_{FA}$.

To compare two potential i-state variables in a three-way table requires the corresponding formulae for multiple and partial associations (analogous to multiple and partial correlation coefficients), which were developed by Gray and Williams (1975). Consider the three-way table with variables age (A), size (S), and fate (F). The multiple association τ_{ASF} measures the extent to which fate is determined by knowledge of *both* age and size. It can be calculated from Eq. (9) by creating a new variable X whose values consist of all combinations of the values of A and S, and then calculating τ_{XF}.

The partial associations $\tau_{AF|S}$ and $\tau_{SF|A}$ measure the importance of age in determining fate, when size is taken into account, and the importance of size in determining fate when age is taken into account. They are obtained from the simple and multiple associations by the relation

$$\tau_{AF|S} = \frac{\tau_{ASF} - \tau_{AS}}{1 - \tau_{AS}} \tag{10}$$

(Gray and Williams 1975), where τ_{AS} is the simple association calculated for the marginal table relating age and size. The formula for $\tau_{SF|A}$ is analogous.

Light and Margolin (1971) and Anderson and Landis (1980) show that the simple, partial, and multiple τ values correspond to simple, partial and multiple R^2 values in regression.

Table 6 gives the simple, multiple, and partial associations between age, size, and fate, for the examples described in the preceeding section, as well as for two of Alm's (1959) experiments (numbered F3B and F3C) on age, size, and maturity in the trout *Salmo trutta*. Alm's data were analyzed twice, once with fate categories mature and immature, and once with fate categories mature males, mature females, and immatures. The results show clearly that the relative importance of age and size in a given species can vary with environmental factors.

Comparison of $\tau_{SF|A}$ and $\tau_{AF|S}$ shows that size is a more important state variable than age in all cases except for the moose (in which size is extremely unimportant), the ramet demography of *Rhododendron*, and one experiment on maturity in trout. For the other species, size is up to 35 times as important as age as a potential i-state variable, even when age has statistically significant effects.

A summary of the relative importance of size and age can be obtained by examining the ratio $(\tau_{SF|A})/(\tau_{AF|S})$. Values greater than 1 indicate that size is more important than age, and vice-versa. This ratio is plotted on a logarithmic scale in Fig. 1. The relative importance of size and age varies smoothly among these data sets. This raises the question of whether size-classified models can be used as approximations when both size and age are important, to which I turn in the next section.

Table 6. Simple, multiple, and partial associations between fate, age, and size for the examples considered in Section 2.2

Name	τ_{AF}	τ_{SF}	τ_{ASF}	$\tau_{SF\|A}$	$\tau_{AF\|S}$	$\tau_{SF\|A}/\tau_{AF\|S}$
Dipsacus sylvestris	0.139	0.390	0.396	0.298	0.010	30.346
Verbascum thapsus	0.013	0.482	0.489	0.482	0.014	35.688
Rhododendron	0.374	0.146	0.424	0.080	0.326	0.245
Alces alces	0.140	0.009	0.147	0.008	0.139	0.058
Porites growth and mortality	0.036	0.151	0.190	0.160	0.046	3.478
Pocillopora growth and mortality	0.026	0.080	0.117	0.093	0.040	2.323
Acropora growth and mortality	0.044	0.109	0.145	0.106	0.040	2.615
Porites mortality	0.014	0.188	0.219	0.208	0.038	5.446
Pocillopora mortality	0.067	0.137	0.212	0.155	0.087	1.788
Acropora mortality	0.003	0.163	0.170	0.168	0.008	20.029
Salmo trutta F3B sex and maturity	0.353	0.361	0.472	0.184	0.174	1.059
Salmo trutta F3C sex and maturity	0.239	0.421	0.483	0.321	0.107	2.994
Salmo trutta F3B maturity	0.552	0.535	0.647	0.212	0.241	0.880
Salmo trutta F3C maturity	0.366	0.624	0.675	0.487	0.136	3.593

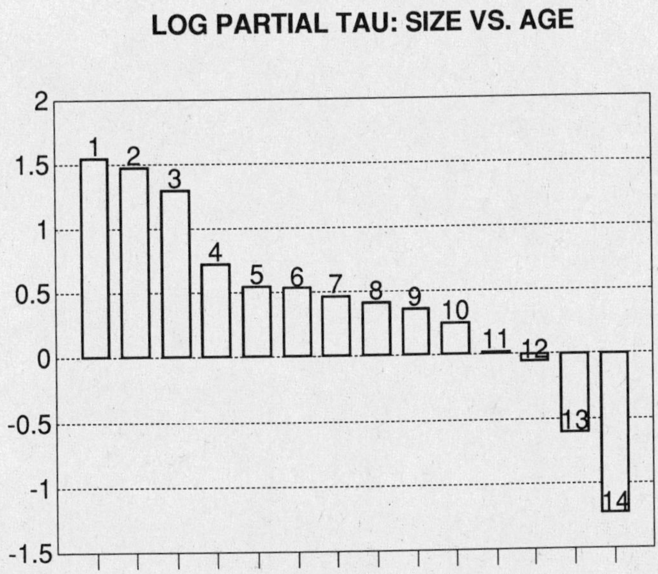

Fig. 1. The relative importance of size and age in determining demographic fate, as measured by $\log_{10}(\tau_{SF|A}/\tau_{AF|S})$; values greater than 0 indicate that size is more important than age and vice versa. Data sources: **1** *Verbascum thapsus*; **2** *Dipsacus sylvestris*; **3** *Acropora* mortality; **4** *Porites* mortality; **5** *Salmo trutta* F3C maturity; **6** *Porites* growth and mortality; **7** *Salmo trutta* F3C sex and maturity; **8** *Acropora* growth and mortality; **9** *Pocillopora* growth and mortality; **10** *Pocillopora* mortality; **11** *Salmo trutta* F3B sex and maturity; **12** *Salmo trutta* F3B maturity; **13** *Rhododendron maximum*; **14** *Alces alces* reproduction

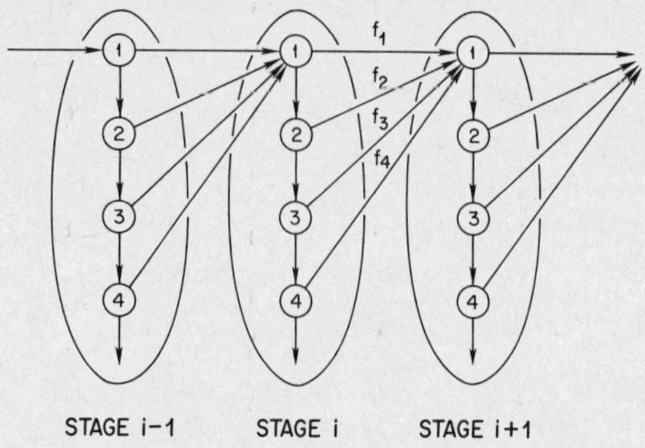

Fig. 2. A portion of the life cycle graph for three successive stages (e.g., instars) with transition (e.g., molting) probabilities determined by the time spent within the stage

3 Size-Classified Models as Approximations

The construction of models using both size and age is possible (Law 1983; Caswell 1983) but may be impractical because of the large number of categories required. An important special case is that in which individuals proceed through a series of developmental stages, with the probability of moving from one stage to the next dependent on the time already spent in the stage, but independent of the time spent in any previous stage. In such a life cyle (Fig. 2), individuals are classified by age *within* stages (Caswell 1983).

Since it is often impossible to measure the age of individuals within a stage, stage-classified models often arise as approximations to this complete model. This requires estimating the parameters, P_i and G_i of the stage-classified model from the stage duration distribution and stage-specific survival. This can be done by estimating γ (the growth probability; for convenience I will refer to the probability of molting) from the stage duration distribution, and combining this with estimates of σ to calculate P_i and G_i.

3.1 The Geometric Distribution

The simplest approximation follows from the fact that if the probability of molting from stage i is a constant γ_i, then among a cohort of individuals which all survive to molt, the stage duration T_i has a geometric distribution with $P(T_i = k) = \gamma_i(1-\gamma_i)^{k-1}, k = 1, 2, 3, \ldots$ with mean $(\bar{T}_i = 1/\gamma_i)$. Thus

$$\gamma_i \approx \bar{T}_i^{-1} . \tag{11}$$

3.2 Fixed Stage Durations

As a second approximation, suppose that T_i is identical for every individual. The proportion of individuals in stage i which molt is then not constant, but depends on the age distribution within the stage. An approximate constant probability γ_i

can be calculated by assuming that the age distribution within the instar has stabilized.

Assume that survival probability is a constant σ_i for all ages within stage i. The stable age distribution within the stage can be written down using the methods of Caswell (1982):

$$\mathbf{w} = \begin{pmatrix} 1 \\ (\sigma_i/\lambda) \\ (\sigma_i/\lambda)^2 \\ \vdots \\ (\sigma_i/\lambda)^{T_i-1} \end{pmatrix} . \tag{12}$$

All individuals in age class T_i molt, so the proportion molting is

$$\gamma_i = \frac{(\sigma_i/\lambda)^{T_i-1}}{1+(\sigma_i/\lambda)+(\sigma_i/\lambda)^2+\ldots+(\sigma_i/\lambda)^{T_i-1}} . \tag{13}$$

Note that if $\sigma_i/\lambda = 1$, this expression reduces to (11).

If $\sigma_i/\lambda < 1$, the series in the denominator of Eq. (13) is

$$\sum_{j=0}^{T_i-1} \left(\frac{\sigma_i}{\lambda}\right)^j = \frac{(\sigma_i/\lambda)^{T_i}-1}{(\sigma_i/\lambda)-1}$$

so that

$$\gamma_i = \frac{(\sigma_i/\lambda)^{T_i}-(\sigma_i/\lambda)^{T_i-1}}{(\sigma_i/\lambda)^{T_i}-1} . \tag{14}$$

This formula is useful when the "stages" are groups of age classes. If a stage consists of individuals aged $k, k+1, \ldots, k+T$, then during $(t, t+1)$ the oldest individuals will graduate to the next stage, and all younger individuals will move up one age class. This model is, in a sense, not "stage classified" at all, since demography is still determined by age, but it is useful when only relatively crude estimates of the survival are available over broad age ranges. Crouse et al. (1987) applied it to estimate stage-classified demographic parameters for the loggerhead sea turtle *Caretta caretta*.

3.3 Variable Stage Durations

Stage duration will, in general, vary among individuals. For each age j within the stage, there is then a probability f_j of molting. The total proportion molting at any time depends on the within-stage age distribution at that time, and by assuming a stable age distribution it is possible to estimate γ_i from information on the stage duration distribution and the stage specific survival probability σ_i (Caswell 1983; what follows here is more complete).

Define f_j as the probability that an individual in age class j molts, given that it survives; g_j as the probability density of age at molting that would be observed in the absence of mortality, and h_j as the probability of not having molted by age j, again in the absence of mortality. The three quantities are related:

$$g_j = f_j \prod_{j=1}^{i-1} (1-f_j) \quad g_1 = f_1 \tag{15}$$

$$h_j = \prod_{j=1}^{i} (1-f_j) \qquad h_0 = 1 \tag{16}$$

$$f_j = g_j/h_{j-1} \tag{17}$$

(If we make an analogy between molting and death, the process of molting is described by a life table, and $h_j = l_x$, $g_j = d_x$, and $f_j = q_x$.)

The stable age distribution within the stage is given by

$$\mathbf{w} = \begin{pmatrix} 1 \\ (1-f_1)(\sigma/\lambda) \\ (1-f_1)(1-f_2)(\sigma/\lambda)^2 \\ \vdots \\ \prod_{k=1}^{j-1}(1-f_k)(\sigma/\lambda)^{j-1} \\ \vdots \end{pmatrix} . \tag{18}$$

The proportion molting is thus

$$\gamma = \frac{\sum_j f_j w_j}{\sum_j w_j} \tag{19}$$

$$= \frac{(\lambda/\sigma) \sum_j g_j (\lambda/\sigma)^{-j}}{(\lambda/\sigma) \sum_j h_j (\lambda/\sigma)^{-j}} . \tag{20}$$

Let $a = \ln(\lambda/\sigma)$. Then we can rewrite this expression as

$$\ln \gamma = \ln \left(\sum_{j=1} g_j e^{-aj} \right) - \ln \left(\sum_{j=0} h_j e^{-aj} \right) . \tag{21}$$

The first term on the right hand side of Eq. (21) is the cumulant generating function[2] of g. The second term differs from a cumulant generating function only because h_j is not a probability density function. However, it can be transformed into one by dividing by the sum of the function h. Define $\hat{h}_i = h_j / \sum_j h_j$. Then Eq. (21) can be rewritten.

$$\ln \gamma = \ln \left(\sum_j g_j e^{-aj} \right) - \ln \left(\sum_j \hat{h}_j e^{-aj} \right) - \ln \bar{T} , \tag{22}$$

where $\bar{T} = \sum_j h_j$ is the expected stage duration, analogous to the life expectancy $e_0 = \sum_x l_x$ of classical demography.

Expanding the cumulant generating functions in Eq. (22) up to terms of second order in a yields

[2] The cumulant generating function of a probability density F is $\psi(a) = \ln \int_0^\infty F(z) e^{-az} dz$. The cumulants K_i are defined by the series expansion $\psi(a) = -aK_1 + \frac{a^2}{2!} K_2 - \frac{a^3}{3!} K_3 + \ldots$ The first cumulant K_1 equals the mean μ; the second cumulant K_2 equals the variance σ^2. Relations between the higher cumulants and the moments of F are given in Kendall and Stuart (1958, p. 71).

$$\ln \gamma \approx -a[K_1(g)-K_1(\hat{h})]+\frac{a^2}{2!}[K_2(g)-K_2(\hat{h})]-\ln \bar{T} \ , \tag{23}$$

Keyfitz (1977, p. 131) shows that

$$K_1(g) = \bar{T}$$

$$K_2(g) = V(T)$$

$$K_1(\hat{h}) = \frac{V(T)+\bar{T}^2}{2\bar{T}}$$

where V(T) denotes the variance of stage duration. Thus, to first order in a, we can write

$$\ln \gamma \approx -a\left[\frac{\bar{T}}{2}-\frac{V(T)}{2\bar{T}}\right] - \ln \bar{T} \ , \tag{24}$$

or

$$\gamma \approx \left(\frac{1}{\bar{T}}\right) \exp\left[-a\left(\frac{\bar{T}}{2}-\frac{V(T)}{2\bar{T}}\right)\right] \ . \tag{25}$$

From Eq. (25) we again obtain the approximation $\gamma \approx \bar{T}^{-1}$, when a is sufficiently small. When $a = \ln(\lambda/\sigma) > 0$, $\gamma \approx \bar{T}^{-1}$ overestimates γ, because it overestimates the abundance of old individuals within the instar by neglecting the effects of mortality and population growth in shifting the age distribution towards young individuals. The smaller the variance V(T) in stage duration, and the larger the average stage duration $\bar{T}$, the greater the overestimate of γ.

3.4 Iterative Calculation

The alert reader will have noted that Eqs. (14) and (25) for γ depend on λ. However, λ is an eigenvalue of the very matrix whose entries are being estimated, and cannot be calculated until the estimation is completed!

One solution is to assume that $\lambda \approx 1$, and ignore it, as in Crouse et al. (1987). An iterative approach, however, may improve the estimates. An initial value of λ is chosen and the entries in **A** estimated. The eigenvalues of **A** yield a second estimate of λ, with which the parameters can be estimated again. If repeated, this process usually (but not always, based on the few calculations I have done) converges to a matrix **A** whose entries are compatible with its own eigenvalues. It is unclear how much difference this makes in practice. For example, when applied to the data of Crouse et al. (1987) on the loggerhead turtle, it changes the estimate of λ from 0.9450 to 0.9644. The uncertainty in the original survival and growth data is certainly greater than this.

3.5 Negative Binomial Stage Durations

The major weakness of the preceding methods is their assumption of a stable within-stage age distribution. This section presents a method which approximates the stage duration distribution by a negative binomial distribution (or, in continuous time, a gamma distribution) with specified mean and variance, without assuming that the age distribution is stable. The negative binomial is a flexible distribution, and can approximate many observed stage duration distributions (see Blythe et al. 1984, who discuss the addition of a lag, representing a minimum age before which no individual molts. Such a lag can easily be added to the life cycle graph, but will not be explored further here.)

Consider a series of k identical stages, each with probabilities $1-\gamma$ of remaining in the stage and γ of moving to the next stage. The time T required to grow through all k stages is the time required for the kth success in a series of identical Bernoulli trials with probability of success γ; this time has a negative binomial distribution

$$P(T=x)=\binom{x-1}{k-1}\gamma^k(1-\gamma)^{x-k} ,$$

with mean and variance

$$\bar{T}=\frac{k}{\gamma} \tag{26}$$

$$V(T)=\frac{k(1-\gamma)}{\gamma^2} . \tag{27}$$

To produce a stage i, the duration of which has a negative binomial distribution, we divide the stage into a series of k identical but invisible "pseudo-stages". These are not ages within the stage, and in fact cannot be identified at all. Their only purpose is to slow down individuals proceeding through the stage in such a way as to produce a distribution of stage durations with a specified mean and variance.

Mortality and fertility are identical for all pseudo-stages within a stage. Thus each transition within the stage is multiplied by the stage-specific survival probability σ_i, and is given the same stage-specific fertility coefficient F_i.

Given the mean $\bar{T}$ and variance $V(T)$ of the stage duration, the parameters specifying the pseudo-stages are given by

$$\gamma_i=\frac{\bar{T}}{V(T)+\bar{T}} \tag{28}$$

$$k_i=\frac{\bar{T}^2}{V(T)+\bar{T}} . \tag{29}$$

3.6 Estimation Methods Compared

Consider the life cycle graph in Fig. 3, which contains an immature stage (n_2) of variable duration. The parameters P_2 and G_2 are determined by the stage dura-

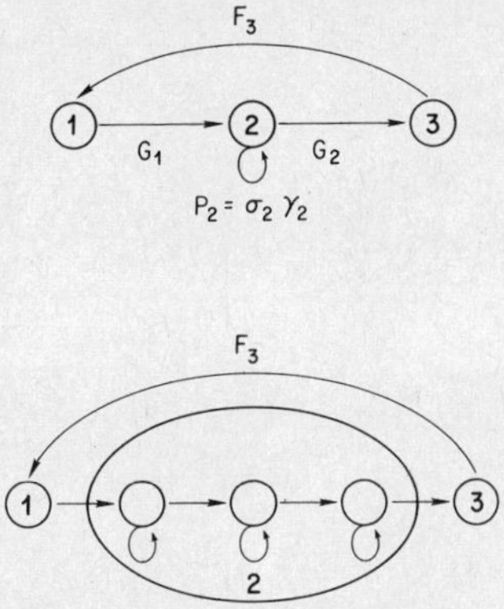

Fig. 3. A life cycle graph (above) for a population with an immature stage n_2 of variable duration, with the corresponding graph (below) for a model assuming that the duration of n_2 follows a negative binomial distribution. The second stage now contains three identical "pseudo-stages"; see text for details

tion; let the other parameters be $G_1 = 0.5$, $F_3 = 10$, $\sigma_2 = 0.75$, and let the mean and variance of stage duration be $\bar{T} = 5$, $V(T) = 3$. Applying the four methods outlined above [using iteration to obtain λ for Eqs. (14) and (25)] produces the following estimates of γ_2, and the resultant matrices and their dominant eigenvalues:

1. Geometric stage duration distribution: $\gamma_2 = 0.2$, $\lambda = 1.1587$,

$$\mathbf{A} = \begin{pmatrix} 0 & 0 & 10 \\ 0.5 & 0.6 & 0 \\ 0 & 0.15 & 0 \end{pmatrix} \tag{30}$$

2. Fixed stage duration: $\gamma_2 = 0.097$, $\lambda = 1.0246$,

$$\mathbf{A} = \begin{pmatrix} 0 & 0 & 10 \\ 0.5 & 0.677 & 0 \\ 0 & 0.073 & 0 \end{pmatrix} \tag{31}$$

3. Variable stage duration, stable age distribution: $\gamma_2 = 0.10$, $\lambda = 1.0289$,

$$\mathbf{A} = \begin{pmatrix} 0 & 0 & 10 \\ 0.5 & 0.675 & 0 \\ 0 & 0.075 & 0 \end{pmatrix} \tag{32}$$

4. Negative binomial stage duration distribution: $\gamma_2 = 0.625$, $k = 3$, $\lambda = 1.0548$,

$$\mathbf{A} = \begin{pmatrix} 0 & 0 & 0 & 0 & 10 \\ 0.5 & 0.2813 & 0 & 0 & 0 \\ 0 & 0.4688 & 0.2813 & 0 & 0 \\ 0 & 0 & 0.4688 & 0.2813 & 0 \\ 0 & 0 & 0 & 0.4688 & 0 \end{pmatrix} \tag{33}$$

In this example, the choice of an approximation makes an appreciable difference in the asymptotic behavior of the resulting stage-classified model. It also affects the transient response. Only the negative binomial model actually delays individuals in their progress through n_2; as a result, it converges to its stable structure much more slowly (and more realistically) than the others.

4 Extracting Age-Specific Information from Size Models

The preceding section considered the development of the standard size-classified model from information on age-related phenomena within a stage. This section addresses the converse problem: how to derive certain inherently age-related statistics (net reproductive rate, generation time) from size-classified models.

4.1 Age-Classified Statistics

The net reproductive rate gives the expected number of offspring by which a newborn individual will be replaced by the end of its life. It is given by

$$R_0 = F_1 + P_1 F_2 + P_1 P_2 F_3 + \ldots \tag{34}$$

$$= \sum_i F_i \prod_{j=1}^{i-1} P_j \; . \tag{35}$$

The cases $R_0 < 1$, $R_0 = 1$, and $R_0 > 1$ correspond to $\lambda_1 < 1$, $\lambda_1 = 1$, and $\lambda_1 > 1$, respectively.

There are several measures of the "length of a generation" in age-classified models. Consider an age-classified Leslie matrix with fertility terms F_i and age-specific survival probabilities P_i. Using Coale's (1972) notation, generation time can be measured by

1. The time T required for the population to increase by a factor of R_0. T satisfies $\lambda_1^T = R_0$, from which

$$T = \ln R_0 / \ln \lambda_1 \tag{36}$$

$$= \ln R_0 / r \; . \tag{37}$$

2. The mean of the net fertility schedule

$$\mu_1 = \frac{\sum_i i F_i \prod_{j=1}^{i-1} P_1}{\sum_i F_i \prod_{j=1}^{i-1} P_1} \; . \tag{38}$$

This gives the mean age of the parents of the offspring produced by a cohort subject to the mortality regime described by the P_i.

3. The mean age of reproduction in a population at the stable age distribution:

$$\bar{A} = \frac{\sum_i i \lambda^{-i} F_i \prod_{j=1}^{i-1} P_j}{\sum_i \lambda^{-i} F_i \prod_{j=1}^{i-1} P_j} \; . \tag{39}$$

This gives the mean age of the parents of the offspring produced by a population at the stable age distribution. The denominator in this case is the characteristic equation, and is thus equal to 1.

Note that μ_1 can be obtained from Eq. (39) by setting $\lambda = 1$. Thus, for a stationary population, μ_1 is equivalent to $\bar{A}$.

4.2 Age-Classified Statistics from Size-Classified Models

At first glance, the calculation of net reproductive rate and generation times for stage classified models seems impossible. These statistics are inherently connected with the *age* of reproducing individuals, and no information on age appears in a size-classified model. Surprisingly, the z-transformed life cycle graph (Hubbell and Werner 1979; Caswell 1982, 1986) technique will, in some cases, permit the extraction of this information.

I assume that the life cycle graph contains, or can be reduced to contain, no disjoint loops[3]. Self-loops are thus permitted, since they can be absorbed in the calculation of transmissions, and in particular the standard size-classified model satisfies this assumption.

The characteristic equation for a life cycle satisfying this assumption is

$$1 = \sum_i L^{(i)} , \tag{40}$$

where $L^{(i)}$ is the ith loop transmission in the z-transformed life cycle graph (Caswell 1982). Each edge in the z-transformed graph, with the self-loops absorbed, has a transmission of the form

$$\frac{a_{ij}\lambda^{-1}}{1-a_{ii}\lambda^{-1}} = a_{ij}(\lambda^{-1}+a_{ii}\lambda^{-2}+a_{ii}^2\lambda^{-3}+\ldots) . \tag{41}$$

Each loop transmission $L^{(i)}$ is a product of terms of this form. Combining terms containing like powers of λ^{-1}, we can rewrite the characteristic Eq. (40) as

$$1 = \sum_{i=1}^{\infty} k_i \lambda^{-i} , \tag{42}$$

where k_i measures the contribution of pathways of length i.

Rewriting Eq. (40) in this form translates the stage-classified model into an age-classified model by including all of the possible age-specific trajectories that an individual might follow as it develops through the stages of the life cycle. The series in Eq. (42) is known to converge, since it is equal to the characteristic Eq. (40).

Once the connection between the two versions of the characteristic equation [Eqs. (40) and (42)] is recognized, the net reproductive rate R_0 can be calculated by direct analogy with the age-classified case as

$$R_0 = \sum_i L^{(i)} \Big|_{\lambda=1} . \tag{43}$$

[3] Thus there is at least one stage through which all individuals must pass. This corresponds to the assumption that "all individuals are born identical" used by Metz and Diekmann (1986, Chap. IV) and Metz et al. (this Vol.) in their discussion of related issues in continuous time.

The generation time indices T, μ_1, and $\bar{A}$ are also easily calculated. T is calculated from R_0 and λ, using Eqs. (36) and (43). The mean age of reproduction in the stable population ($\bar{A}$) is calculated by noting that

$$\sum_i ik_i\lambda^{-i} = -\lambda^{-1}\sum_i \frac{\partial(k_i\lambda^{-i})}{\partial\lambda}. \tag{44}$$

Thus

$$\bar{A} = -\lambda^{-1}\sum_i \left.\frac{\partial L^{(i)}}{\partial\lambda}\right|_{\lambda=\lambda_1}, \tag{45}$$

The mean of the net fertility function (μ_1) is obtained by evaluating this partial derivative at $\lambda = 1$, and dividing by R_0:

$$\mu_1 = \frac{-\lambda^{-1}\sum_i(\partial L^{(i)}/\partial\lambda)|_{\lambda=1}}{\sum_i L^{(i)}|_{\lambda=1}}. \tag{46}$$

Examples. The basic size-classified model is a suitable first approximation for the genet demography of clonal or modular organisms, in which genets are classified by the number of ramets or modules they have accumulated (Caswell 1985). The survival probability of large clones of such organisms can be very high. Indeed, some of the oldest living organisms are large clones of vegetatively reproducing shrubs and ferns (Cook 1983).

Potts (1983, 1984) has raised an interesting question concerning the generation times of clonal organisms in the context of the evolutionary dynamics of Indo-Pacific reef corals. About half of these reefs lie on continental shelves which have been subject to high frequency sea level fluctuations since the Pliocene-Quaternary glaciations. He estimates that any given bathymetric level in this region has remained in the depth zone of active coral growth for an average of only 3200 years. He suggests that if large clones of corals were nearly immortal, only a few generations could be completed in this period and these species would never have had a chance to reach evolutionary equilibrium.

To evaluate this hypothesis adequately would require size-classified demographic data on the appropriate corals, which is not available. However, size-classified matrices are available for tree populations. Trees are also long-lived modular organisms, and the survival probabilities of large individuals are very high. Generation times from these matrices may tell us something about Pott's hypothesis.

Consider the standard size-classified matrix [Eq. (6)]. We wish to obtain the derivatives of the characteristic equation. Define

$$U_i = \frac{F_i}{\lambda - P_1} \tag{47}$$

$$W_i = \frac{G_{i-1}}{\lambda - P_i} \tag{48}$$

and use these quantities to calculate

Table 7. The rate of increase λ, the net reproductive rate R_0 and the generation time measures T, $\bar{A}$ and μ_1 (in years) for size-classified tree populations

Species	R_0	λ	T	$\bar{A}$	μ_1
Avicennia marina	∞	1.227	∞	8.05	∞
Pentaclethra macroloba	1.2	1.002	112.9	110.84	115
Astrocaryum mexicanum	8.0	1.004	495.1	287.30	1107
A. mexicanum 2	17.3	1.011	251.3	115.22	1017
A. mexicanum 3	81.3	1.019	229.3	100.30	1179
A. mexicanum 4	0.5	0.993	650.6	–	93
A. mexicanum 5	1433.0	1.040	185.6	82.36	1137
A. mexicanum 6	281.6	1.023	251.3	98.59	2471

$$V_1 = 1 \tag{49}$$

$$V_i = V_{i-1} W_i \ . \tag{50}$$

The loop transmissions $L^{(i)}$ are then given by

$$L^{(i)} = V_i U_i \ . \tag{51}$$

The generation times $\bar{A}$ and μ_1 are obtained by applying Eqs. (45) and (46) to this expression for the loop transmissions. Letting primes denote differentiation with respect to λ,

$$V_1' = 0 \tag{52}$$

$$V_i' = W_i V_{i-1}' - \frac{V_i}{\lambda - P_i} \tag{53}$$

$$L^{(i)\prime} = V_i' U_i - \frac{L_i}{\lambda - P_1} \ . \tag{54}$$

The results of applying this analysis to matrices for the mangrove *Avicenna marina* (Burns and Ogden 1985), for *Pentaclethra macroloba* (Hartshorn 1975), and for six populations of *Astrocaryum mexicanum* (Pinero et al. 1984) are shown in Table 7.

It is apparent that generation times for these long-lived size-classified populations can indeed be very long, although not as astronomical as Potts (1983, 1984) suggested for corals. The crucial factor, of course, is that the average generation time reflects the structure of the entire population, not just the extreme age of the oldest large individuals.

R_0, T, and μ_1 are all extremely sensitive to the diagonal elements of A when those elements are close to 1, because of terms of the form $\lambda - P_i$, evaluated at $\lambda = 1$, in the denominator. In many studies, estimates of P_i for the largest size classes are very crude; sample sizes for such large individuals are usually small, and mortality rates may be too small to detect. The mean age in the stable population ($\bar{A}$) is much less sensitive to changes in the P_i and is probably the preferred measure of generation time in such cases. However, for chronically disturbed situations such as that considered by Potts, the cohort generation time μ_1 may be the more relevant index. In such cases, its value and interpretation will depend heavily on estimation of the survival probability of large clones.

5 Conclusions

The standard size-classified model provides one means to investigate the consequences of growth, survival and reproduction in size-classified populations. The methods outlined here extend that usefulness, by providing powerful tests of the relative importance of age and size in determining demographic fate, by permitting the incorporation of information on the distribution of stage durations, and by extracting information on generation times from the size-classified model.

Of course, the standard size-classified model is (purposely) simple. For many organisms, more complex life cycle graphs will be necessary, especially as more information on development is incorporated into demographic models. The methods presented here may also shed some light on these situations.

Acknowledgments. This research was supported by National Science Foundation grants OCE 85-16177, BSR 86-9395, and BSR 87-4936. I would like to thank K. Gross, T. P. Hughes, J. B. McGraw, and P. A. Werner for providing data. The comments of participants in the Symposium, especially Hans Metz, Andre de Roos, and Bo Ebenman, were very helpful. Woods Hole Oceanographic Institution Contribution 6666.

6 References

Alm G (1959) Connection between maturity, size, and age in fishes. Rep Drottningholm Inst Freshwater Res 40:5–145

Anderson RJ, Landis JR (1980) CATANOVA for multidimensional contingency tables: nominal-scale response. Commun Stat Theor Meth A9:1191–1206

Bishop YMM, Fienberg SE, Holland PW (1975) Discrete multivariate analysis: theory and practice. MIT, Cambridge, Mass

Blythe SP, Nisbet RM, Gurney WSC (1984) The dynamics of population models with distributed maturation periods. Theor Popul Biol 25:289–311

Burns BR, Ogden J (1985) The demography of the temperate mangrove [*Avicennia marina* (Forsk.) Vierh.] at its southern limit in Zew Zealand. Aust J Ecol 10:125–133

Caswell H (1978) A general formula for the sensitivity of population growth rate to changes in life history parameters. Theor Popul Biol 14:215–2320

Caswell H (1982) Stable population structure and reproductive value for populations with complex life cycles. Ecology 63:1223–1231

Caswell H (1983) Phenotypic plasticity in life history traits: demographic effects and evolutionary consequences. Am Zoologist 23:35–46

Caswell H (1984) Optimal life histories and the costs of reproduction: two extensions. J Theor Biol 107:169–172

Caswell H (1985) The evolutionary demography of clonal reproduction. In: Jackson JBC, Buss LW, Cook, RE (eds) Population biology and evolution of clonal organisms. Yale Univ Press

Caswell H (1986) Life cycle models for plants. Lect Math Life Sci 18:171–233

Caswell H (1988) Matrix population models: construction, analysis and interpretation. (in preparation)

Caswell H, Koenig H, Resh J, Ross Q (1972) An introduction to systems analysis for ecologists. In: Patten BC (ed) Systems analysis and simulation in ecology, vol 2. Academic Press, New York London

Caswell H, Naiman RJ, Morin R (1984) Evaluating the consequences of reproduction in complex salmonid life cycles. Aquaculture 43:123–134

Coale AJ (1972) The growth and structure of human populations: a mathematical approach. Univ Press, Princeton

Cook RE (1983) Clonal plant populations. Am Sci 71:244–253

Crouse DT, Crowder LB, Caswell H (1987) A stage-based population model for loggerhead sea turtles and implications for conservation. Ecology 68:1412–1423

Fienberg SE (1977) The analysis of cross-classified categorical data. MIT, Cambridge, Mass

Fingleton B (1984) Models of category counts. Univ Press, Cambridge

Goodman LA, Kruskal WH (1954) Measures of association for cross classifications. J Am Stat Assoc 49:732–764

Gray LN, Williams JS (1975) Goodman and Kruskal's tau b: multiple and partial analogs. Proc Soc Stat Sect Am Stat Assoc 444–448

Gross KL (1981) Predictions of fate from rosette size in four "biennial" plant species: *Verbascum thapsus, Oenothera biennis, Daucus carota*, and *Tragopogon dubius*. Oecologia (Berlin) 48:209–213

Hartshorn GS (1975) A matrix model of tree population dynamics. In: Golley FB, Medina E (eds) Tropical ecological systems. Springer, Berlin Heidelberg New York, pp 41–51

Hubbell SP, Werner PA (1979) On measuring the intrinsic rate of increase of populations with heterogeneous life histories. Am Nat 113:277–293

Hughes TP, Connell JH (1987) Population dynamics based on size or age? A reef-coral analysis. Am Nat 129:818–829

Kendall MG, Stuart A (1958) The advanced theory of statistics, vol 1. Hafner, New York

Keyfitz N (1977) Applied mathematical demography. John Wiley & Sons, New York

Lande R (1982a) A quantitative genetic theory of life history evolution. Ecology 63:607–615

Lande R (1982b) Elements of a quantitative genetic model of life history evolution. In: Dingle H, Hegmann JP (eds) Evolution and genetics of life histories. Springer, Berlin Heidelberg New York

Law R (1983) A model for the dynamics of a plant population containing individuals classified by age and size. Ecology 64:224–230

Liebetrau AM (1983) Measures of association. Quantitative applications in the social sciences 32. Sage, Beverly Hills

Light RJ, Margolin BH (1971) An analysis of variance for categorical data. J Am Stat Assoc 66:534–544

McGraw JB (1987) Age- and size-specific shoot life histories and shoot population growth in *Rhododendron maximum* (submitted)

Metz JAJ, Diekmann O (1986) The dynamics of physiologically structured populations. Springer, Berlin Heidelberg New York

Pinero D, Martinez-Ramos M, Sarukhan J (1984) A population model of *Astrocaryum mexicanum* and a sensitivity analysis of its finite rate of increase. J Ecol 72:977–991

Potts DC (1983) Evolutionary disequilibrium among indo-pacific corals. Bull Mar Sci 33:619–632

Potts DC (1984) Generation times and the Quaternary evolution of reef-building corals. Paleobiology 10:48–58

Robertson A (1968) The spectrum of genetic variation. In: Lewontin RC (ed) Population biology and evolution. Univ Press, Syracuse

Saether BE, Haagenrud H (1983) Life history of the moose (*Alces alces*): fecundity rates in relation to age and carcass weight. J Mammal 64:226–232

Werner PA (1975) Prediction of fate from rosette size in teasel (*Dipsacus fullonum* L.). Oecologia (Berlin) 20:197–201

Zadeh LA (1969) The concepts of system, aggregate, and state in system theory. In: Zadeh LA, Polak E (eds) System theory. McGraw-Hill, New York, pp 3–42

Population Models Incorporating Physiological Structure: A Quick Survey of the Basic Concepts and an Application to Size-Structured Population Dynamics in Waterfleas

J. A. J. METZ, A. M. DE ROOS and F. VAN DEN BOSCH

1 Introduction

In this chapter we give a survey of the modeling methodology for physiologically structured populations developed in Metz and Diekmann (1986) with a stress on simple techniques and applications. The main application considered is a size- (and age-) structured model for *Daphnia* population dynamics based on the ideas in Kooijman and Metz (1984). Although admittedly considerably simplified, this model appears to be able to generate all three types of observed *Daphnia* population behavior distinguished by Murdoch and McCauley (1985), a feat not (yet) reproduced by any of its competitors.

1.1 Why Structure?

Any mathematical model, just as any experiment, necessarily entails a good deal of simplification. The idea is to study mechanisms in their bare essentials, unobscured by all the wonderful detail present in nature. The ultimate aim is to help unravel the natural network of cause and effect by elucidating the general principles pertaining to its links. Therefore any simplifications made by the modeler, or rather any complexity retained, should directly bear on the principles he or she is searching for; and model and observation should be confronted judiciously, taking each for what it is worth.

In the classical lumped models of population dynamics, all individuals are treated as if they were the same. This is a useful ploy when the essentials of the population process under study can be captured in a few empirically meaningful parameters that do describe the gross interactions among individuals.

Usually the individuals making up a population are far from the same, and quantities like per capita birth and death rates are but "population properties", i.e., averages over the whole population, instead of direct reflections of chances common to all individuals. This often precludes the in-depth quantitative confrontation of lumped models with real data.

A second limitation of lumped models is that the essence of many mechanisms determining population behavior is directly related to precisely those individual differences referred to above, as, for example, when food scarcity af-

Institute of Theoretical Biology, University of Leiden, Groenhovenstraat 5, 2311 BT Leiden, The Netherlands

Size-Structured Populations
B. Ebenman and L. Persson (Eds.)

fects reproduction mainly indirectly, by keeping individuals longer in the non-reproductive size range.

Apparently there is a need for modeling strategies, and corresponding mathematical toolboxes, that not only take into account the total size of a population but also its structure, where "structure" refers to any subdivision on the basis of one or more traits of the individuals. In this paper we shall concentrate on physiological structure, where the adjective "physiological" serves to single out traits such as age or size, which reflect dynamical processes in the individual.

1.2 State Representations

Below we shall often need corresponding concepts on the individual and the population levels. The prefixes i- and p- will be used as distinguishing qualifiers.

1.2.1 The i-State

The fact that we are looking for models with a mechanistic pretension means that we cannot just use any collection of i-traits: the traits considered should together qualify as i-state.

Loosely speaking, a collection of variables qualifies as the state of a system if (1) the behavior (output) of the system at time t is fully determined by the values of these variables plus the condition of the environment (input) at t, and (2) the values of these variables at $t+\tau$ are fully determined by their values at t plus the intervening environmental history. Here "fully determined" should be interpreted in a stochastic sense: taking account of the past does not improve the prediction, all its influence is channeled through the present state.

Below we shall allow i-outputs to be stochastic, but we shall consider only deterministic i-state movements. We shall moreover assume that we may characterize the i-state by a finite list of continuous variables; in other words, the i-state will be a finite dimensional vector. The components of the i-state vector we call i-state variables.

Example: If the probability of producing offspring as well as offspring number, the probability of dying, and the feeding rate of an individual (i.e., the population-dynamically relevant i-behavior) all depend only on its size and the current food conditions, and if growth only depends on the current size and the current food conditions, size qualifies as i-state.

i-State representations can be constructed using either mechanistic, i.e., physiological, or purely empirical, input-output-based, considerations. Examples of state variables of a physiological provenance are size and energy reserves; examples of state variables constructed from input-output data are (physiological) age and development index.

1.2.2 The p-State

The tenet of the physiologically structured population methodology as set out in Metz and Diekmann (1986) is that we should take the p-state to be a frequency distribution over the space of possible i-states Ω. This frequency distribution serves as the mathematical counterpart of the real population present in nature. The dynamics of the p-state takes the form of a continual shifting of this frequency distribution due to the continuous movement of individuals through Ω in the course of their lives, combined with local decreases due to deaths, and local increases as a result of births. Any p-outputs, such as for example total population size or food consumption, can be calculated as integrals of this frequency distribution with some weighting function.

Implicit in this picture of a population is the assumption that all individuals experience the same environmental inputs, such as food availability or predator density. Biologically this is equivalent to the individuals (or the environment) mixing homogeneously on the time scale of the processes under study.

A final assumption will be that we are dealing with large total numbers of individuals, so that we can phrase our models in a deterministic fashion. This also entails that we can represent the p-state by a density function n, where $n(t, x)\,dx$ is the number (or rather number per unit of area of volume) of individuals in a little Ω-volume of size dx around the i-state x. (If you are not familiar with vector notation, please imagine that we are dealing with scalar i-states only. There really is not a great difference.)

1.3 Concluding Remarks

The argument so far may be summarized in the following recipe for the construction of a physiologically structured population model:

1. Track down an appropriate set of i-state variables (e.g., size and age) and determine the corresponding i-state space Ω.
2. Find expressions for the deterministic motion of the i-state (growing, "proceeding of time"), in dependence on the i-state itself and the environmental conditions.
3. Find expressions for the rate of occurrence of the stochastic jump contributions (deaths, births), again in dependence on i-state and environment.

To this we have to add:

4. Write down equations for the p-state incorporating those features, and, last but not least,
5. analyze the resulting p-equations to distil relevant biological information.

Step (4) will be covered in general terms in Section 2. There we shall also consider two equivalent alternative equations which can be of considerable help when it comes to step (5). If that section seems unduly abstract we ask you to bear with us, as the examples in the next sections probably will clarify the issue (and if it is really too much: Subsections 2.2 and 2.3 can, if necessary, be skipped on first reading).

As yet there are few general procedures for dealing with step (5), but neither are there for lumped population models of any complexity. In Section 3 we shall give some indication of what can be done. The example in that section has size as its main i-state variable in accordance with the theme of this volume. However, the reason why we worked so hard on studying this example to begin with, is that size usually exerts a quite drastic influence on the population dynamical behavior of ectotherms, making it an i-state variable worthy of a great deal of attention.

2 Population Equations

2.1 The p-Equation

In this subsection we shall discuss in general terms the basic differential equation for the p-state, expressing the balance of all the infinitesimal contributions to the change in $n(t,x)$. For more details you are referred to Metz and Diekmann (1986) and the references therein.

To simplify the discussion, we shall start by assuming that the i-state is one-dimensional, so that we can make the imagery of a shifting frequency distribution somewhat more concrete by visualizing the population as a heap of sand lying on a conveyor belt, with x the distance from the start of the belt, as depicted in Fig. 1. Only this conveyor does not move everywhere with the same velocity: at time t and position x the local velocity equals the speed of movement of the i-state, $v[E(t),x]$ where $E(t)$ is the condition of the environment at time t. If you wish you may think of x as size, and of $v[E(t),x]$ as the growth rate of individuals of size x at food availability $E(t)$.

If we neglect births and deaths, we can calculate the change in the number of individuals between x and $x+dx$, $\partial n(t,x)dx/\partial t$, by subtracting the outstream at $x+dx$, $v[E(t),x+dx]n(t,x+dx)$, from the instream $v[E(t),x]n(t,x)$ at x. Next we divide through by dx. Adding births and subtracting deaths finally gives

$$\frac{\partial n(t,x)}{\partial t} = -\frac{\partial v[E(t),x]n(t,x)}{\partial x} + b_I(t,x) - \mu_I[E(t),x]n(t,x) \;, \tag{1a}$$

where the index I is added to indicate that we are dealing with a birth rate and a per-capita death rate pertaining to the interior of Ω.

The fact that v depends on x has an interesting consequence. A conveyor belt can only move at different speeds at different places if it contracts wherever it slows down and expands wherever it speeds up. Where it contracts the sand cumulates, as the same amount of sand has to be accommodated on a smaller amount of length, and where it expands the sand spreads out. To see how this effect is accounted for in Eq. (1a) we write

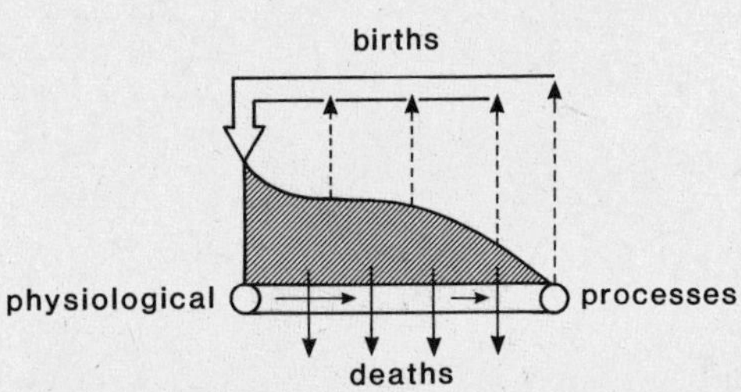

Fig. 1. Conveyor belt representation of a physiologically structured population with a one-dimensional i-state space

$$\frac{\partial vn}{\partial x} = v\,\frac{\partial n}{\partial x} + \frac{\partial v}{\partial x}\,n\,.$$

If n had represented the heights of a collection of rigid sticks riding along with the conveyor belt, as depicted in Fig. 2A, only the term $v\partial n/\partial x$ would have appeared in our equation (we would then of course also miss out the birth and death terms). The term $\partial v/\partial x$ n takes care of mass conservation; $-\partial v/\partial x$ is positive where the conveyor belt contracts and negative where it expands.

Apart from the deaths represented by $\mu_I n$, some more deaths may be implicit in the combination of our assumptions about Ω and Eq. (1a). If the conveyor belt is of finite length, say that it ends at x_{max}, and $v[E(t), x_{max}]\,n(t, x_{max}) > 0$, then there is a net loss of sand at the end. This occurs for example when x represents age, and we decide to idealize a near constant duration of life by letting everybody die at exactly x_{max} [a mathematical alternative is formally to extend Ω beyond x_{max}, and to set $n(t, x)$ equal to zero for $x > x_{max}$].

Our balance law expressed in (1a) has to be completed by indicating what happens at the start of the conveyor belt. We shall denote the corresponding i-state as x_b, $x_b < x_{max}$, where the b symbolizes that this usually is the place where newborns start their lives [the main exception are purely size-dependent models

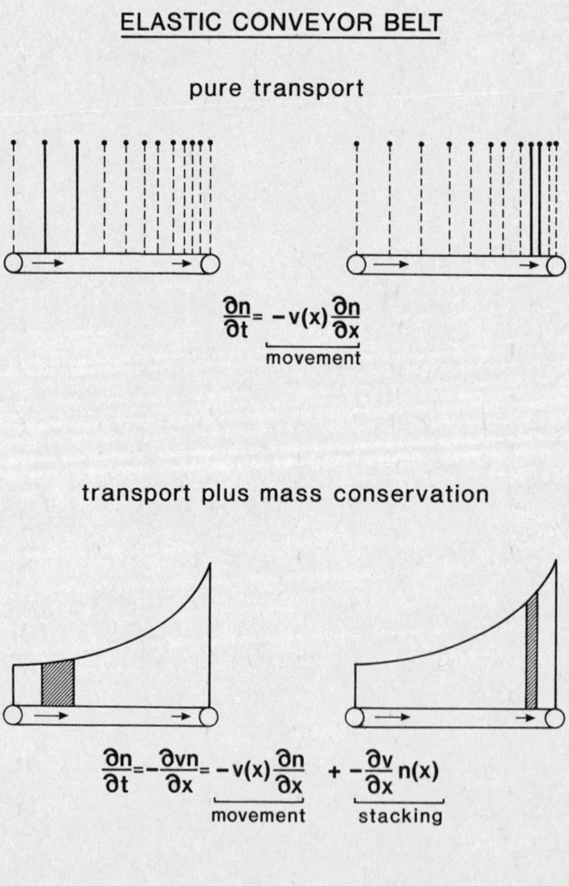

Fig. 2. The effect of local speed changes of the conveyor belt

in which individuals multiply by dividing; it is to accommodate such models as well, that we introduced the term b_I in Eq. (1 a)]. Equating the rate at which individuals enter Ω past x_b to the birth rate at the boundary, b_B, gives

$$v[E(t), x_b]\, n(t, x_b) = b_B(t, x_b) \ . \tag{1 b}$$

[Note that the second argument in $b_B(t, x_b)$ takes a fixed value; it is introduced only as a preparation for the case of higher dimensional Ω.]

The p-equation is completed by giving recipes for calculating the birth rates by summing the births coming from all different sorts of parents

$$b_B(t, x_b) = \int_\Omega \beta_B[E(t), x_b, y]\, n(t, y)\, dy \ ,$$

$$b_I(t, x) \ = \int_\Omega \beta_I[E(t), x, y]\, n(t, y)\, dy \ , \tag{1 c}$$

where β_B and β_I are the per capita birth rates (usually either β_I or β_B is zero), and by giving recipes for calculating the relevant p-outputs F_i, $i = 1, \ldots, k$. The latter are necessarily of the form

$$F_i(t) = \int_\Omega f_i[E(t), x]\, n(t, x)\, dx \ . \tag{1 d}$$

An example is $k = 2$, F_1 the population feeding rate, and F_2 the population size, in which case we should choose f_1 to be the food-availability and i-state-dependent i-feeding rate, and $f_2[E(t), x] = 1$.

All the quantities v, μ_I, β_B, β_I, and f_i in theory at least may depend on the condition of the environment. Whether this also is the case in practice depends on the particular biological problem under consideration. For a given course of the environment Eq. (1) is linear, which is an important mathematical asset. However, we may also make the environment dependent on the p-output, either indirectly, for example by letting v depend on food availability and writing down an equation for the food dynamics which contains the p-feeding rate as one of its terms, or directly, for example by letting the death rate μ_I include a component due to cannibalism, thus making the coupled equations nonlinear. It is especially this possibility of incorporating feedbacks through the environment which allows us to study all sorts of interesting biological questions.

The generalization of Eq. (1 a) to higher dimensional Ω (see Fig. 3) is

$$\frac{\partial n(t, x)}{\partial t} = -\sum_i \frac{\partial v_i[E(t), x]\, n(t, x)}{\partial x_i} + b_I(t, x) - \mu_I[E(t), x]\, n(t, x) \ , \tag{2 a}$$

where the v_i denote the components of the i-state velocity in the direction of the coordinate axes. The generalization of Eq. (1 b) is

$$\langle h(x_b), v[E(t), x_b]\rangle\, n(t, x_b) = b_B(t, x_b) \quad \text{for} \quad x_b \in \partial_+ \Omega \ , \tag{2 b}$$

where h denotes the so-called "inward normal" at the instream boundary $\partial_+ \Omega$ (i.e., a vector of unit length orthogonal to $\partial_+ \Omega$ pointing into Ω) and $\langle .,. \rangle$ denotes the inner product. The corresponding Eqs. (2 c) and (2 d) look exactly the same as their one-dimensional counterparts.

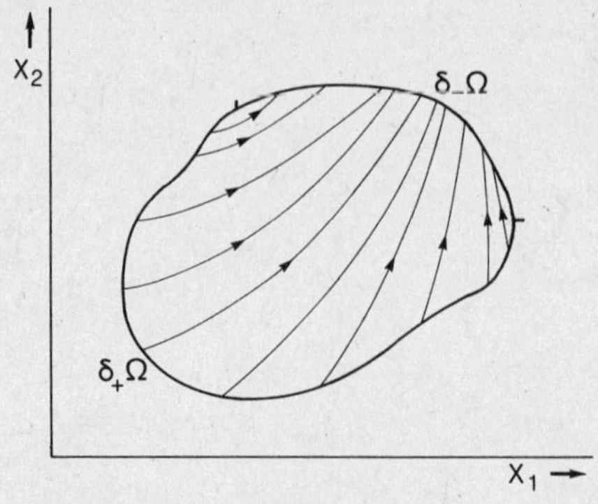

Fig. 3. A hypothetical two dimensional i-state space Ω

2.2 The Cohort Representation

We can also look at a population as a collection of cohorts. The cohort representation and the representation as a frequency distribution over an i-state space coincide only in the purely age-dependent case.

In the general case it is not sufficient to distinguish cohorts by how long ago they were born, a (from age). We also have to distinguish between individuals born with different i-states x_b. Moreover we have to keep track of the i-state of the individuals making up a cohort, $x(t, a, x_b)$. As the cohorts themselves are necessarily infinitesimally small, we introduce a density function m such that $m(t, a, x_b)\,dx_b\,da$ equals the number of individuals having ages in an age bracket of size da around a and birth states in a small Ω_b-volume of size dx_b around x_b, where Ω_b is the set of possible birth states. By the same argument as in the previous section (we just have many age conveyor belts moving in parallel, one for each value of x_b) we find that m satisfies

$$\frac{\partial m(t, a, x_b)}{\partial t} = -\frac{\partial m(t, a, x_b)}{\partial a} - \mu_I[E(t), x(t, a, x_b)]\, m(t, a, x_b) \ , \tag{3a}$$

together with

$$m(t, 0, x_b) = b(t, x_b) = \int_0^t \int_{\Omega_b} \beta[E(t), x_b, x(t, a, y_b)]\, m(t, a, y_b)\, dy_b\, da + b_0(t, x_b) \ , \tag{3b}$$

where b is the birth rate, split up according to birth state x_b, β is equal to either β_B or β_I depending on which of them is nonzero, and b_0 represents the births from parents present at the start of the experiment; as these parents may be of mixed origins, and hence may have experienced different environmental histories, we cannot assume a simple relation between their i-state and their age plus state at birth even when the latter quantities are known.

To arrive at an equation for $x(t, a, x_b)$ we first consider the case where x is size so that we may visualize x as the heights of a collection of growing sticks hitching a ride on the same conveyor belts as the cohorts. This picture has already made its appearance in the previous subsection, except that there the sticks did stay the same length. To account for the growth of the sticks we just add a growth term $v[E(t), x(t, a, x_b)]$ at the right hand side:

$$\frac{\partial x(t,a,x_b)}{\partial t} = -\frac{\partial x(t,a,x_b)}{\partial a} + v[E(t), x(t,a,x_b)] \; , \tag{3c}$$

$$x(t,0,x_b) = x_b \; . \tag{3d}$$

For the general case we only have to interpret v, x and x_b as vectors.

To calculate b_0 we split up the parent population into "cohorts" characterized by their i-state y at $t = 0$. The cohort i-states as a function of time and initial cohort i-state we shall denote as x_0, and the corresponding density function representing the cohort sizes as m_0. Then

$$b_0(t,x_b) = \int_{\Omega} \beta[E(t), x_b, x_0(t,y)]\, m_0(t,y)\, dy \; . \tag{3e}$$

Moreover m_0 and x_0 satisfy the equations

$$\frac{\partial m_0(t,y)}{\partial t} = -\mu_I[E(t), x_0(t,y)]\, m_0(t,y) \; , \qquad m_0(0,y) = n(0,y) \; , \tag{3f}$$

and

$$\frac{\partial x_0(t,y)}{\partial t} = v[E(t), x_0(t,y)] \; , \qquad x_0(0,y) = y \; . \tag{3g}$$

We complete the cohort representation by indicating how we should calculate the p-outputs:

$$F_i(t) = \int_0^t \int_{\Omega_b} f_i[E(t), x(t,a,x_b)]\, m(t,a,x_b)\, dx_b\, da + F_{0i}(t) \; , \tag{3h}$$

where $F_{0i}(t)$ denotes the contribution by the survivors from the parent population,

$$F_{0i}(t) = \int_{\Omega} f_i[E(t), x_0(t,y)]\, m_0(t,y)\, dy \; . \tag{3i}$$

2.3 An Integral Equation for the Birth Rate

Equations (2) and (3) may both rightfully be considered generalizations of the McKendrick (Von Foerster) partial differential equation of age dependent population theory. The counterpart of Lotka's integral equation is

$$b(t,x_b) = \int_0^t \int_{\Omega_b} B[E(t), a, x_b, y_b]\, b(t-a, y_b)\, dy_b\, da + b_0(t,x_b) \; , \tag{4a}$$

where $B[E(t), a, x_b, y_b]\, dx_b$ is the rate at which individuals born a time units ago in i-state y_b produce offspring in an Ω_b-volume of size dx_b around x_b. b_0 has already been defined above. This equation has to be completed with a number of equations from which B can be calculated. We first express B in the per-capita birth rate β and the probability $\mathscr{F}$ that a prospective parent survived from its birth a time units ago to the present:

$$B[E(t), a, x_b, y_b] = \beta[E(t), x_b, x(t, a, y_b)]\, \mathscr{F}(t, a, y_b) \ . \tag{4b}$$

x satisfies Eq. (3c), for future reference here renamed into (4c), just as before, together with

$$x(t, 0, y_b) = y_b \ , \tag{4d}$$

and $\mathscr{F}$ satifies

$$\frac{\partial \mathscr{F}(t, a, y_b)}{\partial t} = -\frac{\partial \mathscr{F}(t, a, y_b)}{\partial a} - \mu_I[E(t), x(t, a, y_b)]\, \mathscr{F}(t, a, y_b) \ , \tag{4e}$$

together with

$$\mathscr{F}(t, 0, y_b) = 1 \ . \tag{4f}$$

The p-outputs can be calculated from

$$F_i(t) = \int_0^t \int_{\Omega_b} f_i[E(t), x(t, a, y_b)]\, \mathscr{F}(t, a, y_b)\, b(t-a, y_b)\, dy_b\, da + F_{0i}(t) \ . \tag{4g}$$

For any realistic population model both $b_0(t, x)$ and $F_{0i}(t)$ go to zero for large time. Therefore Eq. (4) suffices for the study of the long-term behavior of the population. If we wish to study an initial value problem we may just add (3e) to (3g) plus (3i).

2.4 Concluding Remarks

Equations (2), (3) and (4) are fully equivalent. Chapter III Section 4 of part A of Metz and Diekmann (1986) gives the formal mathematical machinery for transforming them into each other. However, they are not equivalent in the ease with which they can be used in various applications: some problems are more easily investigated using (2) and others using (3) or (4). For example, in De Roos (1988) an efficient numerical technique is developed for the treatment of physiologically structured population models by what is essentially an unusual discretization of (3) in combination with an expansion procedure applied to (2). The various *Daphnia* simulations below were all performed using this technique. Therefore it is a good idea to keep them all three in mind, and to switch equations, and associated imagery, whenever there seems to be an advantage in doing so.

3 Example: *Daphnia* Population Dynamics

3.1 Introduction

In this section we shall treat in some detail a size- (and age-) structured model for the population dynamics of ectotherms. This model was originally developed with an eye on the waterflea *Daphnia magna.* Its experimental underpinning for different constant food availabilities is discussed in Kooijman and Metz (1984).

Various mathematical and biological fine points in the derivation of the p-equation, as well as further refinements of the model structure are considered in Diekmann et al. (1984), Metz and Diekmann (1986), Heijmans and Metz (1986), and Thieme (1988).

A number of detailed and complicated *Daphnia* models have been proposed (Sinko and Streifer 1969) or are currently under investigation, notably by Lynch (in press), by Nisbet, Gurney, McCauley and Murdoch (in prep), and by Kooijman (1986a, Kooijman, Van der Hoeven and Van der Werf in press). This presence of alternative, competing, frameworks adds a good deal to the scientific interest of the *Daphnia* work. At the two extremes, Lynch's model is mainly a collation of empirical curves pertaining to various aspects of the overall individual physiology, whereas Kooijman's model is to a large extent based on a priori considerations. The models by Nisbet and coworkers, and by Kooijman include energy reserves as a third i-state variable, in addition to size and age. This is especially important at times of rapid changes in food availability. Kooijman's model is a direct extension of the model described below. In contrast to the models by Lynch and by Nisbet and coworkers it was never meant to specifically mimic the known details of *Daphnia* physiology, but to provide insights of a general nature. Various papers by Kooijman (1986b, c, 1988) i.a. attest to its usefulness for collating data on individual growth physiology from all over the animal kingdom.

For the sake of tractability, we shall temporarily neglect the undoubted presence of energy reserves, as well as any other refinements. We just had to learn our trade in easy steps. Moreover, results from the simple model already are of considerable interest in their own right. After all, the search for principles that is science should be guided by a plethora of particulars. A logical next step should be investigating how our present results stand up against the various modification that have been proposed.

3.2 Model Formulation

3.2.1 The i-Model

Below l will denote length and a age. These are the i-state variables. The environment is characterized by only one variable, food availability (i.e., density or concentration) x. A survey of the identifiable parameter combinations together with their interpretation can be found in Table 1.

The model formulation starts from the simplified picture of an animal shown in Fig. 4, plus the assumption that animals of different sizes have similar geometries, so that surface area and (wet) weight are proportional to l^2 and l^3 respectively. Below we shall just identify these quantities; this only affects the values of the various proportionality constants. Food intake is assumed to be proportional to surface area and to depend on food availability through a hyperbolic functional response. The energy gained is channeled either to reproduction or basal metabolism plus growth in fixed proportions $1-\kappa$ and κ. Growth results only after basal metabolism has had its share, which is assumed to be proportional to the animal's weight, with proportionality constant ζ. (All energetic

Table 1. *Daphnia* parameters used in preparing the figures[a]

ξ	$7 \cdot 10^{-6}$ ml cell^{-1}	Shape parameter of the functional response
ν	$1.776 \cdot 10^{6}$ cell mm^{-2} d^{-1} Daphnia^{-1}	Maximum ingestion rate per unit surface area[b]
κ	0.3333	Fraction of assimilated energy not channeled to reproduction
$\underline{l}_b$	0.8 mm	Length at birth
$\underline{l}_J$	2.5 mm	Length at which reproduction starts
$\underline{l}_m = \kappa\nu/\zeta$	6.6 mm	Maximum attainable length
$\gamma = \zeta/(3\eta)$	0.17 d^{-1}	Time constant of approach to asymptotic size under constant food conditions
$R_m = \theta(1-\kappa)\nu l_m^2$	10 d^{-1}	Maximum attainable reproduction[c]
a_{max}	70 d	Maximum attainable age

[a] Based on the figures in Kooijman and Metz (1984) for *Daphnia magna* feeding on *Chlorella*.
[b] This quantity is only used to calculate the uptake of food cells in Eq. (12) and (13). In the *Daphnia* equations it never appears separately, but only in the parameter combinations l_m, γ, R_m determining growth and reproduction.
[c] This quantity is much too low. A realistic estimate is 30 or even higher. However, this is the only value for which we have a complete set of pictures. Making R_m larger does not appear to lead to qualitatively different results. The main quantitative difference is that the subsistence food density, x_c from Sections 3.3 and 3.4 becomes lower.

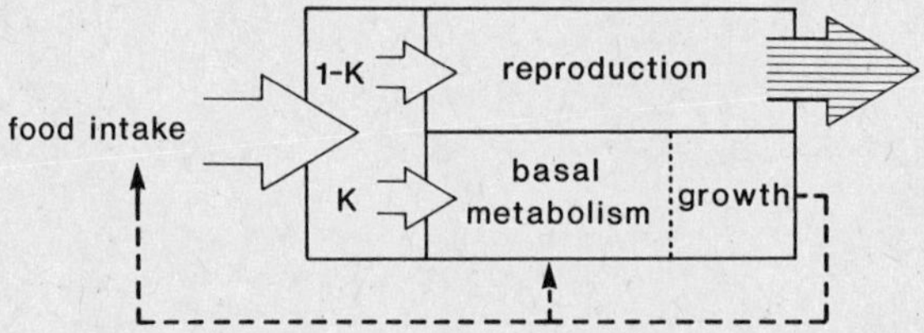

Fig. 4. A schematic representation of the energy channeling in the *Daphnia* model

quantities will be expressed in food equivalents, as this eliminates one more parameter.) The (constant) growth cost per unit weight will be called η. For consistency reasons we moreover add (remember, we did not include reserve energy in the model, and we want to minimize deaths from starvation): If the food availability decreases to the extent that κ times the ingested energy does not suffice for the basal metabolic requirements, the energy-channeling rule changes so that basal metabolism still is catered for, but only just, and growth stops. Together these assumptions lead to the, slightly adapted, Von Bertalanffy growth equation (see, e.g., Beverton and Holt 1957) (we shall from now on refrain from always writing the arguments of any functions in full)

$$\frac{dl}{dt} = g(x,l) = \frac{1}{3\eta}\,[\kappa\nu f(x) - \zeta l]_+ \tag{5a}$$

together with

$$f(x) = \frac{\xi x}{1+\xi x}\,, \tag{5b}$$

where the + signifies that any negative quantity between the brackets is replaced by zero. Size at birth is assumed to be fixed at l_b.

In calculating the birth rate β we shall cheat a bit. Experimentally, waterfleas grown at different food concentrations were found not to produce eggs below a certain size l_J, and to start producing eggs as soon they grew above this size, independent of the age at which this happened. We shall incorporate this by just setting the birth rate equal to zero for $l < l_J$ (as a tongue in cheek explanation we may say that below l_J the reproductive energy flow drains to the production of gonadal tissue and/or differentiation). The energetic cost of one young we shall call $1/\theta$. All together this gives us

$$\begin{aligned} \beta(x,l) &= 0 && \text{for } l < l_J \\ &= \theta(1-\kappa)\nu f(x) l^2 && l_J < l < \underline{l}(x) \\ &= \theta[\nu f(x) l^2 - \zeta l^3] && \underline{l}(x) < l \ , \end{aligned} \tag{6a}$$

where

$$\underline{l}(x) = \kappa \nu f(x)/\zeta \tag{6b}$$

is the size for which at the current food density the + in Eq. (5a) starts to work.

For the death rate we assume

$$\mu(x,a,l) = \phi + \chi(a) + \psi(x,l) \ , \tag{7a}$$

with

$$\begin{aligned} \chi(a) &= 0 \quad \text{for} \quad a < a_{max} && \text{and} \quad \psi(x,l) = 0 \quad \text{for} \quad l < \bar{l}(x) \\ &= \infty \qquad\quad a > a_{max} && \qquad\quad\ \ = \infty \qquad\quad l > \bar{l}(x) \ , \end{aligned} \tag{7b}$$

where

$$\bar{l}(x) = \nu f(x)/\zeta \tag{7c}$$

is the size for which basal metabolism equals energy intake. ϕ represents random deaths due to for example predation. Below we shall treat ϕ as a free parameter, as it is so well amenable to experimental manipulation. χ was introduced first of all to allow us to study any possible effects of aging. Moreover, the presence of a cutoff to the age distribution is quite helpful technically. As it turns out, none of the conclusions discussed below depended noticeably on the choice of a_{max}, as long as both ϕ and a_{max} were given reasonable values. Finally ψ represents death from starvation. However, ψ never has exerted its guillotine action in any of our simulations so far.

Table 1 gives the values of the fixed parameters that were used in preparing the figures.

3.2.2 The p-Equation

As a preliminary assume that $a_{max} = \infty$, so that we can drop a from the i-state description. Equation (5) entails that l can only take values in the size range between l_b and $l_m = \kappa\nu/\zeta$. This defines Ω. Following the lines laid out in Section 2.1 we arrive at the p-equation

$$\frac{\partial n}{\partial t} = -\frac{\partial g n}{\partial l} - \mu n \quad \text{for} \quad l < \bar{l}(x)$$

$$n = 0 \quad \text{for} \quad l > \bar{l}(x)$$

$$g(x, l_b) n(t, l_b) = b(t) \tag{8a}$$

$$b(t) = \int_{l_b}^{l_m} \beta(x, l) n(t, l) dl \ . \tag{8b}$$

In the general case Ω is a subset from the square spanned by the ages from 0 to a_{max} and the lengths from l_b to l_m. The p-equation becomes

$$\frac{\partial n}{\partial t} = -\frac{\partial n}{\partial a} - \frac{\partial g n}{\partial l} - \mu n \quad \text{for} \quad l < \bar{l}(x)$$

$$n = 0 \quad \text{for} \quad l > \bar{l}(x) \tag{9a1}$$

$$n(t, 0, l) = b(t) \delta(l - l_b) \tag{9a2}$$

$$b(t) = \int_{\Omega} \beta(x, l') n(t, a, l') dl' da \ , \tag{9b}$$

where $\delta(y)$ denotes the Dirac delta function, i.e., a "function" which is zero everywhere except for an "infinitely high peak" at $y = 0$ such that the surface area under this peak is exactly one.

The delta function in Eq. (9a2) is due to the fact that all young are born in one point of the one-dimensional instream boundary $a = 0$ only, instead of being spread out over that boundary. Worse still, because all individuals in a cohort experience exactly the same environmental history they stay together. Therefore the density function n develops the same degeneracy, spawned by the delta function on the boundary (see Fig. 5). All the mass is concentrated on just a one dimensional curve in Ω, the current, or period, age-length relation. As a result calculations using Eq. (9), even though still possible, are slightly unpleasant.

The idea of a cohort representation developed in Section 2.2 provides the way out. We just sweep all the mass onto the age axis, and write down the usual equation for the age distribution $n(t, a)$ plus a separate equation which updates the

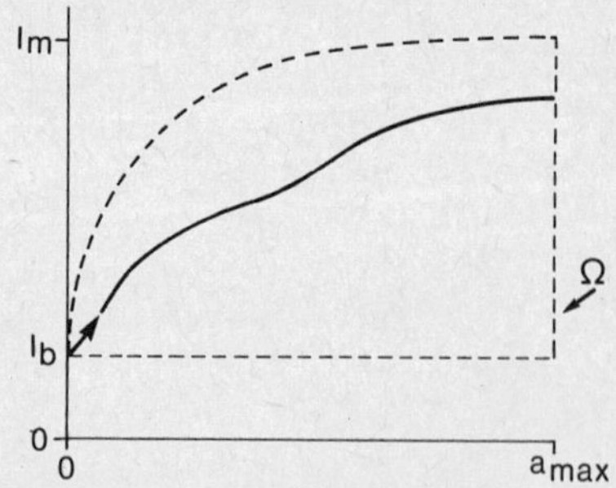

Fig. 5. The *Daphnia* i-state space, and the support of the population distribution (the current, or period, age length relation)

current age length relation l(t, a) (think again of that conveyor belt planted with growing sticks):

$$\frac{\partial n}{\partial t} = -\frac{\partial n}{\partial a} - \mu n \quad \text{for} \quad l(t,a) < \bar{l}(x)$$

$$n = 0 \quad \text{for} \quad l(t,a) > \bar{l}(x)$$

$$n(t,0) = b(t) \tag{10a}$$

$$\frac{\partial l}{\partial t} = -\frac{\partial l}{\partial a} + g(x,l)$$

$$l(t,0) = l_b \tag{10b}$$

$$b(t) = \int_0^{a_{max}} \beta[t, l(t,a)]\, n(t,a)\, da\ . \tag{10c}$$

For the food dynamics we shall make three different choices: We start in Section 3.3 with

$$x(t) = x\ , \quad \text{a parameter}\ . \tag{11}$$

In Section 3.4 we shall consider a chemostat-type food dynamics in which food is fed to the experimental tank at a rate α per unit of volume, and removed again at a relative rate β, for example through decay or washout. The resulting environment equation is

$$\frac{dx}{dt} = \alpha - \beta x - \nu f(x) \int_0^{a_{max}} l^2(t,a)\, n(t,a)\, da\ . \tag{12}$$

In this equation the feeding rate to the thank, α, will be treated as a free parameter. In preparing the figures the relative decay rate β was kept fixed at 0.5 d^{-1}. Finally we shall consider self-renewing food, which we shall represent by the environment equation

$$\frac{dx}{dt} = \alpha x(1 - x/\beta) - \nu f(x) \int_0^{a_{max}} l^2(t,a)\, n(t,a)\, da\ . \tag{13}$$

In this equation we shall treat the carrying capacity β as a free parameter. In preparing the figures the maximum capacity for self-renewal α was set to either 2 d^{-1} (Fig. 10) or 0.5 d^{-1} (Fig. 11).

3.3 Constant Food Availability

When food availability is kept constant, the coupling between the equation for the current age-length relation and the age distribution goes only one way. The biological interpretation as well as Eq. (10b) (think of the conveyor belt interpretation again!) both tell us that in that case the current age-length relation quickly stabilizes to the usual Von Bertalanffy growth curve, dependent only on the, fixed, *Daphnia* parameters and the value of x. This in turn lets Eq. (10a) plus

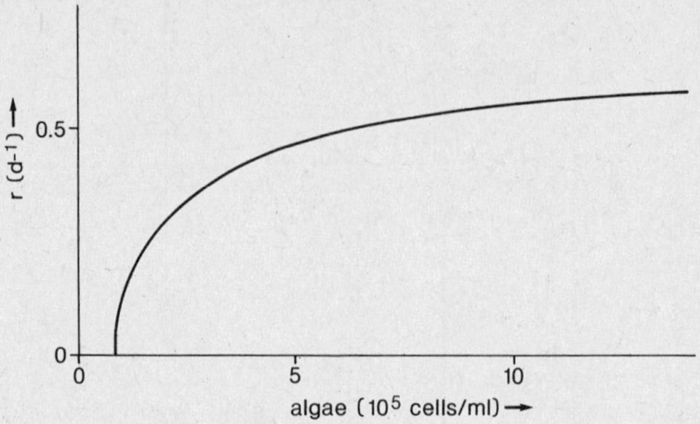

Fig. 6. Intrinsic rate of natural increase as a function of food density for the *Daphnia* model

Eq. (10c) degenerate into the usual equations of age dependent population theory in constant environments.

Figure 6 shows how the intrinsic rate of natural increase r would depend on food availability if there were no random deaths. From this figure one can immediately determine r for more realistic values of ϕ using

$$r(\phi, x) = r(0, x) - \phi \ . \tag{14}$$

Random deaths just act as a constant dilution without affecting the composition of the population.

r is seen to increase with x, in accordance with biological intuition. The critical value of the food concentration above which the population eventually grows and below which it eventually declines, we shall call x_c. To calculate x_c we just set $r = 0$ in the characteristic equation. Alternatively we may go back to the i-equations and use the property that at x_c individuals produce on average exactly one young during their lifetime. (NB: The property of eventual exponential growth or decline under constant environmental conditions holds almost universally for structured population models, but the ability to write down a characteristic equation in just the intrinsic rate of natural increase is tightly linked to the existence of a finite collection of selected states such that any individual has to pass through at least one of these states during its life-time. In our *Daphnia* model this collection consists of the unique state at birth: $a = 0$, $l = l_b$.)

3.4 Dynamical Food

3.4.1 Introduction

The results of the previous subsection could have been obtained just as easily direct from the i-model. The main reason for going through the complicated business of writing down p-equations is that this allows us to study the effect of food limitation.

Since the early work by Pratt (1943) a good number of papers on *Daphnia* population dynamics have been published. A detailed survey of published data on laboratory and (semi-)natural waterflea populations led Murdoch and

McCauley (1985; see also McCauley and Murdoch, 1987) to distinguish three qualitatively different types of dynamical behavior: (i) stable equilibrium, (ii) predator prey oscillations, in which both the *Daphnia* and the food algae fluctuate with the same period, and (iii) pure *Daphnia* oscillations in combination with a barely fluctuating algal population. One of the challenges to the modeler is to reproduce all three types of behavior, including the elusive type (iii) oscillations, by means of just one model for the *Daphnia* population. In the next subsections we shall try to convince you that our *Daphnia* model is at least potentially capable of doing precisely this, plus producing a fourth type of oscillation, in which the *Daphnia* and the food are in antiphase instead of the usual one-quarter phase lag characteristic of predator prey oscillations.

3.4.2 The Stability of the Internal Equilibrium

At equilibrium the *Daphnia* population should neither grow nor decline. Apparently the value of the food density at equilibrium, $\hat{x}$, is set wholly by the *Daphnia* dynamics, making $\hat{x} = x_c$; provided, of course, that $x_c < x_e$, the equilibrium of the food population in the absence of any *Daphnia*. If $x_c > x_e$ *Daphnia* goes extinct.

The equilibrium age-length relation is again a Von Bertalanffy growth curve. Formally this can be derived from Eq. (10b) with $\hat{l}(a)$ substituted for $l(t, a)$, making $\partial l/\partial t = 0$. In the same manner we can calculate $\hat{n}(a)$ from Eq. (10a) after we have also set $b(t) = \hat{b}$. The result is $\hat{n}(a) = \hat{b}\, \mathcal{F}(a)$, where $\mathcal{F}(a) = \exp(-\phi a)$ is the fraction of Daphnids that survives to age a. Substituting both results in either Eq. (12) or (13) finally leads to an explicit, though slightly cumbersome, expression for $\hat{b}$.

Apparently the combination of the food and *Daphnia* dynamics leads to a unique interior equilibrium.

When we try to study the stability of the equilibrium we by necessity enter into a rather shady realm. At the present state of the art we cannot fall back on existing theory to answer even the first questions about existence and uniqueness of solutions to our full population equations. (That such questions are not trivial can be seen from the fact that Thieme (1988) has shown that Eq. (10) can indeed be rigged such that it does not allow a solution; fortunately the combination of conditions leading to this somewhat bizarre phenomenon is such that we feel safe in ignoring it for our present considerations.) However, we shall boldly plunge into the deep, and just assume that all the usual dynamical systems theorems hold good. This also means that the calculations which we describe below in general terms were formal only; the fact that the results of, for example, formal stability calculations and numerical solutions of the full population equations gave consistent results inspires some trust.

We start our study of the internal equilibrium by observing that it exists if and only if the trivial equilibrium, in which the *Daphnia* population is extinct, is unstable. By changing either α or β depending on the food dynamics or ϕ we can let the interior equilibrium disappear or appear at will. When it appears it does so from the trivial equilibrium. The general results from bifurcation theory about the exchange of stability then imply that close to the trivial equilibrium it should be stable. Moreover the absence of any other interior equilibria implies that it can only become unstable by merging with a, stable or unstable, limit cycle, in a so-called Hopf bifurcation.

To study the stability of the internal equilibrium more closely we can linearize Eq. (10) together with either Eq. (12) or (13). After going through the usual motions of substituting trial solutions which depend on time as $\exp(-\lambda t)$ we end up, after a lot of messy algebra, with a characteristic equation in just the one complex variable λ. As in the case of constant x, this is made possible by the fact that all individuals have to pass through a unique i-state at birth. The Hopf bifurcation result cited above implies that we do not have to look for any real λ's crossing zero. Therefore we can set $\lambda = i\omega$, ω real, and look for any parameter combinations which make this happen. Any stability changes are bound to occur at such parameter values. The stability diagrams in the next two subsections were all calculated in this way. The details of the calculations, which are not completely trivial, we hope to publish elsewhere.

3.4.3 Chemostat Food Dynamics

Figure 7 shows the stability diagram for the chemostat food dynamics [Eq. (13)]. It can be seen that the predatory death rate ϕ exerts a fairly dominating influence, especially at high feeding rates to the tank α. Any changes in the food decay rate β change only the size of the peninsula that juts out from the unstable continent into the mare stabilitatis, but not the behavior of the stability boundary at high feeding rates.

Interestingly enough, Fig. 7 does not give the whole dynamical picture, as stability of the equilibrium may go together with the existence of a stable limit cycle. This happens in the area surrounding the peninsula, in particular in the bay separating the peninsula from the continent. As yet we have no idea of the extent of this feature.

Figure 8 shows the pattern of oscillations at high α. Observe that food and *Daphnia* are in clear antiphase. The phase relations between food and *Daphnia*

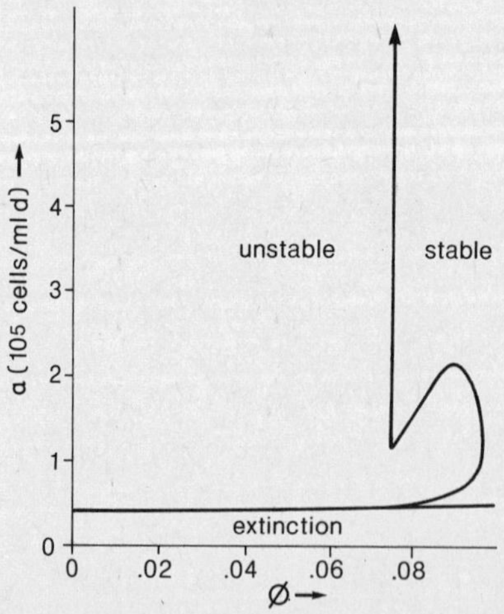

Fig. 7. The stability boundary for the interior equilibrium for the *Daphnia* model with chemostat food dynamics. α feeding rate to the tank, ϕ random death (harvesting) rate

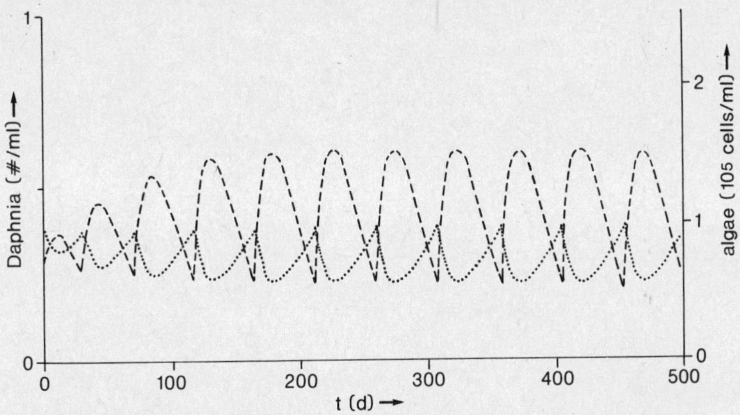

Fig. 8. Oscillations of the *Daphnia* model with chemostat food dynamics at high feeding rate to the tank. – – – Daphnia, ······ algae

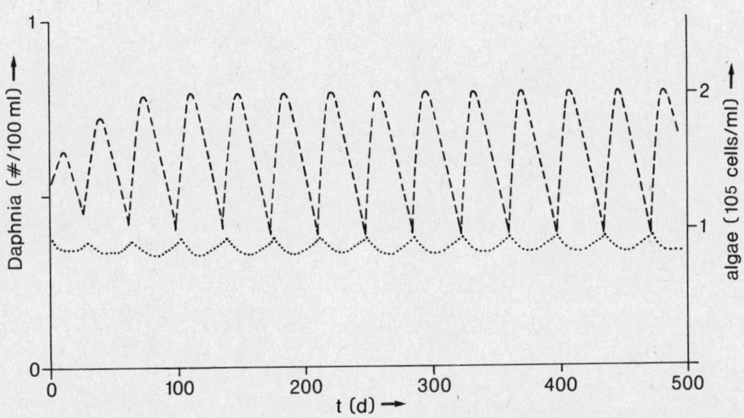

Fig. 9. Oscillations of the *Daphnia* model with chemostat food dynamics at low feeding rate to the tank. – – – Daphnia, ······ algae

also provide a clue to the mechanism driving the oscillations. One feature which is not visible in Fig. 8, is that births occur in sharp peaks, starting with a very steep increase followed by a fairly gentle decline. The drop in food concentration coincides with the birth peak. When the new crop of young ages, their number, and that of their parents, decreases. This allows the food to recover. When the firstborns of a crop have grown large enough they start the new birth peak. So far this is just the variable maturation time scenario of Nisbet and Gurney (1984). However, this is not the whole story. To complete it we have to return to our picture of the conveyor belt. While rolling on it also contracts, pushing all the young from one birth cycle closer and closer together, so that when they reach reproductive size they do so almost simultaneously. This effect strongly contributes to the oscillatory tendency of the model.

At low a the birth peaks become much more gentle and the decrease in the food starts to lag somewhat behind. Moreover, the effect of the deaths on the population feeding rate is largely compensated for by growth. Figure 9 shows an example. It can be observed that the food fluctuates only very mildly, well within the 20% criterion used by Murdoch and McCauley to define their type (iii) oscillations. These oscillations are made possible by the size structure: Even though both *Daphnia* numbers and *Daphnia* biomass do fluctuate considerably, the total *Daphnia* surface area, and therefore the *Daphnia* feeding rate, stays much more constant. Moreover, due to the fact that at low a there are only few *Daphnia* around these fluctuations in feeding rate have only a relatively minor impact on the food population.

3.4.4 Self-Renewing Food

Figure 10A shows the stability diagram for the self-renewing food dynamics [Eq. (13)], and Fig. 10B a corresponding stability diagram for the unstructured predator prey model

$$\frac{dx}{dt} = \alpha x(1-x/\beta) - u f(x) N$$

$$\frac{dN}{dt} = [v f(x) - \phi] N \ . \qquad (15)$$

The maximum capacity for self-renewal of the food a is same in both diagrams, and the same applies to the shape parameter ξ in the functional response. However, the two remaining parameters u and v do not bear any simple relation to the *Daphnia* parameters. They were chosen only for the likeness of the corresponding stability diagrams.

Here too there exist shoals in the mare stabilitatis where the stable limit cycle extends beyond the peninsula.

Finally Fig. 11 shows some typical predator prey cycles in which *Daphnia* lags a quarter period behind its food.

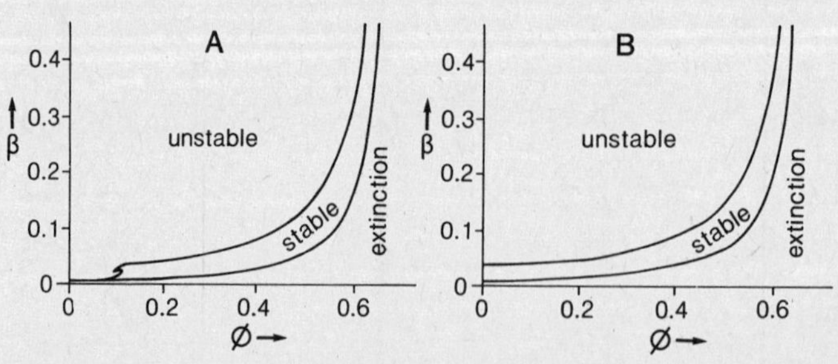

Fig. 10. A The stability boundary for the interior equilibrium for the *Daphnia* model with self-renewing food. **B** The stability boundary for an unstructured model, with the parameters without counterparts in the *Daphnia* model chosen to maximize the likeness to **A**. β carrying capacity of the food in the absence of *Daphnia;* ϕ random death (harvesting) rate

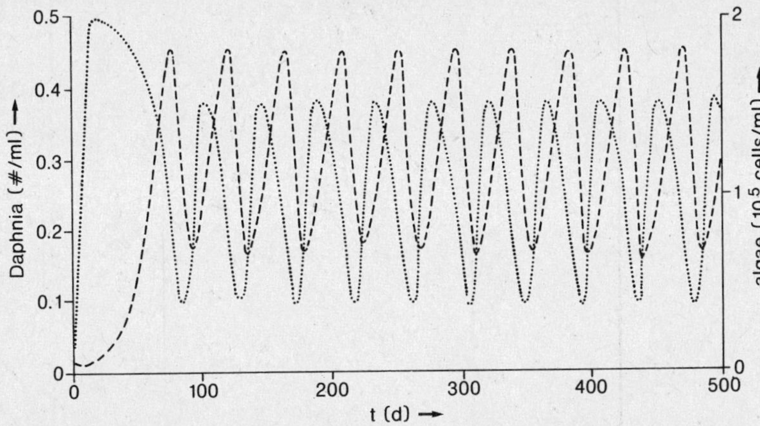

Fig. 11. Predator-prey oscillations in the *Daphnia* model with self-renewing food

Apparently the unstructured model [Eq. (15)] captures the dynamical essentials of an interaction where one species feeds upon an other quite well, at least in a qualitative sense. However, our failure to find any clear relation between the parameters u and v and the parameters characterizing individual behavior is symptomatic of its weaknesses. It is not clear a priori how one should arrive at a quantitative match over a range of different circumstances. It is here that structured models, in which all the parameters have well-defined empirical counterparts, have at least the potential to score higher.

4 Concluding Remarks

We hope that the previous sections have convinced you that one does not need to be a mathematician to deal with physiologically structured population models. What matters is the interplay between reasoning in terms of individual behavior, a good use of the conveyor belt analogy, and some formal mathematical manipulations. Of course there is some quite deep mathematics hidden underneath, but we can safely leave that to the mathematicians.

A second message was that an almost necessary condition for making life simple is that all individuals are bound to pass through one, or a few, selected states. In particular, when everybody is born equal, many calculations reduce to those for the age-dependent problem.

One good reason to replace the more conventional lumped models by their structured counterparts, is that usually only the latter allow a clear mechanistic interpretation of the model parameters. What we are interested in ultimately is mechanistic insight. Models are just a tool.

But structured models may also have interesting dynamical properties special to their kind!

Acknowledgments. We thank Odo Diekmann for giving the Dutch structured population movement its impetus, Bas Kooijman, Roger Nisbet, Bill Gurney and Ed McCauley for many exasperating but fruitful discussions about *Daphnia*, Anneke Leipoldt for calculating stability boundaries for the *Daphnia* model with chemostat food dynamics and for writing the first simulation programs, Erik Evers for calculating stability boundaries for the case of self-renewing food, Peter Hock for drawing the figures, and the organizers of the Falsterbo workshop for a good job and a pleasant week.

5 References

Beverton RJH, Holt SJ (1957) The dynamics of exploited fish populations. H M Stn Off, London

De Roos AM (1988) Numerical methods for structured population models, the escalator boxcar train. Num Meth Part Diff Equ 4(3):173–195

Diekmann O, Metz JAJ, Kooijman SALM, Heijmans HJAM (1974) Continuum population dynamics with an application to *Daphnia magna.* Nieuw Arch Wisk (4) 2:82–109

Heijmans HJAM, Metz JAJ (1986) Small parameters in structured population models and the Trotter-Kato theorem. REP AM-R8603, Cent Math Comput Sci, Amsterdam

Kooijman SALM (1986a) Population dynamics on basis of budgets. In: Metz JAJ, Diekmann O (eds) The dynamics of physiologically structured populations. Springer, Berlin Heidelberg New York, pp 266–297 (Springer lecture notes in biomathematics, vol 68)

Kooijman SALM (1986b) Energy budgets can explain body size relations. J Theor Biol 121:269–282

Kooijman SALM (1986c) What the hen can tell about her eggs: egg development on the basis of energy budgets. J Math Biol 23:163–185

Kooijman SALM (1988) The Von Bertalanffy growth rate as a function of physiological parameters: a comparative analysis. In: Hallam TG, Gross LJ, Levin SA (eds) Mathematical ecology, World Sci, pp 3–45

Kooijman SALM, Metz JAJ (1984) On the dynamics of chemically stressed populations: the deduction of population consequences from effects on individuals. Ecotox Environ Saf 8:254–274

Kooijman SALM, van der Hoeven N, van der Werf DC (in press) Population consequences of a physiological model for individuals. Func Ecol

Lynch M (in press) The life history consequences of resource depression. Ecology

McCauley E, Murdoch WW (1987) Cyclic and stable populations: plancton as a paradigm. Am Nat 129:97–121

Metz JAJ, Diekmann O (eds) (1986) The dynamics of physiologically structured populations. Springer, Berlin Heidelberg New York (Springer lecture notes in biomathematics, vol 68)

Murdoch WW, McCauley E (1985) Three distinct types of dynamic behaviour shown by a single planctonic system. Nature (London) 316:628–630

Nisbet RM, Gurney WSC (1984) "Stage-structure" models of uniform larval competition. In: Levin SA, Hallam TG (eds) Mathematical ecology, Proceedings Trieste 1982. Springer, Berlin Heidelberg New York, pp 97–113 (Springer lecture notes in biomathematics, vol 54)

Pratt DM (1943) Analysis of population development of *Daphnia* at different temperatures. Biol Bull 85:116–140

Sinko JW, Streifer W (1969) Applying models incorporating age-size structure of a population to *Daphnia.* Ecology 50:608–615

Thieme HR (1988) Well-posedness of physiologically structured population models for *Daphnia magna.* J Math Biol 26:299–317

Dynamics of Age- and Size-Structured Populations: Intraspecific Competition

B. EBENMAN

1 Introduction

In most organisms different aged individuals differ in size and morphology. In some taxa the body weight of individuals within species spans 4 or more orders of magnitude (Werner and Gilliam 1984). The radical changes in morphology accompanying metamorphosis in organisms with complex life cycles is well known to all biologists. Such variation within a species will have important evolutionary and population dynamic consequences. In this chapter I will discuss the implications of such differences in size/morphology on the intensity of competition between age classes, and show that competition between age classes can have interesting population dynamic consequences.

In many animals the individuals change their resource use as they develop from newborns to adults. Such niche changes during the life cycle will affect intraspecific competition and hence population dynamics. The strength of competition between different-aged individuals will often depend on their patterns of growth or on the form of their life cycles (Polis this Vol.; Wilbur this Vol.). In most organisms with complex life cycles, like amphibians, holometabolous insects and many marine invertebrates, there is no competition between the juveniles and adults (Wilbur 1980). In organisms with "simple" life cycles (i.e., organisms not undergoing metamorphosis), like reptiles, fishes and hemimetabolous insects, the niche differences between the age classes are in most cases not so pronounced. Instead, individuals in different age classes partly overlap in their use of resources (Polis 1984; Werner and Gilliams 1984). Finally, among birds and mammals – which are almost fullgrown at the age of independence – there is probably seldom any significant ecological segregation between the age classes. This is also often the case for plants (e.g., Goldberg 1987).

With the help of general discrete-time models with two age classes I have investigated how competition between adults and juveniles affects population dynamic behavior. I will first review results from a model of a purely age-structured population (Ebenman 1987 and 1988) and later compare these results with new results from a model that also incorporates size-structure.

2 Age-Structured Populations

There is a rather large literature on the effects of density dependence on the dynamic behavior in age-structured populations (reviewed by Nisbet and Gurney

Department of Biology, Linköping University, 581 83 Linköping, Sweden

Size-Structured Populations
B. Ebenman and L. Persson (Eds.)

1982; Metz and Diekmann 1986; for recent studies see Crowley et al. 1987; Desharnais and Liu 1987; Hastings and Costantino 1987; Liu and Cohen 1987). However, very few studies have directly considered the effects of competition between age classes in a population (May et al. 1974; Bellows 1982a and b; Tschumy 1982; Ebenman 1987 and 1988). It is not my intention to review this literature here. In this paper the focus will be on size-structured populations. However, the purely age-structured case is a useful point of departure and I will therefore briefly review the results from my own recent studies (Ebenman 1987 and 1988). (For a more detailed treatment the reader is referred to these works.)

The model based on age only has the following form:

$$\begin{aligned} N_0(t+1) &= N_1(t)\,Bf\,[N_1(t)+\alpha N_0(t)] \\ N_1(t+1) &= N_0(t)\,Sg\,[N_0(t)+\beta N_1(t)] \end{aligned} \tag{1}$$

Here N_0 and N_1 are the densities of juveniles and adults respectively, B is the maximum per capita birth rate and S is the maximum survival rate (i.e., the vital rates in absence of density dependence), and α and β are the competition coefficients. The functions f and g give the effects of the effective population densities on per capita birth rate and survival rate respectively.

2.1 Density-Dependent Reproduction

First, consider the case when survival is density independent (i.e., $g(N_0+\beta N_1)\equiv 1$). The equilibrium densities are then implicitly defined by

$$N_1^* = SN_0^* \quad \text{and} \quad SBf[(S+\alpha)N_0^*] = 1 \ . \tag{2}$$

The local stability of the equilibrium can be determined by the conventional techniques of linearized stability analysis. When survival is density-independent the condition for stability is

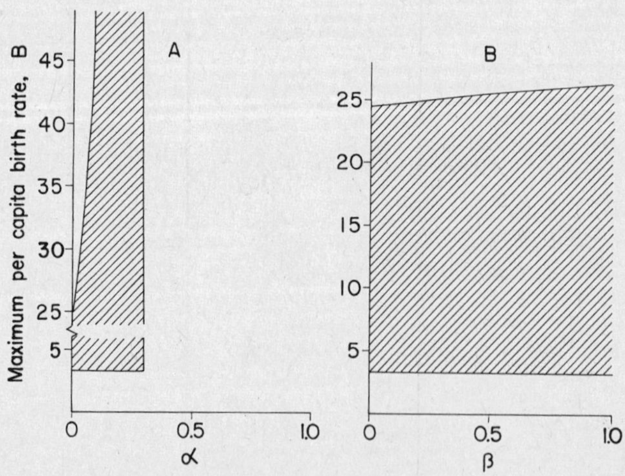

Fig. 1A, B. Stability of equilibrium points of Eq. (1). **A** Reproduction is density-dependent and of the form $f = \exp[-k(N_1+\alpha N_0)]$, S = 0.3. **B** Juvenile survival is density-dependent and of the form $g=\exp[-k'(N_0+\beta N_1)]$, S = 0.3. Shaded areas correspond to stable, positive equilibrium points. (Modified from Figs. 1 and 2 in Ebenman 1988)

$$-N_0^* S B \alpha f'[(S+\alpha)N_0^*] < -N_0^* S^2 B f'[(S+\alpha)N_0^*] < 2 \ , \tag{3}$$

where f′ is the derivative of the function f. The left part of the condition simplifies to

$$\alpha < S \ . \tag{4}$$

The right part of condition (3) is not so straightforward, but implies that increased competition between juveniles and adults has a stabilizing effect as long as α is smaller than S (Ebenman 1988). Figure 1a shows the parameter space that correspond to stable, positive equilibrium points for $f = \exp[-k(N_1 + \alpha N_0)]$ (here k is a measure of the density-dependent feedback on reproduction). It can be seen that an increase in α increases the domain of stability as long as α is smaller than S, but that strong competition has a destabilizing influence.

2.2 Density-Dependent Survival

Now, consider the case when reproduction is density-independent (i.e., $f(N_1 + \alpha N_0) \equiv 1$). Then the equilibrium densities of adults and juveniles are implicitly defined by

$$N_0^* = B N_1^* \quad \text{and} \quad S B g[(B+\beta)N_1^*] = 1 \ . \tag{5}$$

The condition for stability now becomes

$$-N_1^* S B g'[(B+\beta)N_1^*] < -N_1^* S B^2 g'[(B+\beta)N_1^*] < 2 \ , \tag{6}$$

where g′ is the derivative of the function g. The left part of the condition simplifies to

$$\beta < B \ . \tag{7}$$

The right part of condition (6) implies that competition between adults and juveniles has a stabilizing effect as long as β is smaller than B, which in reality is probably almost always the case. Figure 1b shows the parameter space that corresponds to stable, positive equilibrium points for $g = \exp[-k'(N_0 + \beta N_1)]$ (here k′ is a measure of the density dependent feedback on juvenile survival).

When both survival and reproduction are density-dependent, it is not possible to solve analytically for the equilibrium densities of juveniles and adults. However, in accordance with the analytical results, computer simulations of specified systems show that strong competition between the age classes has a stabilizing effect when the density dependent feedback on reproduction is small compared to that on survival, whereas if density dependence acts primarily on reproduction it has a destabilizing effect (Ebenman 1987).

Destabilization of an equilibrium point leads to limit cycles or some other kind of complex dynamic behavior (chaos) about the unstable equilibrium point (Ebenman 1987 and 1988). The outcome may also be competitive exclusion of one cohort by the other. This will always be the case when $\alpha > S$ or $\beta > B$ [i.e., when the left side of the stability conditions (3) and (6) are violated] (Fig. 2).

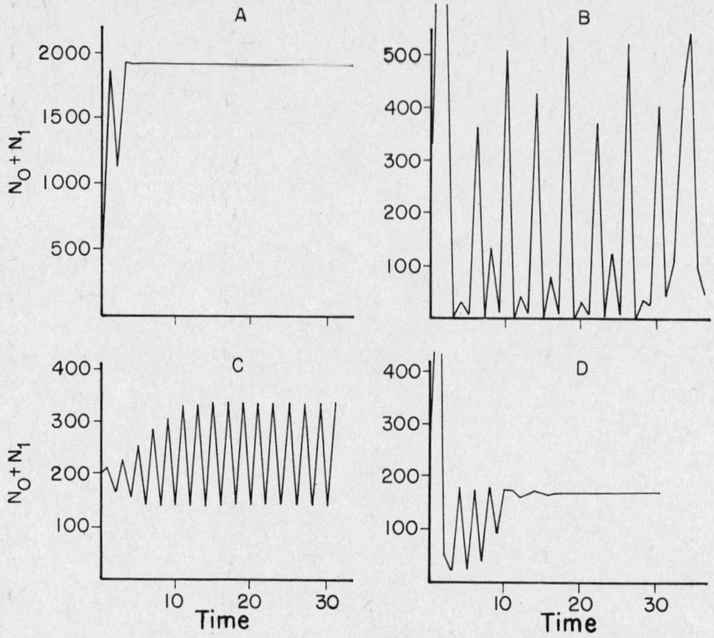

Fig. 2A–D. Dynamics of populations governed by Eq. (1) with $f = \exp[-k(N_1+\alpha N_0)]$ and $g = \exp[-k'(N_0+\beta N_1)]$. **A** Organisms with no competition between age classes ($\alpha = \beta = 0$, e.g., organisms with complex life cycles) and where density dependence is primarily on reproduction ($k > k'$). Parameter values: $B = 50$, $S = 0.3$, $k = 0.01$, $k' = 0.001$ and $\alpha = \beta = 0$. **B** Organisms with no competition between age classes and where density dependence is primarily on juvenile survival ($k < k'$). Parameter values: $B = 50$, $S = 0.3$, $k = 0.001$, $k' = 0.01$ and $\alpha = \beta = 0$. **C** Organisms with competition between age classes ($\alpha > 0, \beta > 0$, e.g., organisms with simple life cycles) and where $k > k'$. Parameter values: $B = 10$, $S = 0.6$, $k = 0.01$, $k' = 0.001$ and $\alpha = \beta = 0.8$. **D** Organisms with competition between age classes and where $k < k'$. Parameter values: $B = 10$, $S = 0.6$, $k = 0.001$, $k' = 0.01$ and $\alpha = \beta = 0.8$. In all cases the starting values are $N_0 = N_1 = 100$. Note the dramatic differences between organisms with no competition and strong competition between age classes: they show completely opposite dynamic responses to density dependence. (From Fig. 3 in Ebenman 1988)

2.3 Conclusions

Strong competition between age classes can be either a stabilizing or destabilizing force depending on the stage of the life cycle – juvenile survival or adult survival/reproduction – at which density dependence occurs. When density dependence acts primarily on juvenile survival, competition has a stabilizing effect, since in reality the intensity of competition (β) is probably always less than the maximum per capita birth rate (B). On the other hand, when density dependence acts mainly on reproduction, strong competition tends to be destabilizing, and will cause the elimination of one of the cohorts.

3 Size- and Age-Structured Populations

An assumption in the above model based on age only is that the competition coefficients are constants. One can say that this follows from an assumption of determinate, density-independent growth of individuals. However, in many organisms, the growth rates of individuals of the same age can differ considerably because of for example differences in the abiotic and/or biotic environment. For example, it is well known from many studies on invertebrates, amphibians, fishes and plants that competition within and among age classes may result in retarded individual growth rates and reduced adult body size (e.g., Underwood 1976; Fletcher 1988; Collins 1980; Prout and McChesney 1985; Wall and Begon 1987; Hart 1987; Van Buskirk 1987; Wilbur and Collins 1973; Botsford 1981; Rubenstein 1981; Jones 1987; Sebens 1987; Weiner 1985; Peet and Christensen 1987; Schmitt et al. 1987; Uchmanski 1985; many contributions is this Vol.). In such species individual growth rates are not known a priori. An individual's size depends not only upon its age, but also upon the history of the population during its past life.

Such a high degree of plasticity in growth can result in a very poor correlation between age and size and hence between age and the probabilities of survival and giving birth. To deal with such organisms, where vital rates depend on size and where growth rates of individuals are plastic and density-dependent, it is important to formulate demographic models that take size into account (Botsford 1981; Murphy 1983; Hughes 1984; Kirkpatrick 1984; Caswell 1985 and this Vol.; Metz and Diekmann 1986; Metz this Vol.).

Since resource use, and hence competition between age classes, generally will be size-dependent, the competition coefficients will not be constants in organisms with plastic and density-dependent growth. Instead, they will vary with the densities and sizes of juveniles and adults. One then needs a model that accounts for both the effect of the size of individuals on the dynamics of the population and the effect of the composition of the population on the growth and size of the individuals.

To my knowledge there has been no analysis of such a model that directly consider competition between age/size classes. I here present results from the analysis of such a model.

3.1 The Model

Consider the case when there is no direct density dependence acting on reproduction. The model then has the following general structure:

$$\begin{aligned} N_0(t+1) &= N_1(t)B(t) \\ N_1(t+1) &= N_0(t)Sg[N_0(t)+\beta(t)N_1(t)] \end{aligned} \tag{8}$$

[with symbols as before; compare with Eq. (1)]. Size structure is introduced in the following way. Let Z_0 be the size of juveniles (which is assumed to be constant) and Z_1 the size of adults. Let the growth rate of the juveniles, σ, be density dependent. That is σ is a function of the effective population density

$$\sigma = \sigma(N_0+\beta N_1)$$

say

$$\sigma = G\exp[-l(N_0+\beta N_1)] \ , \tag{9}$$

where G is the maximum growth rate (i.e., the growth rate in absence of density dependence) and l is a measure of the density dependent feedback on juvenile growth rate. Then the size of adults is

$$Z_1(t+1) = Z_0\{G\exp[-l(N_0(t)+\beta(t)N_1(t))]+1\} \ . \tag{10}$$

Let the magnitude of the competition coefficient β be a function h of the sizes of juveniles and adults

$$\begin{aligned}\beta(t+1) &= h[Z_1(t+1)-Z_0] \\ &= h\{Z_0 G\exp[-l(N_0(t)+\beta(t)N_1(t))]\} \ ,\end{aligned} \tag{11}$$

i.e., β is a function h of the size difference between adults and juveniles, which seems reasonable (an alternative would be to have $\beta = h(\ln Z_1 - \ln Z_0)$, see Roughgarden 1979, pp. 530–533).

There are basically two assumptions that can be made about the birth rate B. I will first assume that birth rate is a constant. The more realistic case is to assume that fecundity depends on body size, or on the size increment, $Z_1 - Z_0$, e.g.,

$$B(t) = bZ_0G\exp\{-l[N_0(t-1)+\beta(t-1)N_1(t-1)]\} \ , \tag{12}$$

where b is a proportionality constant. Thus, there is a delayed effect on reproduction arising from competition when the adults grew up (see Prout and McChesney 1985; Ebenman 1988 for the consequences of such delayed effects in a purely age-structured population). I will discuss this case later.

Finally, let the function g take the following form

$$g = \exp[-k'(N_0+\beta N_1)] \ ,$$

where k' is a measure of the density-dependent feedback on juvenile survival.

Putting together all these functions gives the following system

$$\begin{aligned}N_0(t+1) &= N_1(t)B(t) \\ N_1(t+1) &= N_0(t)S\exp\{-k'[N_0(t)+\beta(t)N_1(t)]\} \\ \beta(t+1) &= h\{Z_0G\exp[-l(N_0(t)+\beta(t)N_1(t))]\}\end{aligned} \tag{13}$$

and

$$Z_1(t+1) = Z_0\{G\exp[-l(N_0(t)+\beta(t)N_1(t))]+1\} \ ,$$

which describes the changes in the densities of juveniles and adults and the changes in the magnitude of the competition coefficient and the size of adults.

Before proceeding with an analysis of this system, I wish to make two points (assumptions) which I think reflect two biologically important characteristics of a size-structured population. First, I will assume that the density-dependent feedbacks on juvenile growth and survival, l and k' respectively, do not vary independently of each other. Besides being dependent on the resource availability they are also functions of the plasticity of growth, P,

$$\begin{aligned}k' &= k'(P) \ , \quad dk'/dP<0 \\ l &= l(P) \ , \quad dl/dP>0\end{aligned}$$

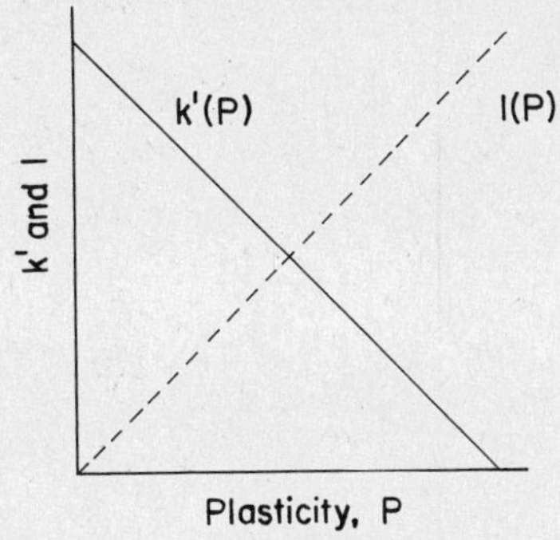

Fig. 3. The density-dependent feedbacks on survival, k′, and growth, l, as functions of plasticity, P. In this particular case $k' = 0.01 - 0.001\,P$ and $l = 0.001\,P$. These formulas are also used in the figures that follow

(Ebenman, Ekstam, Landin and Wennergren, unpubl.). One measure of plasticity may be the average variance in size for individuals of a given age and genotype over a range of environments. Another measure may be the average modulus of the slope of the norm of reaction for individuals of a given age and genotype with respect to size over a range of environments. Growth rate is highly plastic if the variance (slope) is large (steep). On the other hand, if growth rate is primarily under genetic control the variance (slope) will be small (flat). When plasticity is high competition will act primarily to reduce juvenile growth rate, not survival. In the limiting case when P equals zero l becomes zero and the model reduces to the original model without size structure (i.e., β becomes a constant) (Fig. 3).

The second point concerns the competition function h. The intensity of competition between juveniles and adults is a function of their sizes. Size will influence the amounts and kinds of resources used. Large individuals consume more resources per time unit than small ones. This leads to asymmetry. Moreover, the form of the competition function will be different for organisms that change their niche as they grow and those that do not. Most animals change their resource use as they grow. In plants however the extent of differentiation in resource use between different sized individuals is often negligible. All individuals require the same few resources and there are few opportunities for resource partitioning. I have chosen the following competition functions to describe these two cases:

$$\begin{aligned}\beta_{An} &= \exp\,[\kappa(Z_1 - Z_0)]\ \exp\,[-(Z_1 - Z_0)^2/4] \quad \text{and} \\ \beta_{Pl} &= \exp\,[\kappa(Z_1 - Z_0)]\end{aligned} \tag{14}$$

for organisms that change their resource use as they grow and for organisms that use the same resources during the whole life cycle respectively.

The term κ measures the strength of the asymmetry. A large κ-value implies that resource consumption increases rapidly with size. Increasing the asymmetry leads to an increase of both β_{An} and β_{Pl}, i.e., $\partial\beta_{An}/\partial\kappa > 0$ and $\partial\beta_{Pl}/\partial\kappa > 0$.

The first term of β_{An} reflects the fact that large individuals consume more resources per time unit than small ones, increasing the intensity of competition. The second term reflects that different-sized individuals use different resources (e.g., different-sized prey) reducing the intensity of competition. It is assumed that the resources can be ordered along a resource axis, i.e., there is a one-dimensional niche (in the case when the adults use a completely different set of resources than the juveniles the second term becomes zero, i.e., $\beta_{An} = 0$). When plasticity, P increases the size of adults, Z_1, will decrease. Then the first term decreases and the second increases. The net effect on β_{An} of a change in P will depend on the values of κ, G and Z_0. When G is large compared to κ, increasing the plasticity, P, will lead to increased competition between adults and juveniles, i.e., $\partial\beta_{An}/\partial P > 0$. Increasing P in organisms that do not change their resource use as they grow will always result in decreased competition, i.e., $\partial\beta_{Pl}/\partial P < 0$.

3.2 *Results and Discussion*

I have analysed how plasticity and asymmetry affect population dynamic behavior. The equilibrium of system (13) has been solved analytically. However, it has not been possible to specify the conditions for stability of the system in terms of the parameters (as in the age-structured case) in a meaningful way. Instead, the eigenvalues of the gradient matrix have been determined with the help of a computer.

Increasing the asymmetry, κ, tends to have a stabilizing effect up to a certain limit (Figs. 4 and 5). Beyond this limit competitive exclusion occurs and one cohort is eliminated. An increase of the asymmetry always leads to an increase of the strength of competition, in both organisms that change their resource use as they grow and those that do not. This result is therefore in accordance with the result from the purely age-based model, where it was found that increased competition between adults and juveniles has a stabilizing effect up to a certain limit (= B). These results also hold for the case when maximum birth rate, B, is size-dependent (Figs. 7 and 8).

The effect of a change in the degree of plasticity, P, is more dependent on other characteristics of the organism in question. Increasing the plasticity will de-

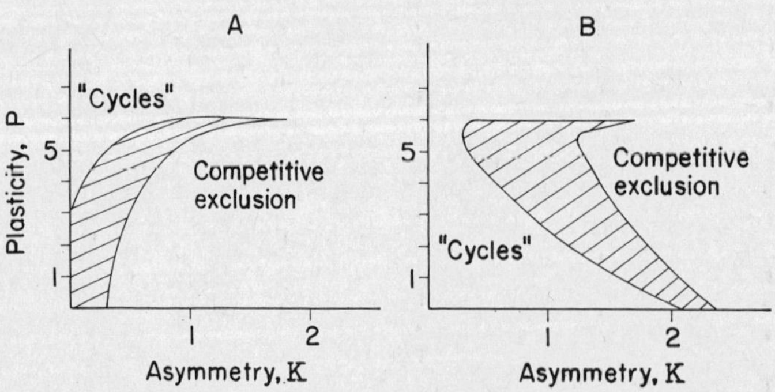

Fig. 4A, B. Dynamic behavior as a function of plasticity and asymmetry. Maximum birth rate, *B*, is a constant and maximum growth rate, *G* is relatively high. **A** Individuals use the same resources during their whole life cycle, $\beta = \exp[\kappa(Z_1 - Z_0)]$. **B** Individuals change their resource use as they grow, $\beta = \exp[\kappa(Z_1 - Z_0)]\exp[-(Z_1 - Z_0)^2/4]$. *Shaded areas* correspond to stable equilibrium points. Parameter values: B = 17, S = 0.5, G = 4, $Z_0 = 2$

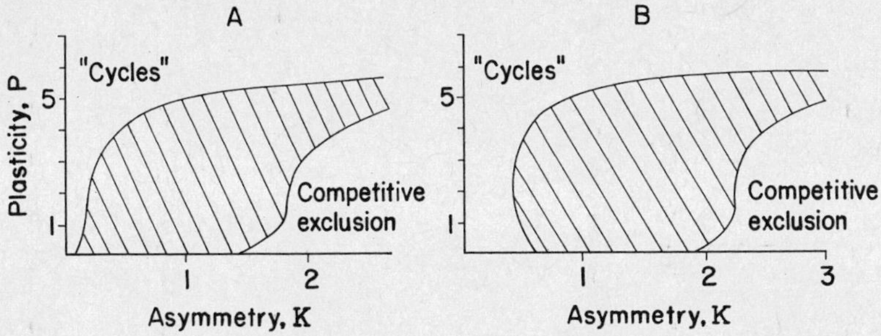

Fig. 5 A, B. Dynamic behavior as a function of plasticity and asymmetry. Maximum growth rate, G, is relatively low. Otherwise see legend to Fig. 4. Parameter values: $B = 17$, $S = 0.5$, $G = 2$, $Z_0 = 1$

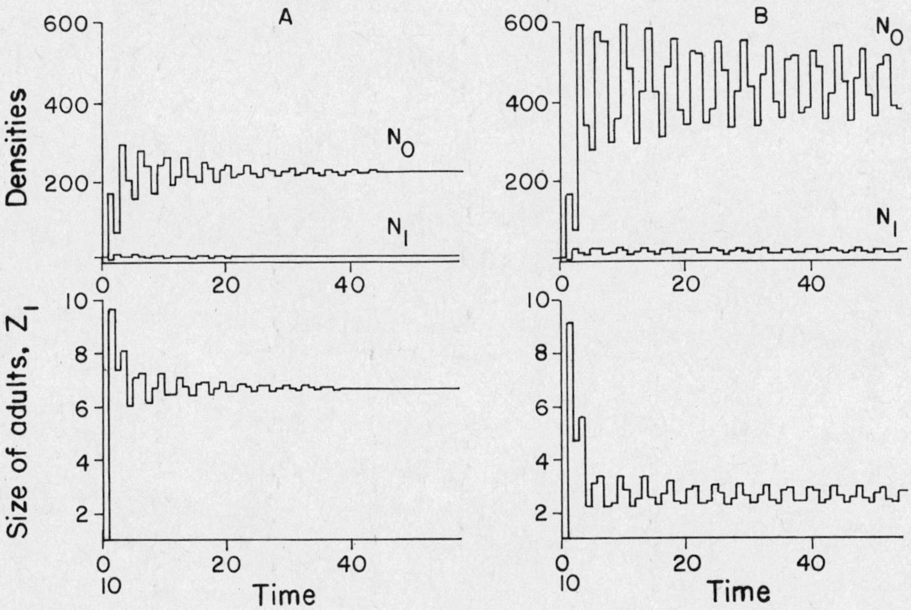

Fig. 6 A, B. Dynamic behavior of populations in which individuals use the same resources during their whole life cycle. Asymmetry, κ, is relatively small. **A** Low degree of plasticity, $P = 2$. **B** Relatively high degree of plasticity, $P = 5.5$. Parameter values: $B = 17$, $S = 0.5$, $G = 4$, $Z_0 = 2$, $\kappa = 0.25$

crease competition between age classes in organisms that use the same resources during their whole life cycle (e.g., most plants). This will tend to destabilize the system when asymmetry is low, leading to a fluctuating size distribution, and stabilize it when asymmetry is large (Figs. 4a, 5a, and 6). For organisms that change their niche as they grow, like many animals, the situation may be the reverse: increased plasticity may have a stabilizing effect when asymmetry is low and a destabilizing influence when asymmetry is large (Fig. 4b). This will be the case when maximum growth rate, G, is large. The reason for this is that under such circumstances increased plasticity will tend to result in increased competi-

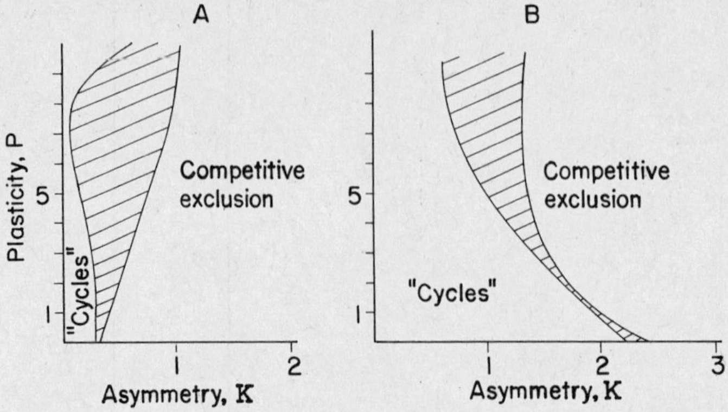

Fig. 7 A, B. Dynamic behavior as a function of plasticity and asymmetry. Maximum birth rate is a function of size, $B = b(Z_1 - Z_0)$, and maximum growth rate, G, is relatively large. **A** Individuals use the same resources during their whole life cycle. **B** Individuals change their resource use as they grow. *Shaded areas* correspond to stable equilibrium points. Parameter values: $b = 3$, $S = 0.5$, $G = 4$, $Z_0 = 2$

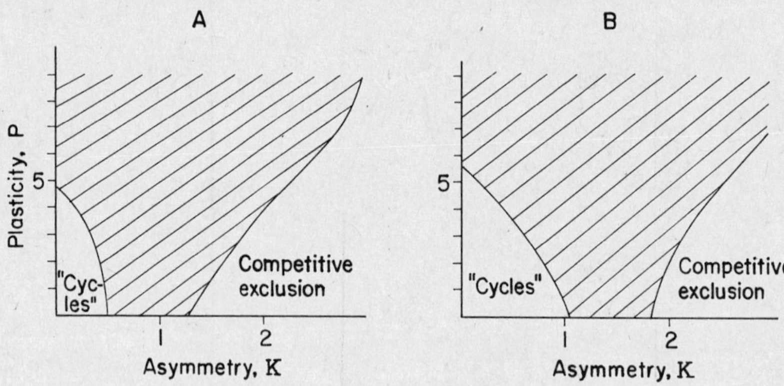

Fig. 8 A, B. Dynamic behavior as a function of plasticity and asymmetry. Maximum growth rate, G, is relatively low. Otherwise see legend of Fig. 7. Parameter values: $b = 8$, $S = 0.5$, $G = 2.5$, $Z_0 = 1$

tion, β_{An}. When G is small, the effect of changing the plasticity will be the same on β_{An} as on β_{Pl}, because then increased plasticity will tend to reduce the strength of competition (Fig. 5b).

The effect of changing the plasticity when maximum birth rate, B, is size-dependent is more complicated, because, plasticity influences both the intensity of competition and the magnitude of the delayed density-dependent feedback on reproduction. However, as can be seen in Figs. 7 and 8, increased plasticity tends to have a stabilizing effect at the level of the population.

The results imply that in organisms that do not change their resource use as they grow (like many plants) high plasticity allows for greater asymmetry before competitive exclusion occurs (Figs. 4a, 5a, 7a and 8a). Plants are actually very plastic in their growth, and competition is often very asymmetric, with large individuals severely depressing the growth of small ones. Perhaps plasticity has evolved, at least partly, to reduce the strength of intra- and inter-specific competition. (How selection operates on plasticity is a very interesting question in itself, which lies outside the scope of this contribution). The same applies to organisms where the individuals change their niche during the life cycle and where individual maximum growth rates, G, are relatively low (Figs. 5b and 8b). On the other hand, in organisms where the individuals change their resource use as they grow and where individual growth rates, G, are relatively large, low plasticity allows for greater asymmetry before competitive exclusion occurs (Figs. 4b and 7b).

The results also have implications for the coexistence of species with size-structure, because in the present model competition between adults and juveniles is analogous to competition between size-structured species whose life cycles are displaced in time (asynchronous). Thus, the results suggest that coexistence between, for example, plant species becomes easier if the growth rates of the individuals are highly plastic.

The results may also be relevant to consider in connection with explanations of synchronized reproduction. The periodical cicadas are well known for their extremely synchronized reproduction. The findings from the present analysis indicate that asymmetric competition between age classes and the potential competitive elimination of a cohort out of phase may act as a strong selective force for the maintenance of synchronized reproduction (see also Bulmer 1977).

Multiple equilibria are not possible in the present models. However, under some circumstances (e.g., when κ is large) a sufficient disturbance may destabilize an initially stable system, driving one of the cohorts to extinction. Thus, a locally stable equilibrium is not always globally stable.

One limitation of the present models is that they have the property that there are two isolated cohorts. Iteroparity or size-dependent semelparity normally allows mixing among cohorts. It is not known to what extent the results of this analysis depend on this non-mixing property. However, analysis of size-classified models that include iteroparity suggests that increased plasticity in growth of individuals will have a stabilizing effect at the level of the population (Ebenman, Ekstam, Landin and Wennergren, unpublished).

4 Conclusions

The analysis of purely age-based demographic models and models that also incorporate size gave concordant results with respect to the effects of intraspecific competition on population stability. The great advantage of a size-based demography as compared to a purely age-based one, is that it provides a mean of explicitly including ecologically important characteristics of individuals – like plasticity in growth – in explanations of population behavior. Many biologically interesting questions – like, for example, the effects of plastic growth of individuals on population dynamic behavior or the effects of the growth patterns of individuals on limiting similarity – can simply not be dealt with from a purely age-specific point of view. Especially, the interplay between growth of individuals and growth of populations can be illuminated with the aid of a demography that includes size.

Models like the ones in this paper, that account for the interplay between growth of individuals and growth of populations, and which are also extended to include iteroparity (more than one reproducing age/size class) may shed new light on some of ecology's most general empirical patterns like the self-thinning law of plant populations (Norberg this Vol.; Weller 1987; Begon et al. 1986; Hara 1988).

Acknowledgments. I wish to thank my colleagues Börje Ekstam, Göran Forsling, Jan Landin, Per Weinerfeld and Uno Wennergren at Linköping University for stimulating discussions and valuable help and Hal Caswell and Richard Law for helpful comments on the manuscript. I am especially grateful to Uno Wennergren for help with the programming. This research was supported by the Swedish Natural Science Research Council.

5 References

Begon M, Firbank L, Wall R (1986) Is there a self-thinning rule for animal populations? Oikos 46:122–124

Bellows TS (1982a) Analytical models for laboratory populations of *Callosobruchus chinensis* and *C. maculatus* (Coleoptera, Bruchidae). J Anim Ecol 51:263–287

Bellows TS (1982b) Simulation models for laboratory populations of *Callosobruchus chinensis* and *C. maculatus.* J Anim Ecol 51:597–623

Botsford LW (1981) The effects of increased individual growth rates on depressed population size. Am Nat 117:38–63

Bulmer MG (1977) Periodical insects. Am Nat 111:1099–1117

Caswell H (1985) The evolutionary demography of clonal reproduction. In: Jackson JBC, Buss LW, Cook RE (eds) Population biology and evolution of clonal organisms. Yale Univ Press, New Haven, pp 187–224

Collins NC (1980) Developmental responses to food limitation as indicators of environmental conditions for *Ephydra cinerea* (Diptera). Ecology 61:650–661

Crowley PH, Nisbet RM, Gurney WSC, Lawton JH (1987) Population regulation in animals with complex life-histories: Formulation and analysis of a damselfly model. Adv Ecol Res 17:1–59

Desharnais RA, Liu L (1987) Stable demographic limit cycles in laboratory populations of *Tribolium castaneum.* J Anim Ecol 56:885–906

Ebenman B (1987) Niche differences between age classes and intraspecific competition in age-structured populations. J Theor Biol 124:25–33

Ebenman B (1988) Competition between age classes and population dynamics. J Theor Biol 131:389–400

Fletcher WJ (1988) Intraspecific interactions between adults and juveniles of the subtidal limpet, *Patelloida mufria.* Oecologia (Berlin) 75:272–277

Goldberg DE (1987) Neighborhood competition in an old-field plant community. Ecology 68:1211–1223

Hara T (1988) Dynamics of size structure in plant populations. Tree 3:129–133

Hart DD (1987) Experimental studies of exploitative competition in a grazing stream insect. Oecologia (Berlin) 73:41–47

Hastings A, Costantino RF (1987) Cannibalistic egg-larva interactions in *Tribolium:* An explanation for the oscillations in population numbers. Am Nat 130:36–52

Hughes TP (1984) Population dynamics based on individual size rather than age: A general model with a reef coral example. Am Nat 123:778–795

Jones GP (1987) Competitive interactions among adults and juveniles in a coral reef fish. Ecology 68:1534–1547

Kirkpatrick M (1984) Demographic models based on size, not age, for organisms with indeterminate growth. Ecology 65:1874–1884

Liu L, Cohen JE (1987) Equilibrium and local stability in a logistic matrix model for age-structured populations. J Math Biol 25:73–88

May RM, Conway GR, Hassell MP, Southwood TRE (1974) Time delays, density-dependence and single-species oscillations. J Anim Ecol 43:747–770

Metz JAJ, Diekmann O (1986) The dynamics of physiologically structured populations. Springer, Berlin Heidelberg New York (Lecture notes in biomathematics, vol 68)

Murphy LF (1983) A nonlinear growth mechanism in size structured population dynamics. J Theor Biol 104:493–506

Nisbet RM, Gurney WSC (1982) Modelling fluctuating populations. John Wiley & Sons, Chichester New York

Peet RK, Christensen NL (1987) Competition and tree death. BioScience 37:586–595

Polis GA (1984) Age structure component of niche width and intraspecific resource partitioning: can age groups function as ecological species? Am Nat 123:541–564

Prout T, McChesney F (1985) Competition among immatures affects their adult fertility: Population dynamics. Am Nat 126:521–558

Roughgarden J (1979) Theory of population genetics and evolutionary ecology: An introduction. Macmillan, New York

Rubenstein DI (1981) Individual variation and competition in the everglade pygmy sunfish. J Anim Ecol 50:337–350

Schmitt J, Eccleston J, Ehrhardt DW (1987) Dominance and suppression, size-dependent growth and self-thinning in a natural *Impatients capensis* population. J Ecol 75:651–665

Sebens KP (1987) The ecology of indeterminate growth in animals. Annu Rev Ecol Syst 18:371–407

Tschumy WO (1982) Competition between juveniles and adults in age-structured populations. Theor Popul Biol 21:255–268

Uchmanski J (1985) Differentiation and frequency distributions of body weights in plants and animals. Philos Trans R Soc London Ser B 310:1–75

Underwood AJ (1976) Food competition between age classes in the intertidal neritacean *Nerita atramentosa* (Gastropoda: Prosobranchia). J Exp Mar Biol Ecol 23:145–154

Van Buskirk J (1987) Density-dependent population dynamics in larvae of the dragonfly *Pachydiplax longipennis:* a field experiment. Oecologia (Berlin) 72:221–225

Wall R, Begon M (1987) Individual variation and the effects of population density in the grasshopper *Chorthippus brunneus.* Oikos 49:15–27

Weiner J (1985) Size hierarchies in experimental populations of annual plants. Ecology 66:743–752

Weller DE (1987) A reevaluation of the $-3/2$ power rule of plant self-thinning. Ecol Monogr 57:23–43

Werner EE, Gilliam JF (1984) The ontogenetic niche and species interactions in size-structured populations. Annu Rev Ecol Syst 15:393–425

Wilbur HM (1980) Complex life cycles. Annu Rev Ecol Syst 1:67–93

Wilbur HM, Collins JP (1973) Ecological aspects of amphibian metamorphosis. Science 182:1305–1314

Maximum Sustainable Yields and the Self-Renewal of Exploited Populations with Age-Dependent Vital Rates

R. LAW[1] and D.R. GREY[2]

1 Introduction

This paper examines the relationship between two problems in the theory of harvesting self-renewing populations with age-dependent vital rates. The first problem is to find a method of harvesting such that the yield sustainable from a population is maximized. Because of the obvious practical importance of this problem, it has been the subject of many investigations (e.g., Beverton and Holt 1957; Beddington and Taylor 1973; Doubleday 1975; Reed 1980; Botsford 1981; Horwood and Whittle 1986). The second problem is to find a method of harvesting which will maximize the capacity of a population to renew itself, given that a certain minimum crop must be taken. At first sight this problem looks relatively unimportant, and it is perhaps understandable that it has received relatively little attention.

Intuition suggests, however, that the two problems are related. Removal of individuals which would otherwise make a large reproductive contribution to renewal of the population should have the consequence that there are relatively few offspring available for harvesting in the future. This was first suggested by MacArthur (1960), who conjectured that the yield would be greatest when the reproductive value of culled individuals relative to their value of the harvester is at its minimum. Slobodkin (1968) argued that a prudent predator would cull individuals at an age of high population efficiency, which he envisaged as being approximately the inverse of the reproductive value. It was pointed out by Law (1979) that the measure of reproductive value would not be fixed for a given life history, but would depend on the rate of population increase under harvesting. Botsford (1981) also emphasized the balance between the current value of an individual and its future value (from future fishing, reproduction etc.) in determining the optimal culling policy. The effect of culling on renewal is particularly critical in the management of endangered species, a problem which has been considered both in a deterministic and stochastic context (Goodman 1980; Grey 1984).

It can be shown that, under certain conditions, the pattern of culling which maximizes the yield is the same as that which maximizes renewal of the population given that a crop greater than a certain minimum value must be taken (Grey and Law 1987). Here we amplify the result of Grey and Law (1987) which was framed in a density-independent setting, and extend the analysis to populations

[1] Department of Biology, University of York, York YO1 5DD, England
[2] Department of Probability and Statistics, The University, Sheffield S3 7RH, England

Size-Structured Populations
B. Ebenman and L. Persson (Eds.)

with density-dependent recruitment. Size dependence enters into the analysis through the weight of culled individuals. Although it would be desirable to allow size-dependent reproduction and mortality as well (see chapters by Caswell; Ebenman; Metz et al. this Vol.), this would entail loss of linearity which makes solutions to the problems straightforward. Further work will be needed to clarify the effects of size-dependent vital rates.

2 Model

We consider a population in a seasonal environment with age-dependent vital rates (following Reed 1980). For the time being, we concentrate on a density-independent framework; the modifications needed for density dependence are introduced later, when needed.

Individuals are classified by age ($i = 1,2 \ldots, k$) at the time of census each year; at census t, there are $n_i(t)$ females of age i per unit area. Reproduction occurs as a pulse immediately after the census, the number of female progeny produced by a mother of age i, which are still alive at the next census, being b_i ($i = 1, 2, \ldots, k$), for all t. For the remainder of the year, mortality due to natural causes and harvesting occurs. The proportion of individuals which survive natural causes of mortality between age 1 and age i is denoted by l_i ($i = 1, 2, \ldots, k$), for all t, with $l_1 = 1$. The proportion of individuals aged i harvested is $1-\theta_i$ ($i = 1, 2, \ldots, k$), for all t. The order in which natural mortality and mortality due to culling takes place is not critical; for simplicity, it is assumed here that harvesting mortality occurs first. Size dependence enters into the model through the weight of an individual at the time of culling; for an individual aged i this is denoted by w_i ($i = 1, 2, \ldots, k$), for all t.

It is assumed that, in the absence of harvesting, the asymptotic rate of population increase (λ) is greater than unity. All harvesting regimes are required to bring this rate of increase down to unity, so that the population neither increases nor decreases under harvesting; the resulting yields are evaluated after the population has achieved steady state (see the Euler constraint, below). In general, however, the rate of increase under harvesting can be set to any value in the range $[0, \lambda]$. Because population density is not defined in this density-independent context, the total yield is unbounded, and it is appropriate to consider the yield per individual recruited at first census (H). The contribution of age i to this yield is

$$w_i l_i \left(\prod_{j=1}^{i-1} \theta_j \right) (1-\theta_i) \ .$$

As it stands, the yield is a non-linear function of the harvest pattern imposed on the population (the θs). The analysis can, however, be simplified by adopting the transformation $\eta_i = \left(\prod_{1}^{i-1} \theta_j \right) (1-\theta_i)$, because the equations and constraints then become linear functions of the harvest pattern, i.e., the vector $\eta = (\eta_1, \ldots, \eta_k)$ (Reed 1980). Thus the yield per recruit is

$$H = \gamma_1 \eta_1 + \ldots + \gamma_k \eta_k \ , \tag{1}$$

where $\gamma_i = w_i l_i$.

The dynamics of the population at steady state must satisfy the Euler equation (Charlesworth 1980: 25), which takes the form

$$1 = \sum_{i=1}^{k} l_i b_i \left(\prod_{j=1}^{i-1} \theta_j \right)$$

because the rate of population increase is unity under harvesting. Although this equation appears to be the same as that defining the net reproductive rate, the resemblance is misleading, because it applies only when the rate of increase under harvesting is set at unity. In terms of $\boldsymbol{\eta}$, the Euler equation is

$$1 = \varepsilon_0 - \varepsilon_1 \eta_1 - \ldots - \varepsilon_{k-1} \eta_{k-1} , \tag{2}$$

where $\varepsilon_i = \sum_{j=i+1}^{k} l_j b_j$. In addition, the harvest pattern must be subject to the constraint

$$1 \geqslant \eta_1 + \ldots + \eta_k \tag{3}$$

to be biologically feasible. This places an upper limit on the total mortality which can be expended in the process of harvesting; when it reaches unity, no further individuals remain to be caught. (Clearly, $\eta_i \geqslant 0$ for all i.)

3 Problem 1 – Maximum Sustainable Yield

The problem of maximizing the yield per recruit is to choose the harvest pattern ($\boldsymbol{\eta}$) which solves the following linear programming problem (Reed 1980):

$$\begin{array}{ll} \text{maximize} & \gamma_1 \eta_1 + \ldots + \gamma_k \eta_k \\ \text{subject to} & \varepsilon_1 \eta_1 + \ldots + \varepsilon_{k-1} \eta_{k-1} = \varepsilon_0 - 1 \\ \text{and} & \eta_1 + \ldots + \eta_k \leqslant 1 \\ \text{with} & \eta_i \geqslant 0 \quad \text{for} \quad i = 1, 2, \ldots, k . \end{array}$$

Thus the problem is defined in k dimensional space, the axes corresponding to the η_is. The Euler constraint [Eq. (2)] constitutes a surface in this space on which the solution must lie, and the harvest-mortality constraint [inequality (3)] is an upper boundary. It follows as a direct consequence of the existence of two linear constraints that the solution must require cropping of no more than two ages (Gass 1969: 55; Reed 1980). This property of the optimal harvest pattern has also been shown in earlier analyses (Beddington and Taylor 1973; Rorres and Fair 1975).

Although a numerical solution of Problem 1 tells us which age(s) to harvest, it does not lead to an understanding of why these are the best ages. This is partly because the high dimensionality of the problem makes it difficult to visualize. We can, however, make use of a fundamental theorem of linear programming to write down another problem with only two variables, the dual problem of the primal above (Gass 1969). The dual, which was first described by Reed (1980), is

$$\begin{aligned}
\text{minimize}\quad & (\varepsilon_0 - 1)X_1 + X_2 \\
\text{subject to}\quad & \varepsilon_1 X_1 + X_2 \geqslant \gamma_1 \\
& \vdots \\
& \varepsilon_{k-1} X_1 + X_2 \geqslant \gamma_{k-1} \\
& X_2 \geqslant \gamma_k \\
\text{with}\quad & X_1 \text{ unconstrained in sign} \\
\text{and}\quad & X_2 \geqslant 0 \;.
\end{aligned}$$

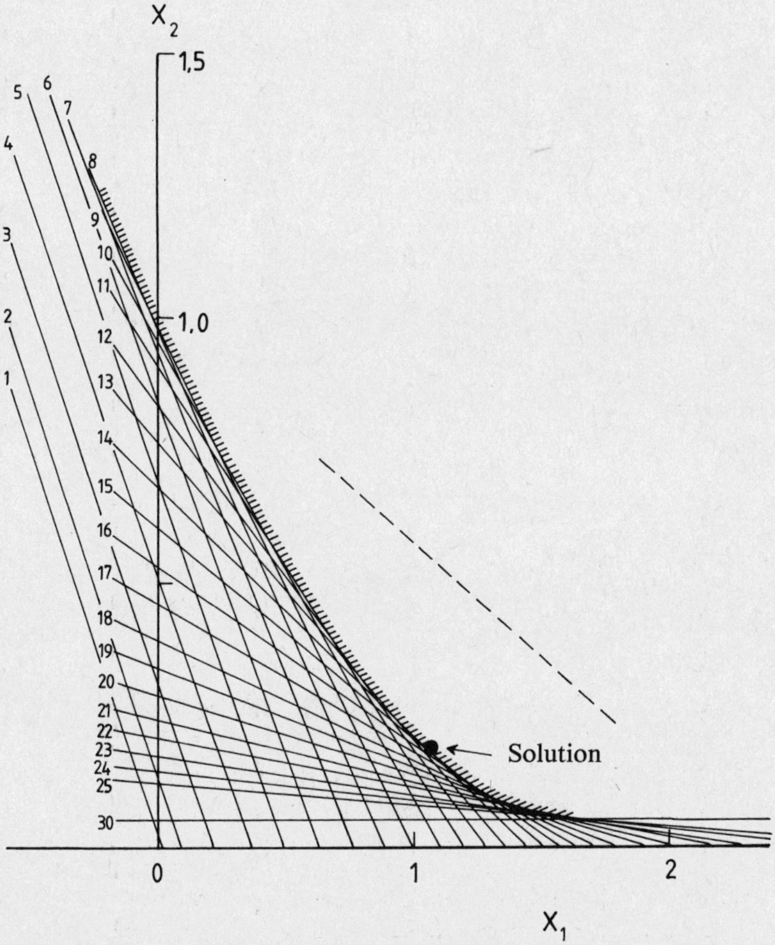

Fig. 1. Graphical representation of the dual of Problem 1 for the Arcto-Norwegian cod. The axes X_1 and X_2 represent the sensitivity of the harvest to changes in the primal constraints. Each line is a constraint corresponding to the numbered age; for clarity, the constraints for ages 26 to 29 years are not shown. The lower boundary of the feasible set of X_1 and X_2 is *hatched.* The gradient of the objective function is indicated by the *discontinuous line,* and the function is minimized at the *point indicated on the graph,* at the intersection of constraints 14 and 15

It is defined in two-dimensional space, X_1 (respectively X_2) being the sensitivity of the harvest to changes in the first (respectively second) primal constraint, while holding the second (respectively first) primal constraint in operation. It has k constraints, one corresponding to each age. The dual has, by definition, the same optimal value as the primal, but being defined in two dimensional space it has a geometry which is simple to understand.

Figure 1 illustrates the dual problem, based on the life history of the Arcto-Norwegian stock of cod described in the Appendix. (It is assumed that the rate of survival from egg to recruitment at age 1 year is 5×10^{-7}. In reality, matters are more complicated than this because survivorship depends on the density of the cohort; we return to this later.) There are 30 constraints, one corresponding to each age. Ages with constraints forming the boundary of the feasible set of X_1 and X_2 are candidates for the optimal harvest pattern, and the solution of the dual here lies at the intersection of the constraints corresponding to ages 14 years and 15 years.

It is worth noting that the value of X_1 at the intersection of constraint i with the X_1 axis is γ_i/ε_i. The biological significance of this is that

$$\frac{\varepsilon_i}{\gamma_i} = \frac{1}{w_i l_i} \sum_{i+1}^{k} l_j b_j \ ,$$

the future contribution to reproduction of a unit weight of age i. Thus, if the solution lies on the X_1 axis, it involves removal of individuals with minimum ε_i/γ_i. This provides some justification for the conjecture made by MacArthur (1960), that a harvester should remove first those individuals of smallest reproductive value relative to their value to the harvester. However, the solution can only lie on this axis if females beyond the final age of reproduction have zero value to the harvester. In general, the solution does not lie on the X_1 axis, and the appealing simplicity of this interpretation is lost. The solution now minimizes the sensitivity of the harvest to changes in the second primal constraint (3) as well as the Euler constraint (2), and is not framed in terms of reproductive values. [Reed (pers. comm.) has also noted the role of the reproductive value in this dual.]

4 Problem 2 – Maximum Renewal Under Harvesting

Although the optimal choice of ages for harvesting is not solely determined by their subsequent reproductive potential, intuition still suggests that, in some sense, the yield should be at its maximum when the self-renewal of the culled population is at its greatest (given that a certain minimum crop has to be taken). Consider the problem of maximizing the lifetime's reproductive contribution to future generations, the reproductive value at first census (v_1), in the presence of harvesting. This can be given as another problem in linear programming in which a harvest pattern (η) is chosen to

$$\text{maximize} \quad \varepsilon_0 - \varepsilon_1 \eta_1 - \ldots - \varepsilon_{k-1} \eta_{k-1}$$

$$\text{subject to} \quad \gamma_1 \eta_1 + \ldots + \gamma_k \eta_k \geqslant H_{max}$$

and $\eta_1 + \ldots + \eta_k \leqslant 1$

with $\eta_i \geqslant 0$ for $i = 1, 2, \ldots, k$,

where H_{max} is the minimum allowable crop, taken to be the solution of the harvest-maximization problem. It can be seen that this problem closely resembles the previous one, being based on the same three surfaces in the same k dimensional space. They differ only in that the objective function and first constraint have been interchanged. Inspection of the surfaces indicates that the solutions of the two problems usually lie at the same point, so that they are solved by the same harvest patterns, as will become clear in the next section. There are, however, exceptions to this which are also considered in the next section. (Note that the resemblance of the objective function above to the net reproductive rate is no more than superficial, and disappears if the rate of increase under harvesting is set to any value other than unity.)

5 Problems 1 and 2 Combined – The Set $\{(H, v_1)\}$

The relationship between Problems 1 and 2 becomes clearer when envisaged in terms of the feasible set of H and v_1. By 'feasible set' we mean the values they take when associated with all harvest patterns satisfying inequality (3) and $\boldsymbol{\eta} \geqslant 0$. This set supresses the dependence of H and v_1 on the harvest pattern, while retaining the relationship between H and v_1. It highlights the geometry of the problems, and in passing simplifies them to a degree which allows 'back-of-an-envelope' solutions. Gilliam and Fraser (this Vol.) have also made use of this geometric approach to optimization to gain understanding of foraging behavior.

The biological motivation of the set is as follows. Consider an arbitrary, feasible, harvest pattern. Associated with this pattern is a value of H and v_1, which we represent as a point in (H, v_1) space. Repeat this until all feasible harvest patterns are included; the space incorporating all of these points is the set $\{(H, v_1)\}$. In practice, construction of the set is very much simpler than this because of the linear dependence of H and v_1 on the harvest pattern. All that is required is the (H, v_1) corresponding to complete removal of each age (for age i, the harvest pattern is $\eta_i = 1$ and $\eta_j = 0$ for $j \neq i$). With complete removal of age i, we have

$$H = \gamma_i \quad = w_i l_i$$

$$v_1 = \varepsilon_0 - \varepsilon_i = \sum_1^i l_j b_j \ .$$

The set $\{(H, v_1)\}$ is then the convex hull of these k points, together with the point $(0, \varepsilon_0)$ (i.e., the smallest convex set containing all of these points). See Grey and Law (1987) for an explanation of this graphical method.

Figure 2 provides an illustration of an $\{(H, v_1)\}$ set. The data on which it is based have been taken from the Arcto-Norwegian stock of cod, as described in the Appendix; it is assumed for the time being that the rate of survival from egg to recruitment at age 1 year is 5×10^{-7}. (In reality, matters are more complicated than this, because survivorship depends on the density of the cohort; we return

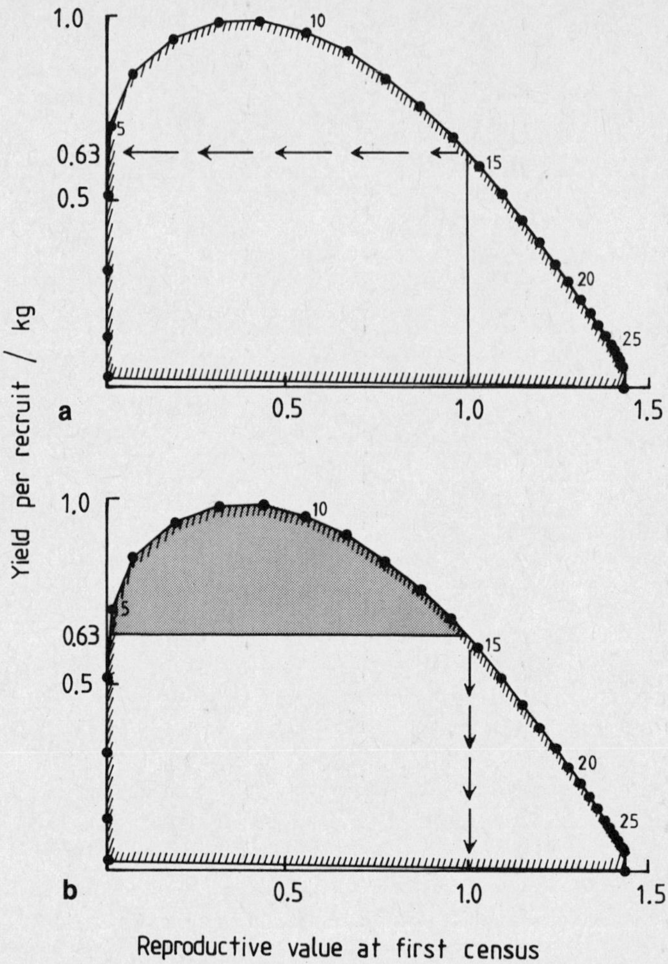

Fig. 2a, b. The set of yields per recruit (H) and reproductive values at first census (v_1) corresponding to all feasible harvest patterns (η) for the Arcto-Norwegian cod, assuming pre-recruitment survivorship 5×10^{-7}. *Circles* represent points corresponding to complete removal of single ages; ages are numbered in multiples of 5 years. **a** Gives the solution of Problem 1, maximization of the yield per recruit. **b** Gives the solution of Problem 2, maximization of the reproductive value at first census

to this in the following sections.) The boundary of the set is completely defined by the vital rates (ls and bs) and the weights (ws), and the set can be envisaged as a device for describing a life history in a way which is simple and convenient for analysis of harvesting. Although all points corresponding to complete removal of single ages lie on the boundary of the set in the case of the Arcto-Norwegian cod, this is not a general feature of $\{(H, v_1)\}$ sets; more often some of these points lie in the interior. Since $\varepsilon_0 - \varepsilon_i$ increases monotonically with age when reproduction takes place, the age of corners of the set increases with v_1 during the adult phase of life. Ages before the onset of reproduction lie on the H-axis, and ages of no value to the harvester (i.e., $w_i = 0$) lie on the v_1 axis. Thus, if there are pre-reproductive ages of no value to the harvester, the origin forms a corner of the set. The shape of the set depends on size at each age (ws); an increase in size at some age would cause the corresponding corner to move up parallel to the H axis. The point $(0, \varepsilon_0)$ forms a right-hand corner of the set.

The relationship between Problems 1 and 2 can be seen by comparing Figs. 2a and 2b. Problem 1 seeks the maximum value of H on the line $v_1 = 1$ in the set $\{(H, v_1)\}$ (Fig. 2a). Problem 2 seeks the maximum value of v_1 in the stippled region $H \geqslant H_{max}$ (Fig. 2b). Clearly the solution of both problems lies at the same point in the set. It should be understood, however, that the solutions coincide only if the upper boundary of the set is a decreasing function of v_1 at $v_1 = 1$; if this boundary is increasing at $v_1 = 1$, then v_1 is at its minimum.

The harvest pattern (η) required to solve the two problems is found by inspection of the set $\{(H, v_1)\}$. It involves removal of the two ages corresponding to the two corners of the boundary adjacent to its intersection with the line $v_1 = 1$ (14 and 15 years in Fig. 2). The η_i of the younger age decreases linearly from 1 to 0 along the line joining the two ages; thus $\eta_{14} = 0.41$ in Fig. 2. On the boundary $\sum \eta_i = 1$; so $\eta_{15} = 0.59$ in Fig. 2. In fact, this means that the older age is completely removed; thus, in terms of θs, the solution of the problem in Fig. 2 is at $\theta_{14} = 0.59$ and $\theta_{15} = 0$.

6 Density-Dependent Recruitment

Extension of the analysis to certain kinds of density-dependent dynamics presents few additional problems. We will assume that mortality between birth and recruitment at first census is a function of the density of the newborn cohort ('nursery competition'); as a first approximation, this applies quite widely in fisheries, the major natural systems exploited by man (Cushing and Horwood 1977). Further, the analysis will assume that there exists a stable equilibrium population under harvesting due to this density dependence; see Reed (1980) for the condition for stability. It is then possible to determine the total yield at equilibrium (H_T), as well as the yield per recruit (H); all the results given here refer to a population at equilibrium under harvesting.

The modifications to the model required to place it in this density-dependent framework are as follows. Offspring are counted at birth, instead of at first census; b_i' denotes the number of female offspring of a mother aged i, counted at the time of reproduction. The average number of female offspring produced by a mother over the course of her life is then

$$\varepsilon = \varepsilon_0' - \varepsilon_1' \eta_1 - \ldots - \varepsilon_{k-1}' \eta_{k-1} \ , \tag{4}$$

where $\varepsilon_i' = \sum_{j=i+1}^{k} l_j b_j'$. At equilibrium, the density of the newborn cohort is $2\hat{n}_1 \varepsilon$, assuming that the sex ratio at birth is 1:1, where $\hat{n}_1$ is the equilibrium density of females at age 1. The population dynamics at equilibrium are described by the balance relation

$$\hat{n}_1 = \hat{n}_1 \cdot \varepsilon \cdot g(2\hat{n}_1 \varepsilon) \ , \tag{5}$$

where g is the density-dependent, per-capita survival rate to first census, assumed to be a monotonic decreasing function of $2\hat{n}_1\varepsilon$, independent of sex. This relation defines $\hat{n}_1$ as an implicit function of ε. The total yield at equilibrium (H_T) is then the product of the equilibrium recruitment and the yield per recruit

$$H_T = 2\hat{n}_1 H \ ,$$

where we assume, for simplicity only, that males have the same ηs, ls and ws as females.

7 Problem 1 Revisited – Maximum Sustainable Yield

To maximize the total yield at equilibrium in the presence of density-dependent recruitment, we must choose a harvest pattern (η) which solves the nonlinear problem

$$\begin{array}{ll} \text{maximize} & 2\hat{n}_1 H \\ \text{subject to} & \eta_1 + \ldots + \eta_k \leqslant 1 \\ \text{with} & \eta_i \geqslant 0 \quad \text{for} \quad i = 1, 2, \ldots, k \end{array}$$

(Reed 1980). In switching to this density-dependent setting, the constraint on self-renewal has been eliminated and, in principle, the value of ε at the solution could lie at any point in the range $0 \leqslant \varepsilon \leqslant \varepsilon'_0$. Nonetheless, the capacity of the population to renew itself still has a major bearing on the solution, because recruitment at equilibrium ($\hat{n}_1$) is an implicit function of ε. The problem is consequently nonlinear, both H and ε (and hence $\hat{n}_1$) being functions of η.

As before the problem and its solution are clarified by constructing the set of H and ε corresponding to all feasible harvest patterns. The linear dependence of H and ε on η again means that the boundary of the set $\{(H, \varepsilon)\}$ is given by the points corresponding to complete removal of single ages [and the point $(0, \varepsilon'_0)$]; for age i, this is

$$H = \gamma_i \quad = w_i l_i$$

$$\varepsilon = \varepsilon'_0 - \varepsilon'_i = \sum_1^i l_j b'_j \ ;$$

the convex hull of these k+1 points is the set $\{(H, \varepsilon)\}$. The upper boundary of the set, which we will call H_{max}, plays a central role in solving the problem. Firstly, it can be used to reduce the problem to the much more straightforward form (Reed 1980)

$$\text{maximize} \quad 2\hat{n}_1 H_{max} \ .$$

Secondly, the harvest patterns which could potentially solve the problem are immediately apparent from inspection of the ages corresponding to corners of the H_{max} boundary. Properties of the optimal harvest pattern, therefore, depend principally on the linear yield-per-recruit problem; the non-linearities introduced

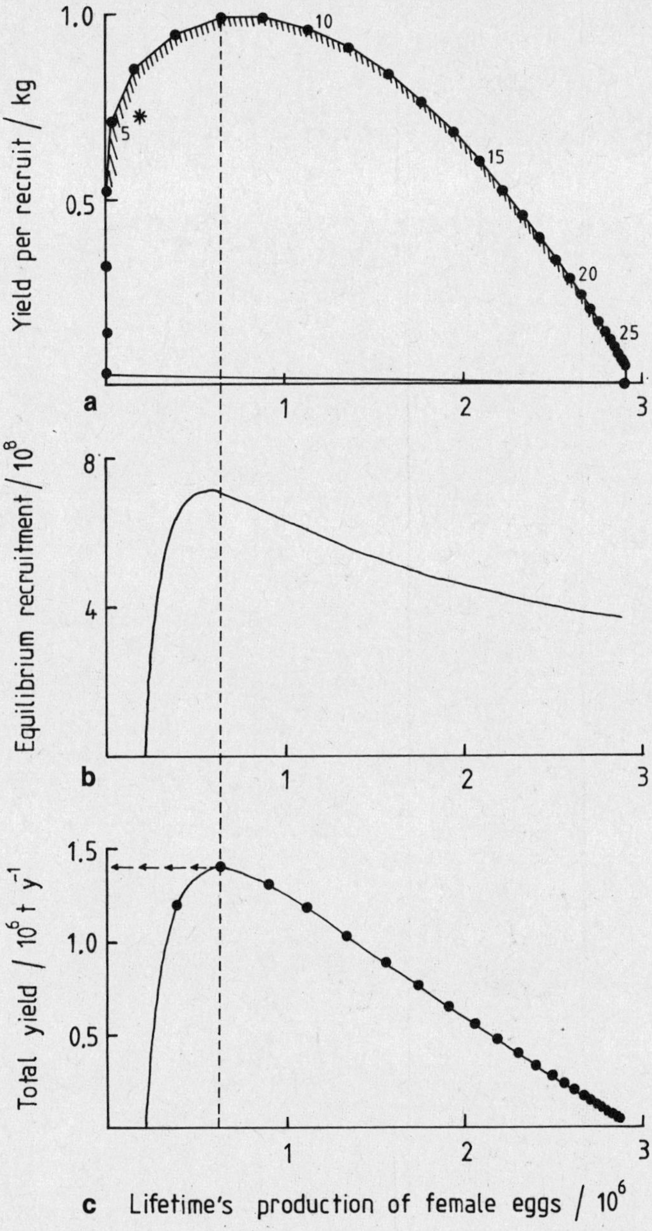

Fig. 3a–c. A graphical solution to the problem of maximizing the total yield at equilibrium for the Arcto-Norwegian cod in the presence of density-dependent recruitment. **a** Gives the set of yields per recruit (H) and lifetime's production of female eggs (ε) corresponding to all feasible harvest patterns (η). *Circles* represent points corresponding to complete removal of single ages, numbered multiples of 5 years. The *hatched boundary* is the maximum yield per recruit (H_{max}) for each ε. The star gives the (H, ε) under current rates of exploitation of the stock. **b** Gives the recruitment of females at equilibrium ($\hat{n}_1$) based on a Ricker curve fitted to post-war stock-recruitment data. **c** Gives the total yield, $2 \cdot \hat{n}_1 \cdot H_{max}$

by density dependence simply tell us where on the H_{max} boundary the solution lies.

Figure 3 illustrates the problem of maximizing the total yield at equilibrium from the Arcto-Norwegian cod, together with its solution. The set $\{(H, \varepsilon)\}$ (Fig. 3a) remains as in Fig. 2; the horizontal axis is, however, rescaled because offspring are counted at birth instead of at first census. H_{max} is given by the upper boundary of the set. To obtain the function $\hat{n}_1(\varepsilon)$ (Fig. 3b), a Ricker curve was fitted to the post-war stock-recruitment data (see Appendix). The total yield (H_T) (Fig. 3c) is given by the product $2\hat{n}_1 H_{max}$, and has kinks corresponding to the corners of H_{max}. Inspection of Fig. 3c indicates that the solution is $H_T = 1.46 \times 10^6\,t\,y^{-1}$, obtained with a lifetime's production of female eggs $\varepsilon = 6.3 \times 10^5$. The harvest pattern responsible for the maximum total yield is found by returning to Fig. 3a and observing the point at which the line $\varepsilon = 6.3 \times 10^5$ intersects the H_{max} boundary. The harvest pattern is $\eta_8 = 1$ (all other $\eta_i = 0$), in other words, complete removal at age 8 y and no harvesting at any other age.

For the sake of comparison, Fig. 3 also gives the yield obtained from the Arcto-Norwegian cod with a harvest pattern based on the current rate of exploitation (see Appendix). This harvest pattern has the unfortunate feature of causing insufficient lifetime's egg production to sustain the population. Thus the total yield at equilibrium is zero. It should, however, be understood that the estimate of the total yield is very sensitive to changes in the stock-recruitment relationship; in view of the wide scatter of data on which it is based, the errors associated with estimates of H_T are large.

8 Problem 2 Revisited – Maximum Renewal Under Harvesting

The equivalence of harvesting to maximize the yield and harvesting to maximize self-renewal, already demonstrated for a density-independent population, suggests that a similar relationship could hold under density-dependent recruitment. One might conjecture that the harvest pattern which maximizes the total yield at equilibrium should also maximize the total production of offspring by the population, given that a certain minimum harvest per recruit must be taken, i.e.,

$$\begin{aligned}
&\text{maximize} && 2\varepsilon\hat{n}_1 \\
&\text{subject to} && H \geqslant H_{max}(\varepsilon^*) \\
&\text{and} && \eta_1 + \ldots + \eta_k \leqslant 1 \\
&\text{with} && \eta_i \geqslant 0 \quad \text{for} \quad i = 1, 2, \ldots, k\ .
\end{aligned}$$

Here, ε^* is the lifetime's production of female offspring at the solution of the previous H_T maximization problem and $H_{max}(\varepsilon^*)$ is the corresponding value of H_{max}. The following line of reasoning indicates that this conjecture is correct, as long as H_{max} is a decreasing function of ε at the solution. Firstly, note that the constraint $H \geqslant H_{max}(\varepsilon^*)$ is equivalent to $\varepsilon \leqslant \varepsilon^*$, if H_{max} decreases with increasing ε. Secondly, note that Eq. (5) can only be satisfied with g monotonic decreasing if the product $\varepsilon\hat{n}_1$ increases with ε. Thus $2\varepsilon\hat{n}_1$ is maximized at ε^*.

In the case of the Arcto-Norwegian cod, however, Problems 1 and 2 are not solved by the same harvest pattern, because H_{max} is an increasing function of ε at the solution of Problem 1. The total egg production is at its minimum at the solution of Problem 1, for harvests in which $H \geqslant H_{max}(\varepsilon^*)$.

9 Concluding Comments

The analysis shows that Problem 1 (maximizing the yield) and Problem 2 (maximizing self-renewal in the presence of harvesting) are equivalent both in a density-independent and density-dependent setting, in the sense that they are solved by the same harvest pattern. The two problems are, in effect, opposite sides of the same coin; when a harvest pattern is chosen to maximize the yield, it also maximizes the renewal of the culled population, given that a yield per recruit at least as great as H_{max} must be taken. This result is, however, contingent upon the shape of the upper boundary of the set $\{(H, v)\}$ or $\{(H, \varepsilon)\}$. The boundary must be a decreasing function of v_1 or ε at the solution; if the boundary is an increasing function, then renewal is minimized when the harvest is maximized.

The sets $\{(H, v_1)\}$ and $\{(H, \varepsilon)\}$, which lie at the heart of the analysis, define the relationship between the yield from an individual and its reproductive contribution to renewal of the population. While this relationship is of interest in its own right, it has the additional feature of providing a bridge between two previously disparate disciplines – the dynamics of exploited populations and life-history evolution. Although not generally accepted at the moment, it is arguable that the heavy widespread culling of exploited populations brings about genetic change in life histories and, as a result, causes evolutionary change in the yield. The relationship between life history and yield described above, in conjunction with results elsewhere in this book (see Lynch and Arnold; Caswell; Kirkpatrick; Lynch), provide a basis for analysis of the evolution of yield in exploited populations.

Acknowledgments. We thank V. M. Terrett and M. H. Williamson for discussions on harvesting problems, D. J. Garrod for information on the Arcto-Norwegian cod, B. Ebenman and J. A. J. Metz for comments on the manuscript, and B. Ebenman and L. Persson for the invitation to attend the meeting on the dynamics of structured populations.

10 References

Beddington JR, Taylor DB (1973) Optimum age-specific harvesting of a population. Biometrics 29:801–809
Beverton RJH, Holt SJ (1957) On the dynamics of exploited fish populations. Fish Invest London Ser II 19:1–533
Borisov VM (1978) The selective effect of fishing on the population structure of species with a long life cycle. J Ichthyol 18:896–904
Botsford LW (1981) Optimal fishery policy for size-specific, density-dependent population models. J Math Biol 12:265–293
Charlesworth B (1980) Evolution in age-structured populations. Univ Press, Cambridge
Cushing DH (1966) The arctic cod. Pergamon, Oxford New York

Cushing DH, Horwood JW (1977) Development of a model of stock and recruitment. In: Steele JH (ed) Fisheries Mathematics. Academic Press, New York London, pp 21–35
Doubleday WG (1975) Harvesting in matrix population models. Biometrics 31:189–200
Garrod DJ (1967) Population dynamics of the Arcto-Norwegian cod. J Fish Res Board Can 24:145–190
Garrod DJ (1977) The North Atlantic cod. In: Gulland JA (ed) Fish population dynamics. John Wiley & Sons, New York London, pp 216–242
Garrod DJ, Jones BW (1974) Stock and recruitment relationship in the Northwest Arctic cod stock. J Con Perm Int Explor Mer 36:35–41
Gass SI (1969) Linear programming methods and applications. McGraw-Hill, New York
Goodman D (1980) Demographic intervention for closely managed populations. In: Soulé ME, Wilcox BA (eds) Conservation biology. Sinauer, Sunderland, Mass, pp 171–195
Grey DR (1984) Non-negative matrices, dynamic programming and a harvesting problem. J Appl. Probab 21:685–694
Grey DR, Law R (1987) Reproductive values and maximum yields. Funct Ecol 1:327–330
Horwood JW, Whittle P (1986) The optimal harvest from a multicohort stock. IMA J Math Appl Med Biol 3:143–155
Law R (1979) Harvest optimization in populations with age distributions. Am Nat 114:250–259
MacArthur RH (1960) On the relation between reproductive value and optimal predation. Proc Natl Acad Sci USA 46:143–145
Reed WJ (1980) Optimum age-specific harvesting in a nonlinear population model. Biometrics 36:579–593
Ricker WE (1954) Stock and recruitment. J Fish Res Board Can 11:559–623
Rollefsen G (1953) Observations on the cod and cod fisheries of Lofoten. Rapp P V Reun Cons Perm Int Explor Mer 136:40–47
Rorres C, Fair W (1975) Optimal harvesting policy for an age-specific population. Math Biosci 24:31–47
Slobodkin LB (1968) How to be a predator. Am Zool 8:43–51

11 Appendix: Life History of the Arcto-Norwegian Cod

The Arcto-Norwegian stock of cod spawns in March off the north west coast of Norway in the vicinity of the Lofoten Islands. See Cushing (1966) for an account of its biology. Ocean currents disperse the eggs to the Barents Sea and the seas around Spitsbergen. The young cod remain in these areas until they reach the age of maturity; thereafter they migrate each year to the spawning grounds. Features of the exploitation and life history are summarized in Table 1, and are based on the calculations described below.

a) Density-Dependent Mortality Before Recruitment at Age 1 Year. The data on stock biomass (S) and subsequent recruitment (R) suggest that a Ricker curve (Ricker 1954) of the form $R = \alpha S \exp(-S/\beta)$ should describe the relationship (Garrod and Jones 1974). The data for 1946 to 1982 have been used to estimate the total number of female eggs produced each year (assuming 250 female eggs g^{-1} body weight, as described below), and the number of female recruits 1 year later, assuming a sex ratio of 0.5 in the eggs and no difference in mortality rates between the sexes. A Ricker curve fitted to these values by least-squares nonlinear regression gives $\alpha = 4.75 \times 10^{-6}$ and $\beta = 4.10 \times 10^{14}$, taking into account the contribution of male as well as female eggs to density dependence. From Eq. (5), the function $\hat{n}_1(\varepsilon)$ is then

Table 1. Data on the life history and exploitation of the Arcto-Norwegian cod. See Appendix for details of calculations

Age (i) Years	l_i Year^{-1}	Females Mature (%)	b'_i $\times 10^6$	w_i kg	F Year^{-1}	η Year^{-1}
1	1	0	0	0.02	0	0
2	0.819	0	0	0.17	0	0
3	0.670	0	0	0.47	0.05	0.049
4	0.579	1	0.00	0.94	0.35	0.281
5	0.449	13	0.05	1.56	0.40	0.221
6	0.368	63	0.36	2.30	0.50	0.177
7	0.301	96	0.75	3.12	0.80	0.150
8	0.247	100	1.00	4.00	0.65	0.059
9	0.202	100	1.22	4.90	0.65	0.031
10	0.165	100	1.45	5.80	0.55	0.014
11	0.135	100	1.67	6.69	0.55	0.008
12	0.111	100	1.89	7.55	0.55	0.005
13	0.091	100	2.09	8.38	0.55	0.003
14	0.074	100	2.29	9.16	0.55	0.002
15	0.061	100	2.47	9.89	0.55	0.001
16	0.050	100	2.64	10.57	0.55	0.001
17	0.041	100	2.80	11.19	0.55	0.000
18	0.033	100	2.94	11.77	0.55	0.000
19	0.027	100	3.07	12.30	0.55	0.000
20	0.022	100	3.20	12.78	0.55	0.000
21	0.018	100	3.30	13.22	0.55	0.000
22	0.015	100	3.40	13.61	0.55	0.000
23	0.012	100	3.49	13.97	0.55	0.000
24	0.010	100	3.57	14.29	0.55	0.000
25	0.008	100	3.65	14.59	0.55	0.000
26	0.007	100	3.71	14.85	0.55	0.000
27	0.006	100	3.77	15.08	0.55	0.000
28	0.005	100	3.82	15.29	0.55	0.000
29	0.004	100	3.87	15.48	0.55	0.000
30	0.003	100	3.91	15.64	0.55	0.000

$$\hat{n}_1 = 4.10\times10^{14}\,\varepsilon^{-1}\ln(4.75\times10^{-6}\,\varepsilon)\ .$$

b) Natural Mortality After Recruitment at Age 1 (l_i). The instantaneous rate of natural mortality is thought to be approximately 0.2 (Garrod pers. comm.); this value has been used to generate the values of l_i in Table 1. It is assumed that no individuals survive beyond age 30 years.

c) Age of Maturity. Accounts of the age at which individuals become sexually mature show a trend towards earlier maturation in recent years (Rollefsen 1953; Garrod 1967, 1977 and pers. comm.; Borisov 1978). The figures of percentage of females mature given in Table 1 refer to the most recent information available, for the calendar year 1985 (Garrod pers. comm.).

d) Number of Female Eggs (b'_i). This has been estimated from the body weight (see below), using Garrod's (1977) figure of 500 eggs g^{-1} body weight, and assuming a sex ratio 1 : 1 in the eggs.

e) Body Weight (w_i). Growth is described by the von Bertalanffy growth curve, with constants $k = 0.12$, $w_\infty = 17$ kg (Garrod 1977) and $t_0 = 0$.

f) Harvest Pattern (η). Instantaneous rates of fishing mortality (F) based on the year 1986–1987, as used by the ICES Arctic Fisheries Working Group, are given in Table 1 (Garrod pers. comm.). These are converted to yearly rates $(1 - \theta_i)$, from which the harvest pattern (η) is obtained.

Interactions in Size-Structured Populations: From Individual Behavior to Ecosystem Dynamics

Interactions Between Growing Predators and Growing Prey

H. M. WILBUR

1 Introduction

Many concepts in community ecology suffer from a typological approach to species interactions in which the ecological relationships between species are classified as competitive, mutualistic or predatory; and then they are represented by −/−, +/+, or −/+ pairs of elements in a community matrix. These relationships are often diagramed by a food web that represents the presumed trophic connections among species in a community. Species within a trophic level are assumed to be potential competitors and predators are considered to have a negative effect on the population dynamics of their prey. This negative effect arises from predators decreasing the rate of increase of prey or lowering the equilibrium density of prey. Recently, the simplicities of this approach have been recognized as ecologists have studied multispecies communities with techniques designed to understand mechanisms that structure natural communities. It is now recognized that the interaction between two species can depend on the context in which it is measured (Boucher 1985; Kerfoot and Sih 1986). For example, two species may have a competitive (−/−) interaction when raised in isolation, but show mutualistic (+/+) interaction when a predator is added to the system because the two species spread the risk of predation both benefiting from the presence of the other.

There is a rich literature on how predators can affect community structure by either elimination of species or promoting the coexistence of prey species (e.g., Darwin 1859; Hrbacek et al. 1961; Paine 1966; Slodbodkin 1962; Yodzis 1976). Predators are generally thought to enhance species diversity by preying differentially on the dominant species in a community, thus permitting the invasion and increase of species that are less successful competitors. Such arguments focus on the numerical effects of predators on their prey. The term predation refers to the act of an individual of one species killing an individual of another species. The consequence of this act on the population dynamics of the prey may be either positive or negative, depending on mechanisms of density-dependent survival or fecundity among the surviving prey. A predator can enhance the success of a population of prey by reducing the intensity of intraspecific competition among the individuals that escape predation. This effect has been well-documented in two systems in which species with complex life cycles are exploiting ephemeral resources: amphibians using temporary ponds (Wilbur 1980, 1987; Wilbur et al. 1983) and dipterans using carrion (Ullyett 1950 and Peschke et al. 1987).

Department of Zoology, Duke University, Durham, North Carolina 27706, USA

Size-Structured Populations
B. Ebenman and L. Persson (Eds.)

The thinning effect of predation, in which a qualitative effect on survivors is emphasized, has been mentioned in discussions of natural history (Errington 1946), fisheries dynamics (Botsford 1981), and "prudent predation" (Odum and Biever 1984; Slobodkin 1962), but it has not often been incorporated in theoretical discussions of species interactions. These two effects of predation, the impact on competing species and the impact on survivors, can both operate in determining the structure of natural communities.

Another context-dependent interaction is one in which the interaction between two individuals depends not on their taxonomic position but on their relative body sizes or life history stage. The outcome of the competitive interaction of two individuals using either interference or exploitative mechanisms of interaction may depend on their relative body sizes (Rubenstein 1981; Wilbur 1980). Similarly the interaction between two generalized predators may be competitive if they are the same size or predatory if they are sufficiently different in size (Polis, this Vol.).

I shall focus on how the body size of individuals may determine their risks of predation, how predation may benefit survivors, then how differences in body size may result in qualitative changes in trophic relationships. My emphasis will be on amphibians that exploit temporary ponds, but I hope much of what I have to say will strike a responsive chord in students of insects, fish, reptiles, and other groups in which growth is associated with changing feeding habits and risks of predation. Some of my conclusions may hold only for species with nonoverlapping generations, or at least ones in which the larval habitat is distinct from the adult habitat. The assumption I make is that density-dependent interactions during the larval stage have implications for the adult phase of the life cycle that may determine the dynamics of the population from one generation to the next. The larval stages of most insects and amphibians are morphologically adapted to take advantage of transient opportunities for rapid growth, such as the high primary productivity often found in temporary ponds (Wilbur 1980). The length of time a pond contains water may be highly variable so that there is a stochastic risk of desiccation. This risk can be reduced by a short larval period and small size at metamorphosis or by a mechanism to sense a deteriorating environment and to initiate metamorphosis within a range of body sizes (Wilbur and Collins 1973). The body size of a larva may also determine its competitive success (Rose 1960) and risk of predation by size-limited (Werner 1974; Paine 1976; Zaret 1980) and size-selective (Brooks and Dodson 1965; Werner and Hall 1974) predators. Once the minimum size necessary for metamorphosis is exceeded, the question of when to terminate the larval period and metamorphose to a juvenile form capable of dispersal and reproduction is a larger question that involves the relative intensities of competition, risks of predation, and opportunities for growth in the aquatic habitat of larvae and the terrestrial habitat of juveniles and adults (Wilbur and Collins 1973; Werner 1986; Werner this Vol.). These costs and benefits change as the environment in the pond and the body size of the larvae change.

The breeding biology of adults determines the spatial and temporal pattern of eggs in the habitat and the initial size distribution of hatchlings, due to maternal effects. Growth rates of larvae are determined by intrinsic factors such as eggs size, yolk quality, genetic factors, and larval feeding behavior as well as extrinsic factors such as temperature, productivity of food resources, and competition. A cohort of siblings with similar sizes at hatching and normally distributed growth rates will develop a lognormal distribution of body sizes (Wilbur and Collins 1973; Wilbur 1984). Populations that have synchronized breeding are expected to have such skewed distributions of body sizes. Populations with asynchronous breeding may also develop skewed size distributions of body sizes as a result of competition, especially if there are strong interference modes of interaction in which a few large individuals suppress the growth of smaller members of the population. Larval anurans are known to interact via both exploitation competition for food and by interference mechanisms involving growth inhibitors that give an advantage to larger members of the population (Rose 1960; Richards 1958; Steinwascher 1979; Wilbur 1977).

2 The Consequences of Predation

2.1 *Effects of Predation on the Growth of Survivors*

Size-selective predators may preferentially feed on either end of a distribution of body sizes. An optimally foraging predator might differentially detect and pursue larger members of the population (Werner 1974; O'Brian et al. 1976). Such a biased risk of predation would tend to make size distributions more symmetrical and perhaps prevent the domination of resources by a few individuals with potent interference mechanisms. The equalizing effect of such size-selective predation might be that the population produced many small metamorphs rather that a few large ones. Vanni (1987) has recently shown that size-selective predation by fish on cladoceran can release smaller individuals from competition. If, however, the prey are in a competitive race to obtain some minimum size required for metamorphosis before their larval habitat deteriorates, then a predator that crops the largest members of the population could have a strongly negative impact if no individuals of the prey are able to metamorphose before the population is exterminated by desiccation (Wilbur 1987).

On the other hand, a predator that is size- or gape-limited might only be able to feed on the smaller members of a population. The prey could be in a race in which obtaining a size-escape from predators is coupled with obtaining a size-threshold for metamorphosis that allows an escape from a deteriorating habitat. Both selective forces would act to maximize larval growth rates. The action of predation could be to remove competitors permitting rapid growth of the survivors (Wilbur 1987). Even random predation can enhance the success of a prey population if it results in competitive release of the survivors (Chambers 1984). This positive, culling effect of predation is most likely in systems dominated by scramble competition (Nicholson 1954) in which even the least successful individuals remove resources from the more successful members of the population. This mode of action appears to be common in larval amphibians (Wilbur 1977, 1980, 1984).

Size-selective and size-limited predation are simple cases of how a trophic relationship is influenced by body sizes of predators and their prey. Aspects of functional, numerical, and developmental responses can vary continuously with body sizes of the interacting species. For example, Thompson (1975) found that parameters of the functional response of an odonate larva changed with sizes of larvae and their cladoceran prey. Persson and Mittelbach et al. have discussed the consequences of age-specific predation in their contributions to this Volume.

Predators may also have an indirect effect on growth of their prey if they cause a behavioral response (McNamara and Houston 1987; Morse 1980; Mittelbach et al. this Vol.; Polis 1980; Sih 1980, 1982, 1984, 1987; Stein 1977; Stein and Magnuson 1976). Several studies have shown that amphibians can detect the presence of predators and then alter their behavior to minimize their risk of predation (Holomuzki 1986; Morin 1986; Roudebush and Taylor 1987; Semlitsch 1987). Predator avoidance may involve spending more time in a refuge or foraging in sites that may have a lower risk of predation, but also a lower rate of resource gain (Cerri and Fraser 1983; Gilliam and Fraser this Vol.). Such behavioral responses to predators may result in smaller sizes at metamorphosis or greater risks of mortality by not obtaining a minimum size necessary for metamorphosis before the larval habitat deteriorates.

Another mechanism for a nonlethal, negative effect of predators on prey is by removal of body parts either by failed predation attempts or by "grazing". Many amphibians are able to tolerate loss of body parts and have remarkable powers of regeneration (Morin 1985). The ecological costs of such losses have not been studied in detail. The impact of loss could be felt as an increased risk of mortality by infection or by future encounters with predators due to loss of speed or maneuverability. In species that are able to rapidly regenerate the lost parts, there may still be a cost in growth rate. Slow growth may delay metamorphosis or maturity or result in a smaller size at maturity, which in turn may entail reduced fecundity or lower survival (Harris MS; Mairorana 1977).

2.2 Effects or Predation on the Dynamics of the Prey Population

The consequences of predation to the population dynamics of the prey depends on whether or not the loss of individuals through the direct action of predation is compensated by an increase in the quality of the survivors. This may depend on both the timing of predation and the ability of prey to respond to a reduction in their density. Predation on a prey population in which growth is limited by the food supply may result in fewer individuals, each with a larger size. The effect of predation on the prey population may be measured in terms of biomass production or the number of individuals that survive to maturity. If the reduction in survival is compensated by an increase in size of individual survivors such that the biomass (B) of the prey population remains constant, the product of the number of survivors (N) and their mean body mass (M) remains constant ($B = NM$) and there will be a linear relationship between the logarithm of M and the logarithm of N, that is $\log(M) = -\log(N) + \log(B)$. In this case, the thinning effect of predation increases the quality of survivors, but does not enhance the biomass productivity of the prey population. This relationship of constant biomass, a slope of -1 for the plot of $\log(M)$ on $\log(N)$, can be compared to the $-3/2$ self-thinning law of plant ecology (Weller 1987; Yoda et al. 1963) (Fig. 1). A $-3/2$

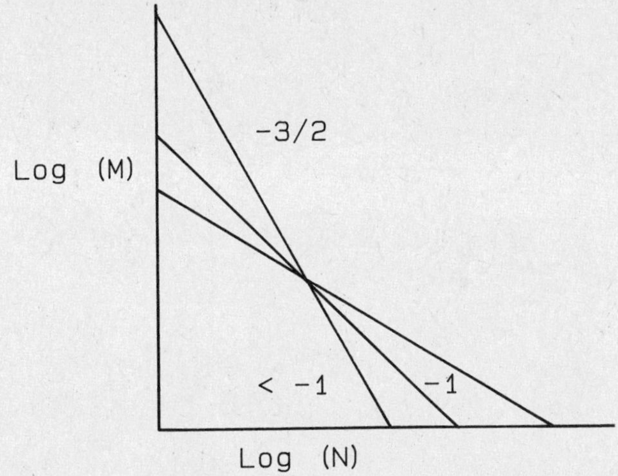

Fig. 1. The relationship between the number of individuals surviving predation (N) and mean mass of individuals (M) according to the law of constant yield (slope = −1) and the −3/2 self-thinning law

thinning law is an empirical result that predicts an increase in total biomass in the population, because as the population density is reduced, individuals increase in size at rate that scales with the −3/2 power of density. These two curves, the law of constant yield (slope of −1) and the self-thinning curve (slope of −3/2), are presented only as lines of reference. Law (this Vol.) has discussed the general problem of yields and the size-specificity of harvest methods. The impact of predation depends on the behavioral and ecological response of the prey individuals and the timing of predation. After predation, the prey population could move above the constant biomass line if individuals can divert resources from interference mechanisms, such as time and energy spent on aggression, to growth. The prey population could move below the biomass line if individuals incur a cost of predation, for example, by spending more time in refuge or foraging in microhabitats with a low rate of resource acquisition.

Biomass production, per se, is irrelevant to models of population dynamics or genetic change in the prey population; these models are concerned with numbers of individuals and their quality. The effects of density on growth and survival of larval amphibians can be used to illustrate how predation may have a positive effect on the population dynamics of prey. The impact of predation depends on the form of the competition curve in the prey population. The probability of metamorphosis and the initial density of eggs are related by the simple exponential model (Fig. 2)

$$N/D = \exp(-sD) \quad \text{or} \quad N = D\exp(-sD)\ , \tag{1}$$

where N is the number of metamorphs, D is the initial density of eggs, and s is an empirically determined coefficient of intraspecific competition (Wilbur 1976, 1984). This curve has been used independently to describe stock-recruitment relationships in fisheries (Harris 1975; Ricker 1954), competition in laboratory populations of insects (Fuji 1965; Nicholson 1954) and, with a constant for adult survival and fecundity, as a discrete-time population model (May 1974). The curve

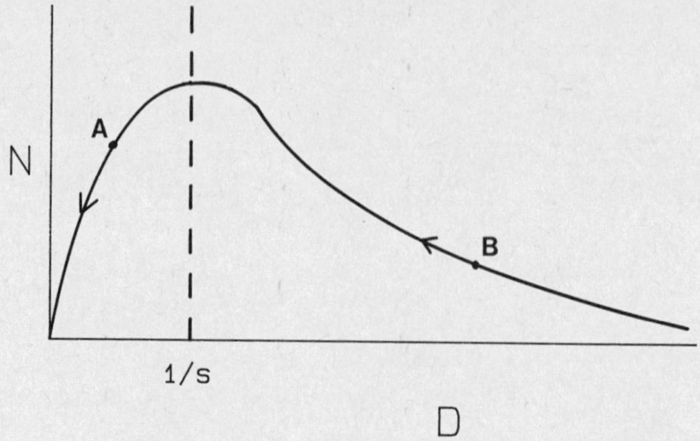

Fig. 2. The competition curve [N = D exp (−sD)] has a maximum number of survivors (N) produced from a population with initial density (*D*) equal to 1/s. *A* Predation on a low density population will result in fewer survivors. *B* Moderate levels of predation on a high density population will result in more survivors

relating N to D (Fig. 2) has a maximum at the initial density of (1/s). If predation acts strictly on eggs or very early in the larval period of populations with densities below (1/s), the number of metamorphs is always reduced. If predation reduces dense populations to sizes closer to (1/s), then predation may increase the number of metamorphs.

The longer-term impact of predation depends not just on the number of metamorphs but also their reproductive success. Reproductive success of a metamorph is likely to be a function of its size at metamorphosis and perhaps its date of metamorphosis. Semlitsch et al. (1988) found that large size at metamorphosis was positively correlated with large size at first reproduction and young age at first reproduction in a salamander (*Ambystoma talpoideum*). Smith (1987) found that recapture rate was correlated with size at metamorphosis in the frog (*Pseudacris triseriata,* Smith 1987). He used cohort marks so the recapture rates are not easily interpreted. His result may be due to an effect of size at metamorphosis on the timing of maturity rather than annual survival rate.

Body size at metamorphosis is related to initial density of the population in the absence of predation as

$$M = b + c \exp(-aD) , \tag{2}$$

where M is mass at metamorphosis, D is initial density of hatchlings, and a, b and c are empirically fitted constants (Wilbur and Collins 1973). The minimum size for metamorphosis is b and the maximum size is (b+c); a is an intraspecific competition coefficient. If predation acts early in the larval period, this relationship can be used to describe the enhanced quality of metamorphs that escape predation (Wilbur 1984). If predation occurs later in the larval period, when competition has had an effect on prey growth, then some other model might be required. I would suggest that a matrix model for the size structure of the population (e.g., Caswell this Vol.; Lefkovitch 1965) with density-dependent transition probabilities and size-dependent predation risks might be a useful approach.

Prout (1986) and Prout and McChesney (1985) have explicitly considered the delayed effects of larval competition on adult fecundity and hence on the dynamics of competitive systems in *Drosophila.* The functional form of the relationship between size at metamorphosis and reproductive rate is known for the salamander *Ambystoma talpoideum* from the data of Semlitsch (1985) and Semlitsch et al. (1988). A useful first approximation of the relationship between the probability of reproducing at age one (p) and size at metamorphosis (M) is

$$p = f - gM^{-h}, \quad (f, g, h > 0) . \tag{3}$$

The constants f, g and h are fitted empirically. The maximum survival rate is f and the minimum size for survival is $(f/g)^{-1/h}$.

The probability of survival as a function of the initial density of the population can be found by substituting M as a function of D [Eq. (2)] into Eq. (3). The product of the number of metamorphs [Eq. (1)] and their probability of survival to maturity [Eq. (3) in terms of initial density] yields the expected number of mature adults (A) from populations of different initial densities,

$$A = D \exp(-sD) \{f - g[b + c \exp(-aD)]^{-h}\} . \tag{4}$$

All of the parameters of this admittedly complex expression can be estimated empirically by experiments in which populations are started at different densities and the number and sizes of metamorphs are determined as a function of initial density and the relationship between size at metamorphosis and survival to maturity is estimated. The expression [Eq. (4)] has a shape similar to that of Eq. (1) (Fig. 3) but the peak is always to the left of the peak of the function for the number of metamorphs, which is at the initial density of 1/s. The displacements of the curves result from the effect of lower larval density on body size of metamorphs. Predation leads to lower densities and larger size at metamorphosis, which in turn leads to higher survival rates to maturity.

Predation can be viewed as having three qualitatively different impacts on the population of prey. If the initial density of prey is below that which yields the highest number of mature adults (region I of Fig. 3), then predation decreases both the number of metamorphs and the number of adults. This is the classical result of predators benefiting from prey and prey suffering from predation. If predation moves the density of prey from the left of the peak of the metamorph curve towards the peak of the adult curve (region II of Fig. 3), then predation decreases the number of metamorphs but survivors have a higher probability of reaching maturity because they are larger. This is a positive effect of predation on the prey population and can be called mutualism. If predation moves the prey population from the right of the peak of the metamorph curve towards the peak (region III of Fig. 3), then predation increases both the number of metamorphs and the number of adults, a clear case for calling the relationship a mutualism.

Size at metamorphosis also influences the timing of maturation and fecundity. As a first step, I have assumed that size at metamorphosis has an impact only on survival; the model can be extended to include effects on fecundity schedules. The beneficial effect of predation would be enhanced if either age at maturity is reduced or fecundity is increased as a consequence of a large size at metamorphosis. If there is insufficient gain in the quality of survivors to compensate for the loss of individuals to predators, then predation has a negative effect on the prey population, even if individuals benefit.

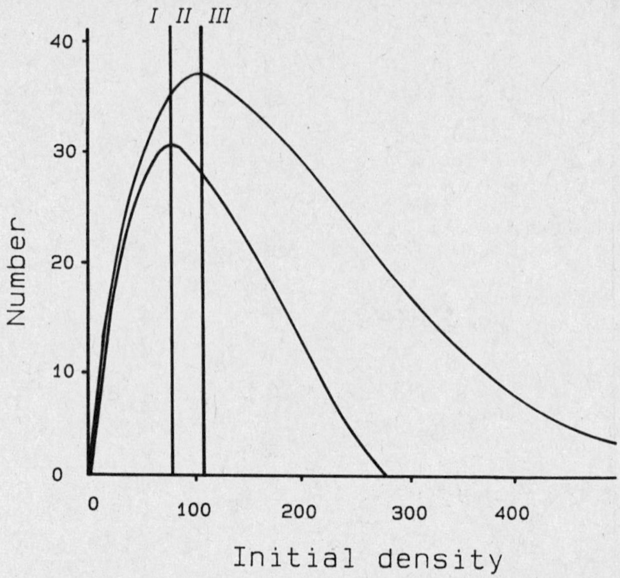

Fig. 3. The number of metamorphs (*upper curve*) and mature adults (*lower curve*) as functions of the initial density of the population with realistic, but arbitrary parameter values (a = 0.01, b = 10, c = 90, f = 1, g = 58, h = 1.5, s = 0.01). If predation acts in the region left of the peak of the curve for adults (*I*), the effect will be negative on both the recruitment of larvae and adults. If predation acts in the region between the peaks of the two curves (*II*), the effect will be negative on juvenile recruitment but positive on adult recruitment. If predation acts in the region to the right of the peak of the curve for metamorphs (*III*), the effect will be positive on both juvenile and adult recruitment. See text additional explanation

3 Cannibalism and Intraguild Predation

Predators may change their feeding behavior and preferences as they grow. Rapidly growing individuals in populations that develop broad or highly skewed size distributions may be able to turn and prey on smaller members of their own population. A predator's developmental response (Murdoch 1971) results in individual predators eating prey, growing larger and eating even more prey. Cannibalism could be viewed as another aspect of the developmental response in which size differences determine the rate of predation. Such a switch in intraspecific interactions from competition to predation (cannibalism) is well documented in some species of insects (Fox 1975; Mordoch and Sih 1978), scorpions (Polis 1980), fish (LeCren et al. 1977; Loadmon et al. 1986; Persson 1983; Werner 1986), frogs (Bragg 1964), and salamanders (Rose and Armentrout 1976). Cannibalism could have a positive effect on a population if the individuals that benefit have a sufficient gain that more than compensates for the loss of individuals as modeled above. I know of no explicit examples in which cannibalism has been clearly shown to benefit the population by this mechanisms, but it is likely to occur in frogs (*Scaphiopus*) and salamanders (*Ambystoma tigrinum*) that breed in temporary ponds. Polis (this Vol.) has reviewed the consequences of cannibalism in detail, but has no evidence that intraguild predation or cannibalism releases scorpions from exploitation competition. His experiments do not reveal food limitation in natural populations.

Predators may also change the nature of their interactions with other species as they grow and alter their feeding behavior. Many species of growing predators may be competitors when they are the same size but prey on each other when they are sufficiently different in size. Mittelbach et al. (this Vol.; Mittelbach and Chesson 1987), Wilbur (1980) and Polis (this Vol.) have all argued that two species may be more similar to one another at equivalent life history stages than they are to other members of their own species at different life stages. This may involve ontogenetic niche shifts such as those documented for centrachid sunfish (Werner and Gilliam 1984; Werner et al. 1983). For example, adults of two species may prey on the young of their own or the other species, but when sizes of the two larvae or two adults are similar, their food habits overlap, and they may compete. Ebenman (this Vol.) has explicitly modeled competition in simple two-lifestage systems. It is also possible for such pairs of species to shift among competition, predation, and perhaps mutualism as a result of different breeding phenologies and growth rates. For example, consider two species of predators. Species A breeds earlier than species B, but it has a slower growth rate (Fig. 4). When species B is young it may be prey for the earlier-breeding and therefore large species A. As the faster-growing species B reaches a size at which it cannot be killed by species A, it may still suffer a period of risk of mutilation or loss of body parts. As the species become even more similar in size and feeding habits thay may be competitors. As the rapidly growing species B becomes larger than species A, it may begin to mutilate species A and then, with additional growth, act as a predator sensu stricto.

The transitions between trophic relationships may not be abrupt as shown in Fig. 4, but rather there may be changing probabilities of the different modes of interaction as the relative body sizes of species shift (Fig. 5). For example, competition for prey may be mixed with mutual mutilation. This mutual mutilation could itself be a result of an aggressive form of interference competition (Polis

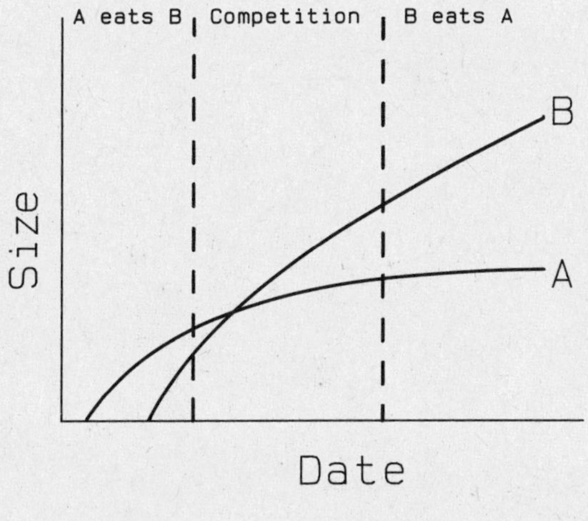

Fig. 4. The trophic relationship between an early breeding, but slowly growing species *A* and the later-breeding, but rapidly growing, species *B* changes as the relative body size of the two species changes through differential growth

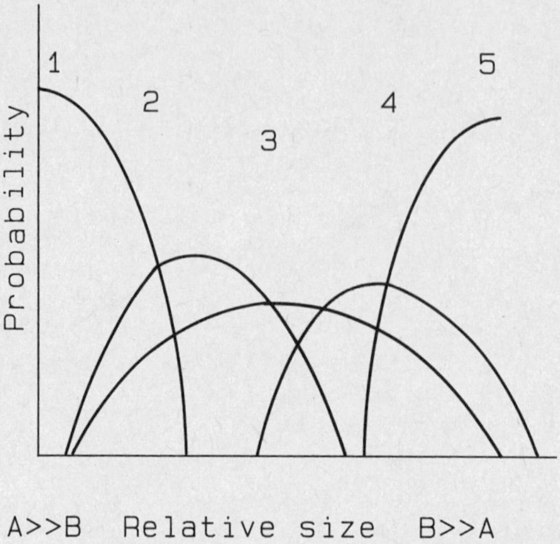

Fig. 5. The probability of the outcome of an encounter between two individuals may shift as the relative body sizes change. *Curve 1* A kills and eats B, *2* A mutilates B, *3* The two species compete for common prey, *4* B mutilates A, *5* B kills and eats A

this Vol.). Such a relationship occurs in salamanders competing for small arthropods with the ever-present risk that a larger salamander may eat a smaller member of its own or another species (Hairston 1986; Holomuzki 1986; Rose and Armentrout 1976; Stenhouse 1985; Stenhouse et al. 1983; Wilbur 1972).

Changes in trophic relationships are likely to be important in many aquatic communities in which larval insects, fish, and amphibians are dominant predators. The species lists of the nekton and macrobenthos of ponds and lakes is dominated by species that are growing as they interact with one another. Insects (odonate larvae, adult hemipterans, and dytiscid beetle larvae) and crayfish are capable of eliminating amphibians from some ponds (Wilbur unpubl.); and yet, these same species may be prey for large salamander larvae and adult frogs. If such changes in trophic position are found to be important, then food webs are not static structures, but rather they are dynamic constructs in which connections are changing and trophic levels become blurred as the outcome of a given interaction shifts between predation, competition, and perhaps mutualism.

4 Examples

Amphibians that breed in temporary ponds are an excellent system to explore the consequences of growth to species interactions. Anuran tadpoles in the southeastern United States are strictly opportunistic filter-feeders on periphyton and phytoplankton. Although they will opportunistically feed on carrion, including dead conspecifics, there is no demonstrated predation in the group. There is a possibility that anuran eggs are eaten incidentally to scraping vegetation and detritus. Anurans that breed in temporary ponds are racing to obtain a minimum size necessary for metamorphosis before the pond dries or the risks of competition or predation outweigh the advantages of growing in the aquatic habitat. The effect of predation on anuran larvae may be positive if predation releases tadpoles

Table 1. The interaction between competition and predation in a drying pond. The number of *Bufo americanus* tadpoles that metamorphosed from experimental ponds that dried in 50 days depended on the presence or absence of the predator, two adult newts (*Notophthalmus viridescens*). The data are the mean number of metamorphs from three replicates (Wilbur 1987)

Initial density of tadpoles	Newts absent	Newts present
200	364	0
800	0	254

from competition thus permitting them to grow rapidly enough to escape the pond before it dries.

Just such a case was demonstrated in an experiment in which density of tadpoles, the presence of a gape-limited predator, and the rate of disappearance of ponds were independently controlled in a factorial design (Wilbur 1987). The toad *Bufo americanus* was unable to complete metamorphosis in crowded ponds that dried rapidly. When predators (the newt *Notophthalmus viridescens*) were added, they reduced the density of tadpoles to levels that permitted rapid growth and escape from ponds that dried early (Table 1). There was also evidence that toad larvae were able to sense the impending loss of their aquatic habitat and were able to accelerate metamorphosis and escape the drying pond by leaving at a small size. The relationship between the populations of newts and *Bufo* in this system was clearly mutualistic when tadpoles were at high densities. When tadpoles were at low densities, newts has a negative effect on *Bufo.* Another anuran, *Scaphiopus holbrooki*, was a dominant competitor, but the preferred prey of newts. Newts had a negative effect on *Scaphiopus* under all experimental conditions. In this experiment the keystone role (Paine 1966) of newts was due both to its direct predation on *Scaphiopus* and its indirect effect on the enhanced growth of surviving *Bufo.*

In another experiment two species of tadpoles and two kinds of predators were raised in all possible combinations of the presence and absence of each (Wilbur in prep.). The anuran tadpoles were *Bufo americanus* and *Rana palustris*, which competed when raised together at moderate densities. The predators were the newt *Notophthalmus viridescens dorsalis*, a gape-limited predator on small tadpoles, and naiads of the dragonfly *Anax junius*, a growing predator than can eat *Bufo* throughout its larval period. When *Bufo* was raised with either predator it had reduced survival when compared to controls without predators (Table 2). The survivors of predation metamorphosed early and at a small size compared to the more dense control populations. A parsimonious explanation is that the *Bufo* tadpoles sensed the presence of the predators and minimized their risk by leaving the ponds – a result that strongly supports the idea of Werner (1986, this Vol.) that tadpoles trade off growth for mortality risk. There may be an additional cost to predator avoidance as tadpoles that metamorphose early are not as mor-

Table 2. Response of *Bufo americanus* tadpoles to the presence of two predators, adult newts *Notophthalmus viridescens* and larval dragonflies *Anax junius*. Menas of four replicate population means are reported

	Control 500 larvae	Newts 2 adults	*Anax* 2 larvae
Number of metamorphs	290	190	121
Mass at metamorphosis (g)	0.155	0.112	0.113
Larval period (days)	37	34	35

Table 3. Response of *Rana palustris* tadpoles after 83 days to the presence of two predators, adult newts *Notophthalmus viridescens* and larval dragonflies *Anax junius*. Means of four replicate populations are reported

	Control 500 tadpoles	Newts 2 adults	*Anax* 2 larvae
Surviving tadpoles	453	352	124
Mass of tadpoles (g)	0.69	0.82	0.95
Number of metamorphs	0	0	15.5

phologically and physiologically well-equipped for terrestrial life as individuals that metamorphose after a longer larval period (Pough and Kamil 1984; Taigen and Pough 1985). The relationship of *Bufo* to both *Notophthalmus* and *Anax* is clearly predation. *Bufo* suffers increased mortality as well as lower quality of those metamorphs that do escape predation.

The response of *Rana* to predation was quite different (Table 3). The experiment was terminated before most *Rana* were able to complete their larval period. The response of *Rana* was measured by survival of tadpoles and their growth rates. Both species of predator reduced the survival of *Rana* compared to the control populations, but those tadpoles that survived predation grew more rapidly than those in the control populations. In populations exposed to a high predation by *Anax* some tadpoles were able to grow fast enough to metamorphose before the experiment was terminated. There was no evidence that individual tadpoles incurred a cost of being exposed to predation unless they were eaten. At the level of the population both predators had a positive effect on *Rana*; the interaction is best called mutualism. Those *Rana* that escaped predation were released from competition and many were able to metamorphose at a large size whereas the control populations remained stunted from the effects of population density on growth.

In these examples the nature of the interaction between the carnivores and the tadpoles depended on the behavioral and developmental response of the tadpoles. *Bufo* modified its size at metamorphosis and incurred a cost beyond the risk of mortality. *Rana* was not observed to modify its developmental pattern or behaviour and individuals that survived predation benefited greatly from the reduced conspecific density that resulted from predation. Initial densities of predators and prey could modify the nature of the relationships in ways such as those illustrated in (Fig. 3).

Morin (1986) raised *Hyla crucifer* tadpoles alone or with adult newts (*Notophthalmus*). Predation did not reduce competition among the survivors because tadpoles avoided predators by a shift in microhabitat distribution, but he could detect no cost to this avoidance behavior.

Semlitsch (1987) compared salamander larvae (*Ambystoma talpoideum* and *A. maculatum*) reared alone and with a predatory fish (*Lepomis macrochirus*) in artificial ponds. The fish had a strong negative effect on both survival and body size of survivors. This double cost of predation was due to behavioral shifts by the salamanders. Larvae restricted activity to the leaf litter of the benthic zone and shifted diet. *Ambystoma talpoideum* also became more nocturnal in the presence of fish.

5 Conclusions

The impact of predation on a prey population can be positive as well as negative, depending on if the survivors of predation are released from competition. The number of individuals that escape the larval habitat and become parents of the next generation may be greater in the presence of predation than in habitats without predators if the thinning effect of predation results in sufficient enhancement of the survivors. Anurans that breed in temporary ponds often begin life at very high densities due to the synchronized breeding and high fecundities of adults. The thinning effect of predation may often enhance the likelihood that a least some members of the population will obtain a sufficient size to initiate metamorphosis before the pond dries, an often unpredictable event.

Species that feed independently of parental care may change their trophic position as they grow. Rapidly growing members of a population of generalized predators may be able to become cannibals on smaller conspecifics. Cannibalism may enhance their chance of reaching maturity by escaping a deteriorating habitat. Trophic connections between species may also change as the result of differential growth. Two species of predators with different growth rates and dates of reproduction may reverse their trophic relationship if the early-breeding species grows more slowly than the later-breeding species. The early species may first be a predator, then become a competitor, and finally become a prey of the later-breeding, but faster-growing, species. Such trophic shifts may occur abruptly or by shifting probabilities of several outcomes of an encounter between the two species. Such shifts may be common in aquatic habitats in which many coexisting predators are fish, salamanders, and insects feeding independently of parents and growing at different seasons and rates. The food web of such communities may be a constantly changing structure with several possible connections between species at any given time.

Acknowledgments. Chris Chambers first got me thinking about the benefits of predation. Jim Gilliam pointed me in the direction of Fig. 3. This paper also benefited from discussions with Joe Pechmann, my other students, and the participants in the symposium, especially Jim Gilliam, Lennart Persson, Gary Polis and Earl Werner. My research on amphibians has been supported by the National Science Foundation and the Duke University Research Council.

6 References

Botsford LW (1981) The effects of increased individual growth rates on depressed population size. Am Nat 117:38–63

Boucher DH (1985) The biology of mutualism. Oxford Univ Press, New York

Bragg AN (1964) Further studies of predation and cannibalism in spadefoot tadpoles. Herpetologica 20:17–24

Brooks JL, Dodson SI (1965) Predation, body size, and composition of plankton. Science 150:28–35

Cerri RD, Fraser DF (1983) Predation and risk in foraging minnows: balancing conflicting demands. Am Nat 121:552–561

Chambers RC (1984) Competition and predation in temporary habitats. Thesis, Duke Univ, N Car

Darwin C (1959) The origin of species. Mentor Edition (1958). New Am Libr World Lit, New York

Errington PL (1946) Predation and vertebrate populations. Q Rev Biol 21:144–177, 221–245

Fox LR (1975) Cannibalism in natural populations. Annu Rev Ecol Syst 6:87–106

Fuji K (1965) A statistical model of the competition curve. Res Popul Ecol 7:118–125

Hairston NG Sr (1986) Species packing in *Desmognathus* salamanders: experimental demonstration of predation and competition. Am Nat 127:266–291

Harris JGK (1975) The effect of density-dependent mortality on the shape of the stock and recruitment curve. J Conserv Int Explor Mer 36:144–149

Harris RN MS Nonlethal injury as a mechanism of population regulation

Holomuzki JR (1986) Predator avoidance and diel patterns of microhabitat use by larval tiger salamanders. Ecology 67:737–748

Hrbacek J, Dvorakova M, Korinek V, Prochazkova L (1961) Demonstrations of the effect of the fish stock on species composition of zooplankton and the intensity of metabolism of the whole plankton association. Verh Int Ver Limnol 14:192–195

Kerfoot WC, Sih A (1987) Predation: direct and indirect impacts on aquatic communities. Univ Press New Engl, Hannover, NH

LeCren ED, Kipling C, McCormack JC (1977) A study on the numbers, biomass and year-class strengths of perch (*Perca fluviatilis* L.) in Windemere from 1941 to 1966. J Anim Ecol 46:281–307

Lefkovitch LP (1965) The study of population growth in organisms grouped by stages. Biometrics 21:1–18

Loadman NL, Moodie GEE, Mathias JA (1986) Significance of cannibalism in larval walleye (*Stizostedion vitreum*). Can J Fish Aquat Sci 43:613–618

McNamara JM, Houston AI (1987) Starvation and predation as factors limiting population size. Ecology 68:1515–1519

Maiorana VC (1977) Tail autotomy, functional conflicts and their resolution by a salamander. Nature (London) 265:533–535

May RM (1974) Ecosystem patterns in randomly fluctuating environments. In: Rosen R, Snell FM (eds) Progress in theoretical biology, vol 3. Academic Press, New York London, pp 1–50

Mittelbach GG, Chesson PL (1987) Predation risk: indirect effects on fish populations. In: Kerfoot WC, Sih A (eds) Predation: direct and indirect impacts on aquatic communities. Univ Press New Engl, Hannover, NH, pp 315–332

Morin PJ (1985) Predation intensity, injury frequency, and prey survival in an amphibian predator-prey interaction. Copeia 1985:638–644

Morin PJ (1986) Interactions between intraspecific competition and predation in an amphibian predator-prey system. Ecology 67:713–720

Morse DH (1980) Behavioral mechanisms in ecology. Harvard Univ Press, Cambridge, Mass

Murdoch WW (1971) The developmental response of predators to changes in prey density. Ecology 52:132–137

Murdoch WW, Sih A (1978) Age-dependent interference in a predatory insect. J Anim Ecol 47:581–592

Nicholson AJ (1954) An outline for the dynamics of animal populations. Aust J Zool 2:9–65

O'Brien WJ, Slade NA, Vinyard GL (1976) Apparent size as the determinant of prey selection by bluegill sunfish (*Lepomis machrochirus*). Ecology 57:1304–1310

Odum EP, Biever LJ (1984) Resource quality, mutualism, and energy partitioning in food chains. Am Nat 124:360–376

Paine RT (1966) Food web complexity and species diversity. Am Nat 100:65–75

Paine RT (1976) Size-limited predation: an observational and experimental approach with the *Mytilus-Pisaster* interaction. Ecology 57:858–873

Peschke K, Krapf D, Fulder D (1987) Ecological separation, functional relationships, and limiting resources in a carrion insect community. Zool Jahrb Syst 114:241–265

Persson L (1983) Food consumption and competition between age classes in a perch (*Perca fluviatilis*) population in a shallow eutrophic lake. Oikos 40:197–207

Polis GA (1980) The significance of cannibalism on the demography and activity in a natural population of desert scorpions. Behav Ecol Sociobiol 7:25–35

Pough FH, Kamel S (1984) Post-metamorphic change in activity metabolism of anurans in relation to life history. Oecologia (Berlin) 65:138–144

Prout T (1986) The delayed effect of fertility of preadult competition: two-species population dynamics. Am Nat 127:809–818

Prout T, McChesney F (1985) Competition among immatures affects their adult fertility: population dynamics. Am Nat 126:521–558

Richards CM (1958) The inhibition of growth in crowded *Rana pipiens* tadpoles. Physiol Zool 31:138–151

Ricker WE (1954) Stock and recruitment. J Fish Res Board Can 11:559–623

Rose FL, Armentrout D (1976) Adaptive strategies of *Ambystoma tigrinum* Green inhabiting the Llano Estacado of West Texas. J Anim Ecol 45:713–729

Rose SM (1960) A feedback mechanism of growth control in tadpoles. Ecology 41:188–199

Roudebush RE, Taylor DH (1987) Behavioral interactions between two desmognathine salamander species: importance of competition and predation. Ecology 68:1435–1458

Rubenstein DI (1981) Individual variation and competition in the Everglades Pygmy Sunfish. J Anim Ecol 50:337–350

Semlitsch RD (1985) Reproductive strategy of a facultatively paedomorphic salamander *Ambystoma talpoideum*. Oecologia (Berlin) 65:305–313

Semlitsch RD (1987) Interactions between fish and salamander larvae: costs of predator avoidance or competition? Oecologia (Berlin) 72:481–486

Semlitsch RD, Scott DE, Pechmann JHK (1988) Time and size at metamorphosis related to adult fitness in *Ambystoma talpoideum*. Ecology 69:184–192

Sih A (1980) Optimal foraging: partial consumption of prey. Am Nat 116:281–290

Sih A (1982) Foraging strategies and the avoidance of predation by an aquatic insect, *Notonecta hoffmanni*. Ecology 63:786–796

Sih A (1984) The behavioral response race between predator and prey. Am Nat 123:143–150

Sih A (1987) Predators and prey life styles: an evolutionary and ecological overview. In: Kerfoot WC, Sih A (ed) Predation: direct and indirect impacts on aquatic communities. Univ Press New Engl, Hanover, NH, pp 203–224

Slodbodkin LB (1962) Growth and regulation of animal populations. Holt, Rinehart & Winston, New York

Smith DS (1987) Adult recruitment in chorus frogs: effects of size and date at metamorphosis. Ecology 68:344–350

Stein RA (1977) Selective predation, optimal foraging and the predator-prey interaction between fish and crayfish. Ecology 58:1237–1253

Stein RA, Magnuson JJ (1976) Behavioral response of crayfish to a fish predator. Ecology 57:751–761

Steinwascher KF (1979) Host-parasite interaction as a potential population regulating mechanism. Ecology 60:884–890

Stenhouse SL (1985) Interdemic variation in predation on salamander larvae. Ecology 66:1706–1717

Stenhouse SL, Hairston NG, Cobey AE (1983) Predation and competition in *Ambystoma* larvae: field and laboratory experiments. J Herpetol 17:210–220

Taigen TL, Pough FH (1985) Metabolic correlates of anuran behavior. Am Zool 25:987–997

Thompson DJ (1975) Towards a predator-prey model incorporating age structure: the effects of predator and prey size on the predation of *Daphnia magna* by *Ischnura elegans*. J Anim Ecol 44:907–916

Ullyett GC (1950) Competition for food and allied phenomena in sheep-blowfly populations. Philos Trans R Soc London Ser B 234:77–174

Vanni MJ (1987) Indirect effects of predators on age-structured prey populations: planktivorous fish and zooplankton. In: Kerfoot WC, Sih A (eds) Predation: direct and indirect impacts on aquatic communities. Univ New Engl Press, Hannover, NH, pp 149–160

Weller DE (1987) A reevaluation of the −3/2 power role of plant self-thinning. Ecol Monogr 57:23–43

Werner EE (1974) The fish size, prey size, handling time relation in several sunfishes and some implications. J Fish Res Board Can 31:1531–1536

Werner EE (1986) Amphibian metamorphosis: growth rate, predation risk, and the optimal size at transformation. Am Nat 128:319–341

Werner EE, Gilliam JF (1984) The ontogenetic niche and species interactions in size-structured populations. Annu Rev Ecol Syst 15:393–425

Werner EE, Hall DJ (1974) Optimal foraging and the size selection of prey by the bluegill sunfish (*Lepomis machrochirus*). Ecology 55:1042–1052

Werner EE, Gilliam JF, Hall DJ, Mittelbach GG (1983) An experimental test of the effects of predation risk on habitat use in fish. Ecology 64:1540–1548

Wilbur HM (1972) Competition, predation and the structure of the *Ambystoma-Rana sylvatica* community. Ecology 53:3–21

Wilbur HM (1976) Density-dependent aspects of metamorphosis in *Ambystoma* and *Rana sylvatica.* Ecology 57:1289–1296

Wilbur HM (1977) Interactions between food level and population density in *Rana sylvatica.* Ecology 58:206–209

Wilbur HM (1980) Complex life cycles. Annu Rev Ecol Syst 11:67–93

Wilbur HM (1984) Complex life cycles and community organization in amphibians. In: Price PW, Slobodchikoff CN, Gaud WS (ed) A new ecology: novel approaches to interactive systems. John Wiley & Sons, New York, pp 195–224

Wilbur HM (1987) Regulation of structure in complex systems: experimental temporary pond communities. Ecology 68:1437–1452

Wilbur HM, Collins JP (1973) Ecological aspects of amphibian metamorphosis. Science 182:1305–1314

Wilbur HM, Morin PJ, Harris RN (1983) Salamander predation and the structure of experimental communities: Anuran responses. Ecology 64:1423–1429

Yoda K, Kira T, Ogawa H, Hozumi K (1963) Self thinning in overcrowded pure stands under cultivated and natural conditions (Intraspecific competition among higher plants XI). J Inst Polytech Osaka Cy Univ Ser D 14:107–129

Yodzis P (1976) The effects of harvesting on competitive systems. Bull Math Biol 38:97–109

Zaret TM (1980) Predation and freshwater communities. Yale Univ Press, New Haven

Resource Depletion and Habitat Segregation by Competitors Under Predation Hazard

J. F. GILLIAM[1] and D. F. FRASER[2]

1 Introduction

Individuals of different sizes in size-structured populations often differ greatly in the use of resources and/or space. Spectacular examples include shifts, within a lifetime, from carnivory to herbivory (e.g., some turtles; Clark and Gibbons 1969), or from herbivory to carnivory (e.g., some copepods; Neill and Peacock 1980), or from aquatic to terrestrial habitats (e.g., many amphibians). Many other species show large shifts in prey size or habitat use as they grow, but yet other species show little difference across sizes (Fraser 1976; Polis 1984; Werner and Gilliam 1984; Persson this Vol.). Thus, patterns of resource and habitat use within a species vary from complete segregation between two given size classes, to partial overlap, to complete overlap. The presence or absence of such diet or habitat segregation between different size classes can greatly affect population structure and dynamics. For example, the degree of overlap between different size classes influences population stability, and the intensity of competition between classes influences, in interesting ways, whether increases in the density of a given class increases or decreases the density of other classes (Tschumy 1982; Mittelbach and Chesson 1987; Ebenman this Vol.).

Some of the shifts in diet or habitat use might be rather inflexible in the ontogeny of a given individual. However, a large literature demonstrates that many mobile animals show rapid behavioral changes in diet or use of space as resources change across time or space (Stephens and Krebs 1986). Further, several recent experimental studies show that alteration of predation hazard alters habitat selection or feeding behavior by mobile animals (e.g., reviews by Helfman 1986; Milinski 1986; Dill 1987; and Sih 1987 for aquatic systems). Flexible behavior by individuals greatly enriches the range of population structures and dynamics that we might expect to observe in nature, because behavioral shifts in habitat use can rapidly alter the degree of size class (or species) overlap, and may also suddenly alter size-specific mortality and growth rates.

A hypothetical example is illustrated in Fig. 1. Mortality rate is usually taken to be an increasing function of predator density, but this need not be true if individuals can shift habitats in response to the hazard of predation (e.g., Iwasa 1982; Werner and Gilliam 1984). In Fig. 1, the mortality rate of a given size class is assumed to increase as a function of predator density over some interval, but

[1] Behavioral Ecology Group, Department of Biological Sciences, State University of New York at Albany, Albany, NY 12222, USA
[2] Department of Biology, Siena College, Loudonville, NY 12211, USA

Size-Structured Populations
B. Ebenman and L. Persson (Eds.)

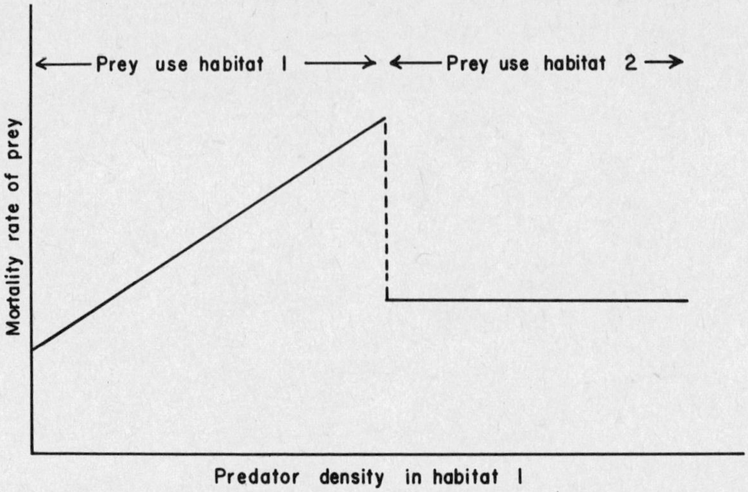

Fig. 1. One reason to consider individual behavior in models of population dynamics. Hypothetical mortality rate of a prey size class or species showing active habitat selection

to show a saltatory drop to a lower level as the predator density reaches a level which induces a discrete shift to a safer habitat (examples of such shifts, and ways of predicting them, are treated below). A corresponding step function in individual growth rate could also be expected, if the habitats differ in resource levels or energetic costs. A discrete habitat shift is depicted, but in some cases the shift away from a hazardous site might be more gradual.

Similarly, appreciation of active, flexible habitat choice can also alter the way we would view genetic responses to biotic or abiotic environmental changes. Natural selection might produce animals adapted to their environment, but by habitat selection animals can choose environments to which they are adapted. For example, the abscissa in Fig. 1 could be relabeled "selection gradient" to indicate the strength of selection on some morphological trait which confers protection from the predator. As the predator increases in density, selection for the trait might increase, but above some threshold the evolution of the trait becomes insensitive to changes in predator density, because the prey just leaves. Such a cusp for a given trait might be accompanied by cusps in other traits which become subject to stronger or weaker selection in the "new" habitat. A question of this nature was addressed, in a preliminary way, by Werner and Gilliam (1984), in the context of morphological response to species introductions.

Below, we first outline experimental studies which demonstrate how the introduction of a predator can cause habitat shifts which increase or decrease segregation between different size classes in fish populations. We then develop a model which addresses two problems. First, if a single size class depletes resources in two sites which differ in the risk of death, how will the difference in predation hazard indirectly affect relative resource levels in those sites? Second, if two or more size classes (or species) have different vulnerabilities to a predator, how will resources be depleted simultaneously by the consumers, and what is the resultant habitat use by each size class?

2 Experimental Studies of Predator-Induced Alterations of Size-Class Overlap

2.1 Predator-Induced Segregation

Experimental work with bluegill sunfish, *Lepomis macrochirus*, has demonstrated how the presence of a predator can cause size-class segregation (Werner et al. 1983 a, b). In the absence of predators in an experimental pond, all bluegill size classes chose the most energetically profitable areas, which were nonvegetated areas. However, in the presence a predator (largemouth bass, *Micropterus salmoides*), small, vulnerable sizes showed a shift to the energetically poorer vegetated areas, but larger sizes showed no such shift. The shift by the smaller bluegills corresponded to a reduced individual growth rate in the presence of the predator, presumably reflecting the poorer foraging rate among vegetation, or possibly increased vigilence in the presence of the predator. Simultaneously, larger bluegills showed higher growth rates than in the absence of predators, apparently reflecting lessened competition from smaller conspecifics. Thus, the effect of the predator was not just to remove individuals; rather, it was to restructure competitive relations among individuals of different sizes. Mittelbach et al. (this Vol.) and Werner and Hall (1988) further discuss this interaction.

Recent experimental work by Schlosser (1987) has shown that predator intimidation can also increase size-class segregation in stream fish, and distributional patterns of size classes often suggest such an effect in fish and other taxa (e.g, Sih 1980; Polis 1981; Stamps 1983; Power 1987; Kneib 1987; Polis and McCormick 1987).

2.2 Predator-Induced Overlap

While the above work shows how behavioral responses to a predator causes size-class segregation, we might also expect to find cases in which the introduction of a predator decreases segregation. Indeed, Power et al. (1985) found this effect in studies of a herbivorous minnow, *Campostoma anomalum.* In the absence of piscine predators in a stream pool, young-of-year minnows (2 – 3 cm standard length) occupied shallow edge areas, and larger conspecifics (4 – 8 cm) mainly used deeper areas. However, the introduction of piscivorous largemouth bass (18 – 23 cm) into the pool resulted in extensive overlap, as the large minnows shifted to shallow areas (Power et al. 1985; Power 1987). Unlike the above bass-bluegill experiment, all size classes of *Campostoma* were vulnerable to the introduced bass.

We have identified a second case in which experimental manipulation of a predator can result in reduction of size-class segregation. In the Northern Range mountains in Trinidad, West Indies, populations of the guppy (*Poecilia reticulata*) in the larger rivers are subject to predation from several effective piscine predators, including the pike cichlid (*Crenicichla alta*) and the guabine (*Hoplias malabaricus*). However, above barrier falls, a weaker predator, Hart's rivulus (*Rivulus hartii*), is often the only piscine predator. Several studies have investigated how upstream and downstream guppy populations differ in behavior, life history, and morphology (e.g., Haskins et al. 1961; Liley and Seghers 1975; Endler 1980; Reznick 1982; Breden and Stoner 1987; Fraser and Gilliam 1987; Houde 1988).

While many sites in the Trinidadian streams fall cleanly into "downstream (strong predation)" or "upstream (weak predation)" categories, others do not. We have been studying such a site, a first- and second-order headwater stream which is a tributary of the Arima River near the William Beebe Tropical Research Station (Simla). The Arima River contains a full complement of predators, including the pike cichlid and the guabine. However, neither the pike cichlid nor the guabine was observed in the tributary during three surveys prior to a survey in August 1987, when guabine appeared in some of the pools. Guabine apparently appear intermittently in the stream, and we asked whether the appearance of guabine might restructure use of space by guppies.

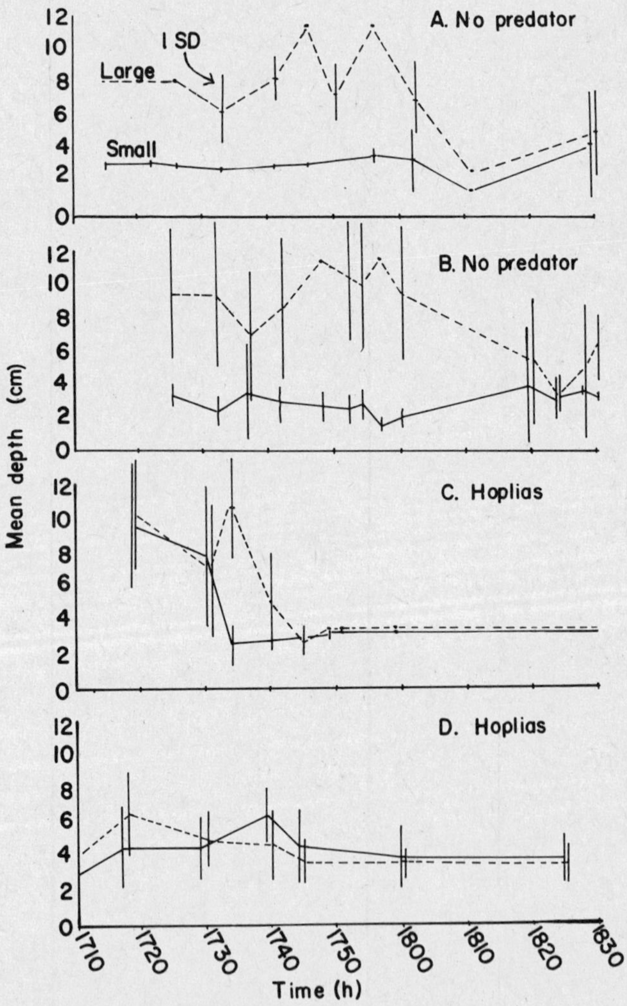

Fig. 2. The effect of a predator, *Hoplias malabaricus*, on the depth distributions of two guppy size classes in a headwater pool

To determine whether the use of space by the guppy population can be altered during periods in which guabine are present, we selected an upstream pool, 2 m wide × 3 m long, averaging 0.16 m deep, and screened the up- and downstream ends with seines. After removing all fish from the pool, we restocked 15 adult (25 – 30 mm TL) and 15 juvenile (15 – 20 mm TL) guppies, approximating natural densities. We then observed the spatial distributions of the two size classes in the late afternoon and through dusk for 2 days, and then introduced one guabine, 120 mm TL, for the following 2 days. Prior to the introduction of the guabine, the two size classes of guppies were spatially separated by depth (Fig. 2, panels A and B). This segregation broke down after sunset (1800 h) when adult guppies shifted to shallow water (Fig. 2A and B). However, following the addition of the predator, the guppies no longer showed segregation (Fig. 2C and D). The guabine spent most of the time in the deeper part of the pool, but it occasionally made forays into shallow areas.

This initial experiment shows that the appearance of guabine can eliminate extensive size-class segregation, at least near dusk (the effect might be quite different in mid-day periods, when guabine are much less active). However, it is not immediately clear whether segregation among the guppies, or the minnows studied by Power et al., might be reestablished over the longer term, as resources are depleted in shallow areas. Below, we explore how resource depletion might result in complete segregation, complete overlap, or some intermediate pattern as resources are depleted.

3 Predicting Spatial Distributions of Size Classes (or Species) Differing in Vulnerability

3.1 Choices with Resources and Predation Hazard Held Constant

It is clear from experimental studies that habitat use is strongly influenced by resources and predation hazard in alternative sites. However, the creation of a practical theoretical basis for predicting such decisions by animals is a difficult endeavor still in its infancy. The difficulty is not only in integrating the different units of energetic gain and mortality hazard into a decision rule, but also in formulating the problem in measurable parameters amenable to field experimentation.

For size-structured populations, one way of reducing the problem to measurable parameters was offered by Werner and Gilliam (1984), who suggested a "minimize mortality rate to growth rate (minimize u/g)" hypothesis. The hypothesis is based on the observation that a juvenile in a purely size-structured population (all properties, including maturation, dependent only on body size) would maximize expected lifetime reproduction by maximizing survival to each size, including size at maturation. This is done by accumulating each bit of mass at the lowest mortality cost; i.e., minimizing u/g at each size, regardless of the relationship between size and mortality. The hypothesis applies in a constant environment, in a constant population with continuous reproduction, but is an approximation otherwise.

We (Gilliam and Fraser 1987) developed an experimental system to test the "minimize u/g" hypothesis. However, in the proccss of developing a way of manipulating growth rates in alternative sites, we realized that we could further simplify the problem. Rather than measuring mortality and growth rates in alternative sites to predict choice, we found that an alternative formulation of the problem would allow predictions of choice on the basis of mortality rates and, more simply, gross foraging rates in alternative sites. The hypothesis we actually tested was a "minimize u/f" hypothesis: the model predicted that foraging fish, when venturing out of a foodless refuge, would prefer the site with the lowest ratio of mortality rate (u) to gross foraging rate (f). The "minimize u/f" hypothesis assumes that: (1) metabolic rates are similar in each site; (2) some time is spent not feeding; and (3) when not feeding, the mortality rate is zero, such as in an absolute refuge. Alteration of these assumptions alters the u/f hypothesis (Gilliam and Fraser 1987).

We tested the utility of the hypothesis in a series of choice experiments by juvenile creek chubs, *Semotilus atromaculatus*, in an experimental field stream. The juvenile chubs could allocate time among a foodless refuge and two foraging areas, in which mortality hazard (density of adult creek chubs) and resource levels (tubificid worms) were manipulated. The experiments established the following (Gilliam and Fraser 1987). First, a higher risk of death in an area does not preclude its use; a sufficiently high energetic return in the more hazardous site can induce a shift to that site. Second, the foragers could assess different levels of hazard (1, 2, or 3 predators) in different sites, as well as different levels of energetic gain; the ability to assess hazard extends beyond detection of presence or absence of predators. Third, choices corresponded closely to those predicted by the "minimize u/f" hypothesis, given independent prior estimates of mortality hazard and intake rate in alternative sites. In particular, it was possible to predict the level of resource enrichment necessary to induce a shift to a more hazardous site.

The above experiments involved resources which were not substantially depleted over the course of the experiment, and a single size class of foragers. Below, we use the "minimize u/f" hypothesis to develop a model which predicts how depletable resources will be structured by a single size class using two sites differing in predation hazard. The model is then extended to multiple size classes.

Throughout, we assume that an individual will choose to forage in a site which minimizes the ratio of mortality to foraging rate (u/f), so the above three assumptions also apply.

3.2 Depletable Resources: Use of Two Sites by a Single Size Class

If feeding animals prefer safer areas, resources should be impacted more severely in such areas. Power (1987) reviews experimental demonstrations of such effects across three trophic levels, in which predators on herbivorous minnows indirectly structure algal communities in streams, by altering use of space by the minnows. The effects on the algae were caused by behavioral shifts by the minnows, rather than numerical reduction of the total minnow population. We also know that the

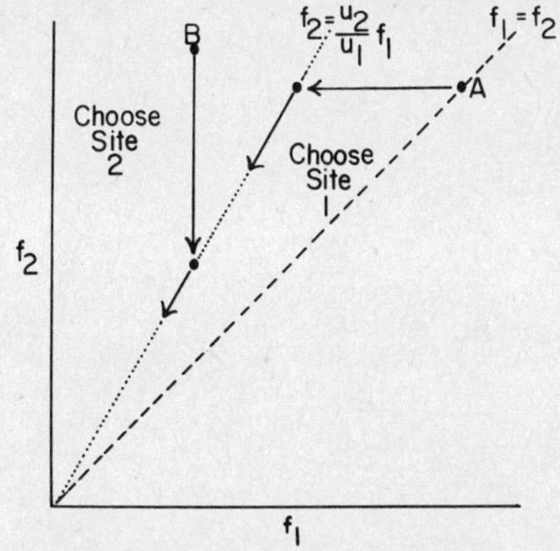

Fig. 3. Predicted trajectories of resource depletion in two sites differing in mortality hazard. *Site 2* is more hazardous than *Site 1*. *f* foraging rate; *u* mortality rate in a site

stream benthic invertebrate community is altered if the local density of juvenile creek chubs is experimentally altered (Gilliam et al. 1989), so the chubs' responses to predation hazard should also tend to structure the invertebrate community. Other processes, such as physical disturbance, could counteract the effect of predation hazard.

The "minimize u/f" hypothesis provides a way to predict the pattern of resource depletion among sites differing in mortality hazard (Fig. 3). Site 1 is safer than Site 2 ($u1 < u2$), and the axes the foraging rates in each site, reflecting resource levels. Given initially equal foraging rates in each of two sites (Point A), a forager is predicted to choose the safer site (Site 1) when feeding, because $u1/f1 < u2/f2$. In the idealized case of perfect choice, resource depletion is depicted by a horizontal vector pointing to the left (in practice, Site 2 will also receive some use, as in Gilliam and Fraser 1987). The foraging rate in Site 1 will be decreased (via resource depletion) until the point representing foraging rates touches the "switching curve" with slope $u2/u1$; along the switching curve, $u1/f1 = u2/f2$, or $f2 = (u2/u1)f1$. Then a hypothetical forager would allocate time to each habitat so as to cause the foraging rate in each to decline along the indicated path. Identical arguments apply to depletable resources started at Point B, except initial use of Site 2 is predicted because $u2/f2 < u1/f1$ at Point B.

Along the switching curve, $f2/f1 = u2/u1$. That is, the hypothetical consumer structures resource levels such that the relative foraging rates equal the relative mortality rates in each site. Identical arguments apply to multiple sites, or to continuous gradients such as distance from shore or cover.

3.3 Simultaneous Depletion by Two or More Consumer Classes

Now consider two size classes (or two species) consuming identical resources in the two sites. One size class is vulnerable to predators, and the other is invulnerable. For convenience, these classes will be called "Small" and "Large," respectively, although in many cases larger individuals need not be less vulnerable.

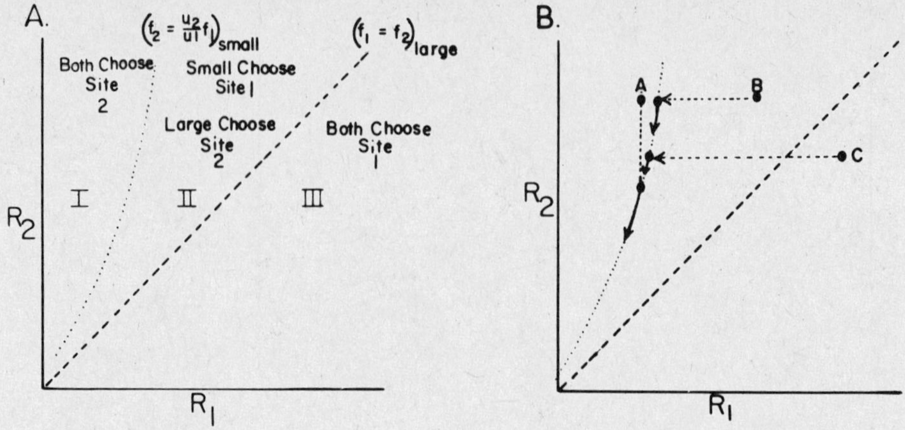

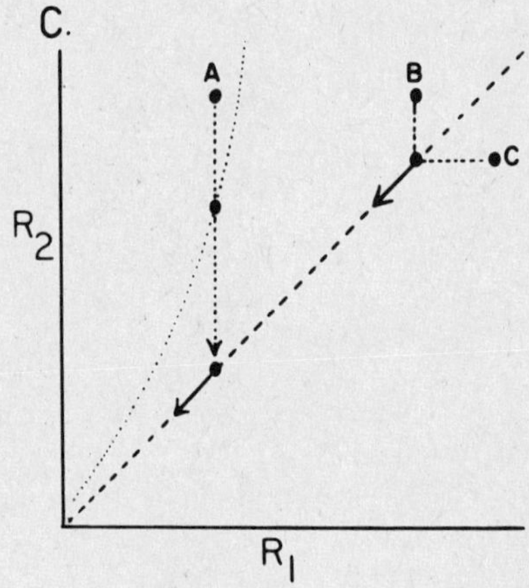

Fig. 4. A Simultaneous habitat choices by Small (vulnerable) and Large (invulnerable) foragers. Site 2 is more hazardous to Small. *R* resource level. **B** Case 1 (Small numerically dominant to Large). Decisions by Small determine resource trajectories. Along Small's switching curve, Large specializes in Site 2, but Small uses both sites. **C** Case 2 (Large numerically dominant to Small). Decisions by Large determine resource trajectories. Along Large's switching curve, Small specializes in Site 1, but Large uses both sites

Exploitative competition occurs in two sites which differ in mortality hazard. If Small tends to produce unequal resource levels in the two sites (as above), but Large tends the equalize resource levels, how does all this settle out? The following approach uses some elements from ideal free models of habitat selection (e.g., Rosenzweig 1981; Parker and Sutherland 1986; Fagan 1987), but also explicitly includes differing mortality hazard in alternative sites.

Hypothetical habitat selection for given resource levels are shown in Fig. 4A. Small is vulnerable with mortality rates u1 und u2 in the safer and more hazardous sites, respectively. Large is invulnerable to predators and is assumed to use the site with the greater resource level. The hypothesized choices by each competitor fall into three regions. In Region I, resources in Site 2 are much higher than in Site 1, and both Small and Large are predicted to choose Site 2. In Region II, Small and Large segregate; Large are hypothesized to still choose Site 2 (there is more food there), but Small chooses Site 1 (because u/f is lower there). In Region III, both Small and Large choose Site 1; it has the highest resources, and is also

safer for Small. Along the boundaries (the "switching curves" of Small and Large) of the Regions, each site is equally attractive. The switching curve of Small is nonlinear because foraging rate is taken to be a nonlinear function of resource level; the curve in Fig. 4A was calculated from data for juvenile creek chubs (Fig. 5 in Gilliam and Fraser 1987). The model assumes purely exploitative competition.

Given initial resource levels, the patterns of habitat shifts and the asymptotic distributions will depend upon the resource trajectories, and hence the densities and relative frequency of Large and Small. Consider two limiting cases: Case 1, in which Small is much more abundant than Large, and Case 2, in which Small is much less abundant than Large.

The trajectories for Case 1 (Fig. 4B) are determined by the decisions of the numerically dominant Small. In the ideal case (perfect omniscient decisions by Small, and negligible effects of Large), the resource trajectories are as indicated. Given initial resource levels indicated by Point A in Fig. 4B, the shifts are as follows. Initially Small and Large both specialize in Site 2, but resources are depleted to a level at which Small generalizes across Site 2 and Site 1, so as to drive the resource level along Small's switching curve; Large remains specialized in Site 2. If resources were started at Point C, the prediction is that initial specialization in Site 1 by both Small and Large will be followed by a period of habitat segregation (Small still in Site 1, but Large in Site 2), and then, as above, generalization by Small but specialization by Large.

Trajectories for Case 2 are shown in Fig. 4C. Here, Large is numerically dominant to Small, and Large tends to equalize resources in the two sites. As before, complete habitat segregation or complete overlap can occur for certain combinations of resource levels. However, the asymptotic distribution, as resources are depleted, is that Small specializes in the safer site (Site 1), and Large allocates time between sites so as to maintain equality of resource levels.

In this example, Large is invulnerable, but different mortality rates for Large in the two sites would just be represented by a different switching curve.

The main predictions for two competing classes are: (1) habitat segregation or complete overlap can result from different, but specifiable, resource levels; (2) if resource levels are "driven" by the consumers, the asymptotic distribution will be characterized by specialization by one competitor class, and generalization (relatively) by the other; and (3) the specializing class will be the one without a dominant demand on resources. Finally, if resource levels are determined by factors other than consumption by Small and Large (e.g. by other consumers in the system, or by changing abiotic conditions), then the resource levels are unlikely to lie on a switching curve at any instant. In that case, segregation (Region II) or congruence (Regions I or III) is predicted at a given time.

4 Conclusion

Our main aim has been to extend the "minimize u/f" hypothesis to begin to predict how competing size classes or species would structure resources when sites differ in mortality hazard, and how patterns of habitat use would result. For illustrative purposes, we have done this by considering just two competitor classes. By introducing a switching curve for each size class or species, the approach is readily extended to multiple consumer classes, or a continuous distribution of sizes. If the switching curves are distinct for each size class (or species), the model predicts specialization by all size classes except possibly one; for a given size class, there is one best place to be, unless the resource levels lie exactly on the switching curve. Parker and Sutherland (1986) make a similar point when considering unequal competitive abilities of different sizes (mortality hazard was not considered). If one size class consumes food x times as fast as another in each of two sites, there are many stable configurations of size classes among the sites (e.g., Fig. 14.3 in Sutherland and Parker 1985). However, if one size class consumes food x times as fast as the other size class in Site 1, but y times as fast in Site 2 ($x \neq y$), then one size class is predicted to specialize in one site, and the other is expected to distribute itself so as to equalize its intake rate in each site. Thus, asymmetries in competitive ability (Parker and Sutherland 1986) or mortality hazard (this chapter) tend to produce specialization, if the asymmetry between consumer classes differs across sites.

The processes hypothesized in Figs. 3 and 4 follow from the "minimize u/f" hypothesis, to which certain assumptions are attached. Notably, the original formulation of the hypothesis (Gilliam and Fraser 1987) assumes use of a foodless refuge when not foraging. This assumption might be met approximately even if no physical refuge is used, if periods of inactivity and extreme vigilance, creating refuge-like conditions temporally, are interspersed with periods of active foraging. However, we have not solved, to our satisfaction, the related problem of ideal free distribution of a single size class between two foraging sites with no refuge or opportunities for safe periods, given that increased food translates into a benefit of faster growth. Abrahams and Dill (1989) offer an empirical approach, which asks the foragers to reveal how they convert energy and death into fitness.

It is clear that foraging fish readily respond to both resource levels and mortality hazard in choosing foraging sites, and some progress has been made in combining theory and field experiments to construct a predictive basis for such shifts. It is also clear that such shifts are important in population dynamics (Ebenman this Vol.; Mittelbach et al. this Vol.). However, a satisfactory link between flexible decisions and population dynamics has not yet been made for size-structured populations. Models of populations without age- or size-structure show how predator-prey dynamics are altered if the model includes flexible behavior of prey responding to predation hazard (Abrams 1984; Sih 1987; Ives and Dobson 1987). In these models, flexible prey behavior often appears to be stabilizing, because responses of prey to high predator density lowers the efficiency of predators in finding prey; the result is analogous to increased interference among predators as predator density increases. Investigations of such questions in size-structured populations is likely to lead to rich dynamical behavior, as different size classes respond differently to hazard, and the effects of behavioral decisions by small individuals linger as the cohort ages. The development of behavioral theory applicable to field experiments is difficult and still in its infancy, but the central roles of resource and habitat use to ecological and evolutionary processes warrants continued attention to flexible behavior in ecological settings.

Acknowledgments. We thank Åke Norberg, Lennart Persson, and Gary Polis for comments on an earlier version of the manuscript. Earlier discussions with others, especially Ronald Pulliam and Andre de Roos, were helpful. We thank Boysie Ramdeen for permission to use his land. Support was from National Science Foundation (USA) grant BSR-8507523.

5 References

Abrahams MV, Dill LM (1989) A determination of the energetic equivalence of the risk of predation. Ecology (in press)

Abrams P (1984) Foraging time optimization and interactions in food webs. Am Nat 124:80–96

Breden F, Stoner G (1987) Male predation risk determines female preference in the Trinidad guppy. Nature (London) 329:831–833

Clark DB, Gibbons JW (1969) Dietary shift in the turtle *Pseudemys scripta* (Schoepff) from youth to maturity. Copeia 1969:704–706

Dill LM (1987) Animal decision making and its ecological consequences: the future of aquatic ecology and behaviour. Can J Zool 65:803–811

Endler J (1980) Natural selection on color patterns in *Poecilia reticulata.* Evolution 34:76–91

Fagan R (1987) A generalized habitat matching rule. Evol Ecol 1:5–10

Fraser DF (1976) Coexistence of salamanders in the genus *Plethodon*: variation of the Santa Rosalia theme. Ecology 57:238–251

Fraser DF, Gilliam JF (1987) Feeding under predation hazard: response of the guppy and Hart's rivulus from sites with contrasting predation hazard. Behav Ecol Sociobiol 21:203–209

Gilliam JF, Fraser DF (1987) Habitat selection under predation hazard: test of a model with foraging minnows. Ecology 68:1856–1862

Gilliam JF, Fraser DF, Sabat AM (1989) Strong effects of foraging minnows on a stream benthic invertebrate community. Ecology (in press)

Haskins CP, Haskins EF, McLaughlin JJA, Hewitt RE (1961) Polymorphism and population structure in *Lebistes reticulata*, a population study. In: Blair WD (ed) Vertebrate speciation. Univ Texas Press, Austin, pp 320–395

Helfman GS (1986) Behavioral responses of prey fishes during predator-prey interactions. In: Feder ME, Lauder GV (eds) Predator-prey relationships: perspectives and approaches from the study of lower vertebrates. Univ Press, Chicago, Ill, pp 135–156

Houde AE (1988) Genetic differences in female choice between two guppy populations. Anim Behav (in press)

Ives AR, Dobson AP (1987) Antipredator behavior and the population dynamics of simple predator-prey systems. Am Nat 130:431–447

Iwasa Y (1982) Vertical migration of zooplankton: a game between predators and prey. Am Nat 120:171–180

Kneib RT (1987) Predation risk and use of intertidal habitats by young fishes and shrimp. Ecology 68:379–386

Liley NR, Seghers BH (1975) Factors affecting the morphology and behavior of guppies in Trinidad. In: Baerends GP, Beer C, Manning A (eds) Function and evolution in behaviour. Oxford Univ Press, pp 92–118

Milinski M (1986) Constraints placed by predators on feeding behaviour. In: Pitcher TJ (ed) The behaviour of teleost fishes. Croom Helm, London, pp 236–252

Mittelbach GG, Chesson PL (1987) Predation risk: indirect effects on fish populations. In: Kerfoot WC, Sih A (eds) Predation: direct and indirect impacts on aquatic communities. Univ New Engl, Hanover, NH, pp 315–332

Neill WE, Peacock A (1980) Breaking the bottleneck: interactions of invertebrate predators and nutrients in oligotrophic lakes. In: Kerfoot WC (ed) Evolution and ecology of zooplankton communities. Univ Press New Engl, Hanover, NH, pp 715–724

Parker GA, Sutherland WJ (1986) Ideal free distributions when individuals differ in competitive ability: phenotype-limited ideal free models. Anim Behav 34:1222–1242

Polis GA (1981) The evolution and dynamics of intraspecific predation. Annu Rev Ecol Syst 12:225–251

Polis GA (1984) Age structure component of niche width and intraspecific resource partitioning: can age groups function as ecological species? Am Nat 123:541–564

Polis GA, McCormick SJ (1987) Intraguild predation and competition among desert scorpions. Ecology 68:332–343

Power ME (1987) Predator avoidance by grazing fishes in temperate streams: importance of stream depth and prey size. In: Kerfoot WC, Sih A (eds) Predation: direct and indirect impacts on aquatic communities. Univ Press New Engl, Hanover, NH, pp 333–351

Power ME, Matthews WJ, Stewart AJ (1985) Grazing minnows, piscivorous bass, and stream algae: dynamics of a strong interaction. Ecology 66:1448–1456

Reznick DN (1982) Genetic determination of offspring size in the guppy (*Poecilia reticulata*). Am Nat 120:181–188

Rosenzweig ML (1981) A theory of habitat selection. Ecology 62:327–335

Schlosser IJ (1987) The role of predation in age- and size-related habitat use by stream fishes. Ecology 68:651–659

Sih A (1980) Optimal behavior: can foragers balance two conflicting demands? Science 210:1041–1043

Sih A (1987) Predator and prey lifestyles: an evolutionary and ecological overview. In: Kerfoot WC, Sih A (eds) Predation: direct and indirect impacts on aquatic communities. Univ Press New Engl, Hanover, NH, pp 203–224

Sih A (1987) Prey refuges and predator-prey stability. Theor Popul Biol 31:1–12

Stamps JA (1983) The relationship between ontogenetic habitat shifts, competition and predator avoidance in a juvenile lizard (*Anolis aeneus*). Behav Ecol Sociobiol 12:19–33

Stephens DW, Krebs JR (1986) Foraging theory. Univ Press, Princeton, NJ

Sutherland WJ, Parker GA (1985) Distribution of unequal predators. In: Sibly RM and Smith RH (eds) Behavioural ecology. Blackwell, London, pp 255–274

Tschumy WO (1982) Competition between juveniles and adults in age-structured populations. Theor Popul Biol 21:255–268

Werner EE, Gilliam JF (1984) The ontogenetic niche and species interactions in size-structured populations. Annu Rev Ecol Syst 15:393–425

Werner EE, Hall DJ (1988) Ontogenetic habitat shifts in the bluegill sunfish (*Lepomis macrochirus*): the foraging rate-predation risk tradeoff. Ecology (in press)

Werner EE, Gilliam JF, Hall DJ, Mittelbach GG (1983a) An experimental test of the effects of predation risk on habitat use in fish. Ecology 64:1540–1548

Werner EE, Mittelbach GG, Hall DJ, Gilliam JF (1983b) Experimental tests of optimal habitat use in fish: the role of relative habitat profitability. Ecology 64:1540–1548

Exploitation Competition and the Evolution of Interference, Cannibalism, and Intraguild Predation in Age/Size-Structured Populations

G. A. POLIS

1 Introduction

There is now recognition that a "population" is actually a complex of phenotypes and age groups that vary in their interactions with the environment. For species that grow slowly through a "wide size range" (Polis 1984a), age/size structure is a major feature and determinant of population dynamics. For these species, the type and intensity of intra- and interspecific interactions depend on size. Interactions may range from neutral to predator-prey or competitive as individuals grow and relative size ratios change (e.g., see Fig. 1).

This paper analyzes population level effects of exploitation and interference competition among and within age/size-structured species of predators displaying a wide size range. The relative strength of these two processes is often a function of size difference: as two entities (= intraspecific age classes or different species) grow closer in size, the potential for exploitation competition (= EC) often (but not always) increases, while the potential for effective interference almost always decreases. Smaller individuals use many of the same resources as large individuals and may therefore compete. Individual competitive effects usually are magnified at the population level: small entities use relatively large amounts of energy

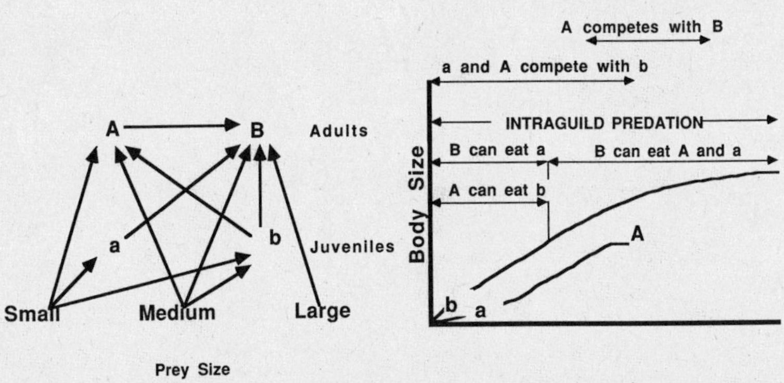

Fig. 1. Simplified food web (*left*) for two species exhibiting age/size structure. *Arrows* indicate that one entity (*base of arrow*) is eaten by a second (*point*). Species potentially compete exploitatively with various degrees of similarity in resource use (e.g., age classes *A*, *b* and *B* all use medium-sized prey). In addition there is (mutual) intraguild predation among age classes (*B* on *a* and *A*; *A* on *b*) and cannibalism (*B* on *b*). Age/size-related changes in interactions (*right*)

Department of Biology, Vanderbilt University, Nashville, Tennessee, USA

Size-Structured Populations
B. Ebenman and L. Persson (Eds.)

because they normally exhibit higher densities and metabolic rates. In some cases, small entities actually use more of jointly exploited resources than larger competitors (this point is controversial – see Sect. 3). Regardless of relative exploitative abilities, dense populations of small entities can exploit great quantities of resources and potentially harm large entities. The possible evolutionary options of large entities living among hordes of smaller entities are assessed. I argue that interference directed by large towards smaller entities is a widespread and frequent evolutionary response that decreases both resource overlap and potential EC.

I focus primarily on two types of predatory interference: intraspecific predation (cannibalism) and intraguild predation. Intraguild predation (= IGP) is the killing and eating of other species that use the same resources and are therefore potential exploitative competitors. These extreme forms of interference are widespread taxonomically and can determine population and/or community structure (Fox 1975; Polis 1981, 1984b; Polis et al. 1989). Cannibalism and IGP are also unique forms of interference that can evolve under less restrictive conditions than other types of interference (see Sect. 6).

Data from long-term studies of scorpions (Polis 1980, 1984a; Polis and McCormick 1986a and b, 1987) are used to illustrate trends general to many wide size range predators. Newborn scorpions are miniature adults in appearance. However, body size, prey use, foraging behavior, home range, microhabitat use, temporal (seasonal and diel) patterns, and intra- and interspecific interactions change greatly during growth (Polis 1979, 1980, 1984a; Polis and McCormick 1986a, 1987; Polis et al. 1985). For example, foraging age classes of the scorpion *Paruroctonus mesaensis* increase 60–80× in weight from 0.03 g (instar 2) to 2.0–2.5 g (nongravid adults); instar 2 individuals eat prey that average 5 mm in length whereas adults use prey averaging three times larger, two-thirds of which are different species. In fact, the differences in body sizes and resource use for the three year classes of *P. mesaensis* are greater than the differences that exist among most real "biological species" (Polis 1984a). This magnitude of size change is typical of other wide size range species (e.g., aquatic and marine invertebrate phyla, arachnids, hemimetabolic insects, larva of holometabolic insects, fish, and reptiles; Polis 1984a).

2 Pertinent Scorpion Biology

This paper analyzes the biology of four species of scorpions inhabiting the desert floor of the Coachella Valley, Riverside County, California. Data were collected intensively from 1973 to 1984 and intermittently since 1985. Size, age and density are presented in Table 1. There are three rare and one abundant species. Most data describe the moderately sized and numerically dominant scorpion, *Paruroctonus mesaensis*. Adults of this species mate from spring through summer. Gestation lasts 10–14 months; first appearance of newborn throughout the population is synchronous and occurred in August for 11 consecutive years. Individuals mature in 19–24 months and growth is determinate. Thus, three age groups exist in the field at any one time: young (<1 yr), intermediate (1–2 yr), and adult (2– >5 yr) animals. Individuals >2 yr are similar in size and ecology. The other species live

Table 1. Bionomics of four species of desert scorpion

Species	Adult mass (g)	Maturity (months)	Longevity (years)	Density (No./ha)	Relative frequency (percent)
Paruroctonus mesaensis	2.53	19–24	4–6	1314–2637	95.7
Paruroctonus luteolus	0.29	19–24	3–4	6–25	0.6
Vaejovis confusus	0.19	19–24	3–4	9–57	3.3
Hadrurus arizonensis	5.53	48–60	20–25	10–31	0.9

4 (*Vaejovis confusus* and *Paruroctonus luteolus*) to >25 years (*Hadrurus arizonensis*) and exhibit two to five distinctly sized age classes.

All scorpions are euryphagous predators, eating whatever prey they can subdue. Since scorpions digest their prey externally in a process that may last hours, it is relatively easy to quantify diet. Insects, spiders, solpugids and other scorpions are eaten primarily. Cannibalism and IGP are readily observed and quantified.

In the following analyses and discussion, both intra- and inter-specific interactions among scorpions are analyzed. Although this tactic may ignore some potentially import differences (e.g., feedback loops between conspecific young and adults, see Schoener 1976), con- and heterospecific size classes exhibit many of the same characteristics (see below and Brown and Maurer 1986).

3 Exploitation Competition

The strength of potential EC for limited food resources is not constant during development and growth. Age related changes in foraging, metabolic rates, and population size characterize most wide size range predators. Such changes alter the degree of resource overlap, energy use, and the overall intensity of potential EC between competing entities.

Prey size generally increases with predator size. Significant positive correlations (Fig. 2) exist for age groups within species (e.g., Wilson 1975; Turner 1979; Polis 1979; Polis and McCormick 1986a) and for differently sized species (Hespenheide 1973; McCormick and Polis 1982; Peters 1983; Vezina 1985; Nentwig and Wissel 1986; Polis and McCormick 1986a). Further, as two (hetero- or conspecific) entities become more similar in size (i.e., body size ratio of larger/smaller predator approaches 1.0), they become more similar in prey size utilization (Fig. 3). However, note that large predators still frequently exploit the same small prey used by smaller entities (Figs. 1 and 2; also see Wilson 1975).

Under the strict conditions necessary for the operation of EC (Polis and McCormick 1987), resource overlap potentially describes the degree of EC between two entities (e.g., Pacala and Roughgarden 1982; it should be noted that overlap may not approximate Alpha, the competition coefficient; see below and Abrams 1980; Alstad 1982). Thus potential EC may be expected to increase as entities approach in size and as prey use converges.

However, other factors modify this simple relationship. Compared to older/larger animals, younger/smaller animals of the same or different species

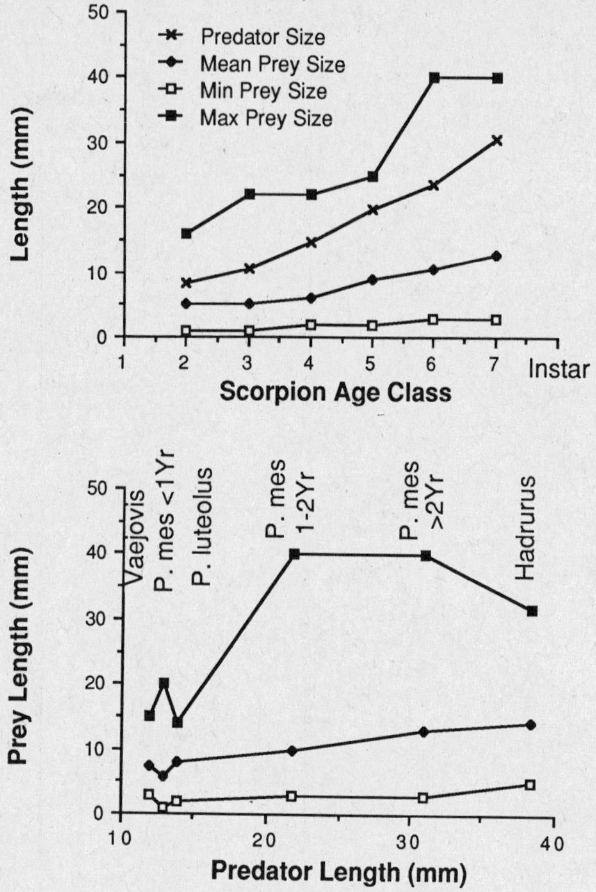

Fig. 2. The relationship between predator size (length) and prey size within the same species of scorpion (*Paruroctonous mesaensis*; top) and for different species of scorpions and *P. mesaensis* year classes (bottom). Note that maximum prey size increases rapidly, mean prey size increases slowly and minimum prey size barely increases with increasing predator size

usually are characterized by: (1) Higher metabolic rates *per gram mass* (U; Peters 1983; Calder 1984; Schmidt-Nielsen 1984: U decreases with body mass, M, so that $U \propto M^{-0.25}$). Thus overall metabolic rate per individual increases as a decreasing function of M ($M^{0.67}$ to $M^{0.75}$) and small individuals proportionately use more energy and food than larger ones. (2) Higher densities (D; Peters 1983; Peters and Wassenberg 1983; Calder 1984: D decreases with increasing M, so that $D \propto M^{-0.98}$ to $M^{-0.75}$). (Exceptions occur intraspecifically if populations are dominated by adults [e.g., some *Tribolium* fish and bivalves; see Polis 1981, p 231]; interspecifically, if smaller species are rare.) Thus, depending on the specifics of size, density, and metabolism, small animals may use more or less energy than larger competitors.

I now present a general method to assess quantitatively potential EC between any two entities. Energy use combined with resource overlap determines the potential competitive effect of smaller on a larger entity. Thus M and U must be used to approximate more closely the intensity of potential EC exerted by one individual on a second. Alpha_{ij}, the per-capita competitive effect of one individual (j) on another (i), is calculated:

Fig. 3. The relationship between similarity in prey use and the size ratio of larger/smaller scorpion entities within the same species of scorpion (*Paruroctonus mesaensis* instars (top) and for different species of scorpions and *P. mesaensis* year classes (bottom). Summation overlap (May 1975) in prey use combined overlap in prey size and prey taxa and is used as a measure of similartiy in prey use. Schoener's (1968) similarity index was used to calculate overlap

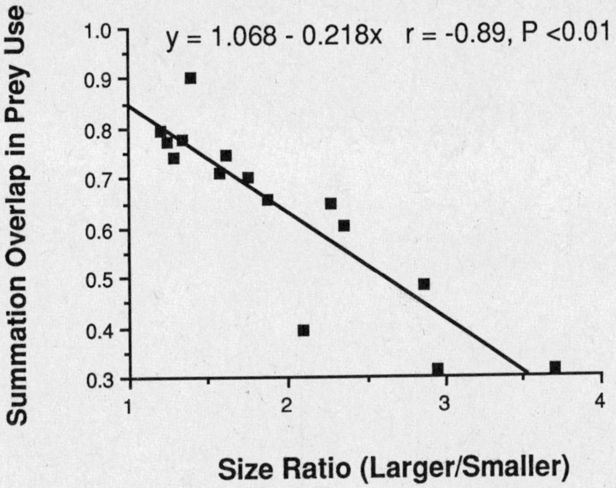

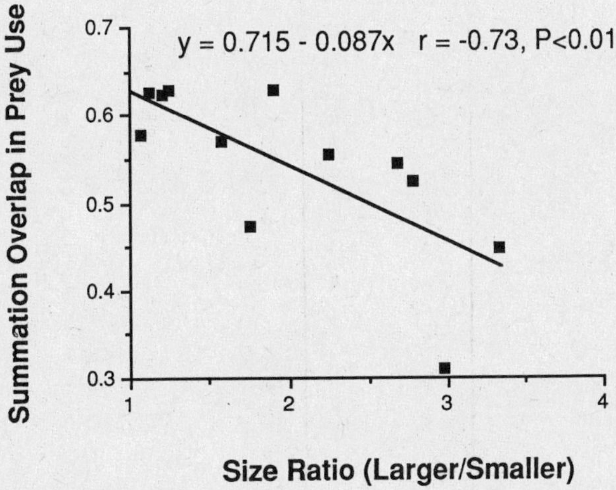

$$Alpha_{ij} = U M O_{ij} ,$$

where O_{ij} is the unitless overlap in resources between the two individuals. $Alpha_{ij}$ is comparable to the competitive coefficient in the Lotka-Volterra equation.

To calculate the potential overall competitive effect of population S (small entity) on population L (large entity) (= A_{LS}), density (D), population biomass ($B_S = D_S M_S$), population metabolism or energy use ($U_S = B_S U$), and resource overlap between S and L (O_{LS}) are combined to calculate A_{LS}:

$$A_{LS} = (U_S O_{LS}) = D_S M_S U O_{LS} .$$

A_{LS} can be viewed as that amount of total energy/prey biomass exploited by small entities that otherwise could be used by larger entities. A_{LS} is then compared with A_{LL} (the energy or prey biomass used by the population of the large entity; $A_{LL} = U_L$ because $O_{LL} = 1.0$) to determine the relative strength of exploitative competition by population S on population L ($= C_{LS}$):

$$C_{LS} = A_{LS}/A_{LL} .$$

The potential harm (measured by use of common resources) to larger competitors caused by EC from populations of smaller entities increases as a function of C_{LS}. In some cases, it is likely that C_{LS} approaches and exceeds 1.0, i.e., smaller entities use relatively more common energy or prey than do larger entities. For example, consider termite and elephant populations eating similar plants in a tropical savanna. Although an individual elephant eats as much as thousands of termites, a population of termites most likely eats more than a population of elephants (termites maintain the highest animal population biomass in many tropical ecosystems and their U is high). This would mean that the competitive impact of termites on elephants is potentially greater than the intraspecific competitive effect of the elephant population of itself (i.e., $C_{LS} > 1.0$). The exact intensity of potential EC among entities can be calculated only if all the necessary individual and population parameters are known.

These parameters have been determined for age classes of the scorpion *P. mesaensis* (Table 2). Energy use was calculated using monthly data on population dynamics, individual growth, age-specific foraging activity/phenology, diet, feeding rates, and field metabolism based on ambient temperatures (exact methods in Polis and Yamashita manuscript). As expected, juveniles (compared with adults) are more dense (3.2× and 2.1×, <1-yr- and 1-2-yr-olds respectively) and maintain greater weight-specific metabolisms (1.8× and 1.5×). Consequently, the total yearly population metabolism ($U_{(1-2)}$) of the 1–2-year class surpasses that of adults. The population of 1–2-year-old (individual mass = 38% of adult mass) maintains a population biomass 97% that of adults, but uses 142% more energy (prey biomass). Greater densities and metabolic rates of the <1-yr-old age class

Table 2. Age-specific energetics of *Paruroctonus mesaenis* year classes. Body mass and biomass are for wet weights. Average metabolism is that at the average annual temperature of 21.4 °C. The population metabolism is based on bimonthly temperature averages. The numbers in parentheses are the fraction of that parameter for the year class as compared to the adult year class

Year class	Average mass (M) (g)	Average density (D) (No./ha)	Population biomass (M_{pop}) (g/ha)	Average metabolism (21.4 °C; U) (ml O_2/g/h)	Year class	Population metabolism (U_{pop}) (kj/ha/yr)	Requirements biomass (M_{pop}) (g/ha/yr)
<1	0.169 (0.07)	1043 (3.20)	175 (0.26)	0.060 (1.77)	<1	12474 (0.55)	1700 (0.55)
1–2	0.971 (0.38)	677 (2.08)	658 (0.97)	0.051 (1.50)	1–2	32219 (1.42)	4391 (1.42)
>2 (2–7)	2.530 (1.0)	326 (1.0)	678 (1.0)	0.034 (1.0)	>2 (2–7)	22757 (1.0)	3101 (1.0)

Table 3. Age-specific exploitation competition among year classes of *Paruroctonus mesaensis*. O_{LS} ($= O_{SL}$) is the summation overlap combining prey size and taxa (see Fig. 4 legend). A_{LS} is the competitive effect at the population level of the smaller population/year class (S) on the larger (L). C_{LS} is the relative strength of competition of S on L and is equal to (A_{SL}/A_{LL}). "Diffuse C" (bottom three combinations) represents the relative strength of competition by two year classes on the third. A_{SL} and C_{SL} are the equivalent parameters of the larger population/year class (L) on the smaller (S). See text for full explanation

Year class combination	O_{LS}/O_{SL}	A_{LS}	A_{SL}	C_{LS}	C_{SL}
$(<1)\times(1-2)$	0.676	8432	21780	0.262	1.746
$(<1)\times(>2)$	0.390	4865	8875	0.214	0.711
$(>2)\times(1-2)$	0.885	28514	20140	1.253	0.625
Diffuse C					
$(1-2)\&(>2)\ \times(<1)$				2.457	
$(<1)\ \&(>2)\ \times(1-2)$				0.887	
$(<1)\ \&(1-2)\times(>2)$				1.467	

cannot compensate for their small size (7% of an adult's mass): $M_{(<1)}$ is only 26% of $M_{(adult)}$ whereas $U_{(<1)}$ is 55% of $U_{(adult)}$.

A_{LS} calculations show that the 1–2-year class eats more prey used in common with the other year classes than do either the <1 or >2-year classes (Table 3). Thus this age class can cause greater competitive damage to the other two year classes than vica versa. For example, they use 28514 Kj/ha/yr [$= A_{(>2)(1-2)}$] of the same prey potentially eaten by adults; conversely, $A_{(1-2)(>2)} = 20140$ Kj/ha/yr of prey that could have been used by 1–2-year-olds. The relative competitive effect of 1–2-year-olds on the adults [$C_{(>2)(1-2)} = 1.253$] is twice as strong as the competitive impact of adults on the 1–2 year class [$C_{(1-2)(>2)} = 0.625$]. This means that the 1–2-year population is equivalent to 125% of the adult population in its exploitation of common prey (the two juvenile age classes combined exert a "diffuse" C_{LS} of 147%; Table 3, bottom).

The <1-year class uses the least quantity of common resources, exerts the least competitive impact and potentially suffers the most from EC. Each C_{LS} of this age class is much lower than the comparable C_{SL}. The diffuse competitive impact on <1-year-olds by all older scorpions indicates that older animals use 2.4× more of commonly exploited prey than does the population of <1-year-olds.

The age-specific population energetics of another scorpion (*Urodacus yaschenkoi*) are slightly different (Shorthouse 1971). The five younger year classes exhibit a greater summed density, population biomass, and energy use than the adults. However, no single year class has a greater population biomass or population metabolism than adults. Although Shorthouse did not present sufficient data to calculate A_{LS}, it would follow the same trends and diffuse C_{LS} by younger year classes on adults is most likely >1.0.

In general, I predict/speculate that age-specific patterns of energy use by many wide size range predators and most other scorpions and arthropod predators (spiders, insects) will be similar to those of *P. mesaensis* or *U. yaschenkoi*. Either a younger age class or the total of all younger classes will use more energy than adults and exert greater potential competitive harm on adults than the reverse.

No clear patterns of relative energy use or competitive abilities emerge from other analyses from population metabolism as a function of size. Some authors (Damuth 1981; Peters 1983; Peters and Wassenberg 1983) calculated that smaller entities use at least the same proportion of resources from their ecosystems as do larger entities. Using different data, Brown and Maurer (1986) showed that large species use a disproportionately high share of resources. The analysis of scorpion energetics provides evidence for both sides: exploitation by a smaller entity (1–2-year class) was higher than that of the larger adults and resultant competition could influence negatively larger adults; however, the smallest/youngest age class (<1 yr) used less energy than either of the two older age classes. In many other cases, smaller entities doubtlessly exert a greater impact on their larger competitors than vica versa (i.e., $C_{LS} > C_{SL}$). In fact, it has been suggested that, in general, populations and/or individuals of smaller entities are better exploitative competitors, especially during periods of food scarcity (Morse 1974, 1980; Persson 1985; but see Schoener 1983, Bengtsson 1987).

Regardless of relative competitive abilities, small and large predators overlap in food/energy use and thus negatively affect each other. This is accentuated during periods of food stress when individuals become less selective, average prey size decreases and overlap increases (Polis 1988 and included references). Such changes increase the potential for EC during critical "crunch" periods of food shortage when competition is expected to be most severe. The important point is that dense populations of small entities potentially harm larger competitors via exploitation. Of course, competitive effects will occur only if there is a limited energy supply and the demand for resources exceeds the supply.

In summary, variable EC occurs in age/size structured populations both intraspecifically (as juveniles grow to adult size) and interspecifically (as growth trajectories intersect and juveniles of one species approximate the adult size of other species). Two general trends describe such dynamic competition. First, as the smaller entity approaches the size of the larger, prey overlap increases but the overall intensity of potential EC (A_{LS}) is a function of population dynamics and energetics. Second, smaller entities often exert greater impact energetically (and potentially competitively) than their small size would suggest. This is because they are usually more dense and less efficient metabolically than larger entities. In some (many?) cases, small entities can cause greater competitive damage than large entities exert either on themselves or on the smaller entity.

4 Interference Competition

Interference is most easily and safely performed when an asymmetry in size exists between the perpetrator and victim. A higher probability of success and a lower

Table 4. A matrix of scorpion-scorpion predatory interference. The entries represent the percent of the diet that each species forms as prey for each of the other four scorpion species. The diagonals represent cannibalism and the off-diagonals represent intraguild predation. (Polis and McCormick 1987)

Predator	Prey				
	Hadrurus arizonensis	*Paruroctonus luteolus*	*Paruroctonus mesaensis*	*Vaejovis confusus*	Total scorpions
H. arizonensis	0.0	6.3	12.5	3.1	21.9
P. luteolus	0.0	33.3	6.7	6.7	46.7
P. mesaensis	0.4	0.5	0.5	8.3	14.2
V. confusus	0.0	0.0	8.0	4.0	12.0

probability of damage-inflicting retaliation occur when the victim is relatively small. For wide size range species, the potential for effective interference is variable, increasing as the size difference between two entities increases. "Incomplete dominance" (Morse 1974) between species may occur when size ratios and relative dominance reverse during growth. Thus small/young individuals of species A may be subordinate to larger individuals of species B whereas larger/older individuals of species A are dominant to all B (e.g., see Fig. 1). Connell (1983) argues that competitive dominance is size/age- rather than species-specific. He predicts that "reversals in competitive rank should be common" (p. 680).

Scorpions illustrate the relationship between relative size and interference. Table 4 is a matrix of scorpion-scorpion predatory interference. The numerically dominant *P. mesaensis* was predator in 94.9% of the 117 cannibalisms and 91.5% of the 141 intraguild predations observed in the field. In particular, adult *P. mesaensis* are the most important predators on other scorpions: smaller species and age classes formed 9–17% and 11–14% respectively (at two sites) of the total adult diet.

Each species, however, was both a predator and a prey because a simple size hierarchy exists: larger scorpions almost always kill and eat smaller individuals, regardless of species. In fact, the victim was larger or equally sized only twice in the 258 observed cases of scorpion-scorpion predation. When two scorpions fight, each grasps the tail of the other with its pincers; the larger individual exerts more leverage with its tail and is able to insert its sting, envenomize, and then paralyze and kill its smaller opponent. Since the outcome of this battle is dependent only on relative size, all species are vulnerable when young and small. Thus individual *H. arizonensis* begin life as prey of *P. mesaensis*, later to become the predator of *P. mesaensis*, all the while eating many of the same prey and potentially competing exploitatively (see Fig. 1). Such age/size-mediated reversals occur among most species combinations (Fig. 4). Reversals may contribute to "age-class bottlenecks". For example, IGP by *P. mesaensis* on younger age classes of *H. arizonensis* partially explains the rarity of this, the largest species in North America (*P. mesaensis* ate > 10% of all observed *H. arizonensis* < 1 month old; Polis and McCormick 1987). Individual adult *H. arizonensis* are predatory dominants to all other scorpions yet they are so rare that IGP by *H. arizonensis* is relatively insignificant.

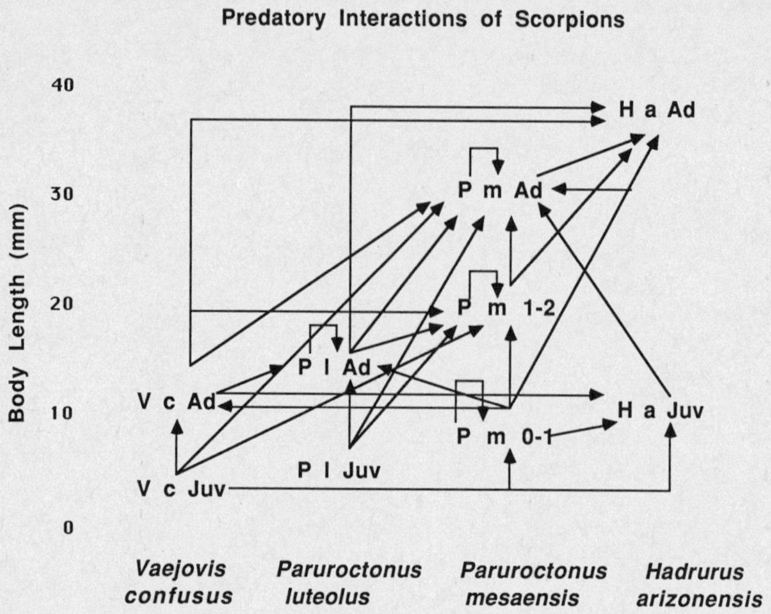

Fig. 4. Observed predatory interactions among scorpions as a function of size (length). *Arrows* indicate that one age class (*base of arrow*) is eaten by a second (*point*). Only two age classes (juveniles and adults) are represented for *Vaejovis confusus* (= *Vc*). *Paruroctonus luteolus* (= *Pl*) and *Hadrurus arizonensis* (= *Ha*); three year classes of *Paruroctonus mesaensis* (= *Pm*) are pictured. An *arrow* returning to the entity signifies cannibalism within the same age class. Predatory reversals (*A* eats *b* but *B* eats *a* and *A*, Fig. 1) are frequent. Such "incomplete dominance" occurs because larger age classes of any species eat smaller age classes of any other species. It is likely that all possible combinations of smaller victim-larger predator occur in nature

Interference is generally directed against smaller entities. Reviews of both cannibalism (Fox 1975; Polis 1981, 1984b) and IGP (Polis et al. 1989) show that the vast majority of victims were markedly smaller and/or younger than their con- or hetero-specific predators. Relative size determines the outcome of predatory interference in a variety of protozoans, rotifers, gastropods, crustaceans, arachnids, insects, fish, amphibians, reptiles and mammals. Thus cannibalism and IGP are usually asymmetric: older/larger individuals inevitably win and are relatively invulnerable to injury and death. Other forms of interference likewise are often asymmetric with larger species dominant (Case and Gilpin 1974; Persson 1985). The victims of murder (e.g., Askew 1971; Stilings and Strong 1983) are always smaller or younger. Larger individuals are the inevitable winner in most aggressive dominance and territorial interactions (Morse 1974, 1980; Connell 1983; Schoener 1983).

5 Coevolution Among Age/Size Groups

In size/age structured populations the strength of both exploitation and interference changes as entities converge in size. What are the possible coevolutionary responses to this dynamic regime? In theory, selection operates to lessen the effects of EC in two basic ways. First competing entities may diverge in resource use (i.e., partition food, space, and time) and/or body size (e.g., Santa Rosalia ratios, limiting similarities, Dyar's constant) thus decreasing potential EC for limited resources (Roughgarden 1979; Polis 1984a). Alternately, (alpha) selection for interference by larger entities should minimize the deleterious effects of EC by smaller competitors (Case and Gilpin 1974; Gill 1974; Morse 1980; Polis 1981; Persson 1985; Polis et al. 1989). These authors maintain that interference is a common, if not the most frequent evolutionary consequence of resource competition. It is the predominant interaction among many competitors (Simberloff 1982; Connell 1983; Schoener 1983; Strong 1983).

I now present hypotheses to distinguish the operation of exploitation from interference; the hypothesis are tested below with data on scorpions. The exact coevolutionary response to resource competition is dependent upon relative size. Consider a regime whereby a population of the smaller entity exploits a higher share of the limiting resources and is the better competitor. If only EC is possible, the larger entity should be selected to lessen competition by diverging from the smaller entity in use of food, space and/or time. Thus, the smaller and more successful exploiter is predicted to occupy better places and times whereas the larger entity should occur in less productive situations.

If interference is possible, the natural asymmetry due to size will favor interference by the larger entity towards the smaller but superior exploiter. In this case, the smaller entity is predicted to diverge (in either ecological or evolutionary time) thus directly decreasing resource related interference and indirectly lessening resource overlap by displacing to less productive situations. The larger and dominant entity will occupy more productive times and places and experience less EC. The smaller entity is also expected to diverge if the larger entity is the superior exploitative competitor. Thus, the distribution of entities relative to resource quality will suggest if larger or smaller entities are better competitors.

These predictions are tested using scorpion data. The relative competitive/exploitative (C) ranking of *P. mesaensis* year classes is: (1 – 2 yr) > (adults) > (< 1 yr). If EC is the major interaction, the 1 – 2-year class are predicted to occur in the most productive situations. The relative ranking of predatory interference is by size: adult *P. mesaensis* are the most important cannibalistic and intraguild predators preying on all scorpions except *H. arizonensis* adults. The prediction based on interference is that *P. mesaensis* (and *H. arizonensis*) adults should occupy best time periods and habitats.

Analysis of foraging periods and microhabitats indicates that adult *P. mesaensis* are the competitive dominant (Table 5 – *H. arizonensis* adults were too rare to test and all age classes of *H. arizonensis* were combined for these analyses). Distributions on different soil types shows that *P. mesaensis* adults primarily occurred in sand, the most productive microhabitat (Table 5). Smaller species and age classes occurred more frequently in significantly less productive soils (Polis and McCormick 1987). Only adult *P. mesaensis* exhibited a significant correlation between surface activity and insect abundance ($r = 0.52$, $p < 0.05$). All other entities showed negative correlations between the times that they emerge to forage and insect abundance, i.e., they were relatively inactive during productive times

Table 5. Differences on soil microhabitats. Each of the three species is significantly different from *P. mesaensis* in its distribution on sand (G-test). The two year classes are significantly different from *P. mesaensis* in their distribution on sand (G-test). IGP by *P. mesaensis* adults occurs significantly more frequently on sand than other soils (G-test). Potential prey are significantly more abundant on sand than other soils (t-test for means). (Data from Polis and McCormick 1986a, 1987)

Distribution of entity	Percent in microhabitat		Comparison and probability
	Sand	Heterogeneous and hardpan	
Vaejovis confusus	15	85	
Hadrurus arizonensis	25	75	(Distribution on soil: each
Paruroctonus luteolus	11	89	spp versus *P. mesaensis*)
Paruroctonus mesaensis	55	45	<0.001
P. mesaensis (<1 yr)	(57)	(43)	(Distribution on soil: >2 yr
P. mesaensis (1–2 yr)	(47)	(53)	versus <1 yr and 1–2 yr)
P. mesaensis (>2 yr)	(65)	(35)	<0.01
Distribution of IGP:			(Distribution on sand: entity versus IGP)
Vaejovis confusus	49	51	<0.001
Paruroctonus luteolus	100	0	<0.001
Prey abundance:			
Mean No. prey/trap	106.1	53.5	(Sand versus other soils)
Standard deviation	(41.3)	(16.3)	<0.001

and active during poor periods (r values between −0.20 and −0.54). Note that the entity potentially best at exploitation (1–2-year class *P. mesaensis*) was significantly less active/abundant in the most productive times/microhabitat as compared with adults. This suggests that EC was relatively unimportant in shaping distribution patterns.

The data indicate the adult *P. mesaensis* are competitively superior to all smaller entities. Polis and McCormick (1987) show that smaller scorpions active during time periods and in microhabitats (sand) primarily occupied by *P. mesaensis* adults suffered significantly higher rates of predation by adult *P. mesaensis* as compared to small scorpions active in other times and microhabitats (see Table 5 for microhabitat data). Moreover scorpion-scorpion predatory interference is an important, if not the major source of mortality for smaller species and age classes (Polis 1980; Polis and McCormick 1987). Data from field experiments indicate that EC either did not exist or was too weak to measure (Polis and McCormick 1987). The strong inference is that frequent (predatory) interference is the major force shaping patterns of distribution and abundance of smaller age classes and species of scorpions. Large *P. mesaensis* dominate the most productive conditions and smaller entities are forced into other times and places.

It is not possible to measure the exact degree that cannibalism and IGP lessened EC. However, it is clear that interference decreased potential EC in at least two distinct ways. First, cannibalism and IGP directly decreased the populations of potential exploitative competitors (smaller scorpions). Second, smaller scorpion entities tended to avoid adult *P. mesaensis* in time and space, thus decreasing overlap and use of common resources. Such niche shifts are analogous to those that occur when prey avoid predators by shifting into "enemy free space" (Jeffries and Lawton 1984). In general cannibalism (Polis 1981, 1984b), IGP (Polis et al. 1989), and other forms of size-related interference (Case and Gilpin 1974; Morse 1974, 1980; Oksanen et al. 1979; Carothers and Jaksic 1984) can produce patterns of food use, spatial and temporal distribution, and body size differences similar to those caused by EC. In these cases, aggression decreases potential EC when smaller entities avoid larger dominants and are thus forced to use other resources.

In closing this section, note that these analyses of competitive effects and energy use share the same fault with most nonexperimental research. Ecologists study post-interactive systems: abundance and the intensity of population energy use, EC and/or interference are themselves products of previous interaction. Interference can limit the density of small entities and exploitation by smaller competitors may limit the density of large entities. Thus current observation and analysis may not provide full insight into the potential intensity of interference and EC.

In summary: (1) Interference (alpha selection) is a frequent coevolutionary response against (competitively superior or inferior) smaller entities whose exploitation of similar resources can harm larger competitors. (2) The differences in size characteristic of wide size range species makes interference by larger entities relatively effective and safe and thus facilitate its possible evolution. (3) Large dominants occupy the most productive times and places; subordinates often coevolve to avoid larger dominants via spatial and temporal displacement to poorer conditions; this response produces niche patterns (i.e., resource partitioning) similar to those predicted under an exploitative regime. (4) Thus interference decreases the potential harm of EC from smaller competitors and may be one explanation why large entities possibly use such a high proportion of environmental resources (Brown and Maurer 1986; but see earlier and Damuth 1981; Peters 1983; Peters and Wassenberg 1983).

6 The Evolution of Interference Competition

What are the possible evolutionary routes open to large entities facing dense populations of potentially damaging, smaller entities that exploit great quantities of resources? Interference is one possible response to lessen the harmful effects of EC (Case and Gilpin 1974; Gill 1974; Polis 1981, 1984b; Roughgarden 1979; Persson 1985). However, interference cannot always evolve in response to EC. For example it cannot be directed indiscriminately against smaller competing entities, no matter how much harm each causes. Cost-benefit considerations constrain the evolution of interference to restricted conditions (Case and Gilpin 1974; Rothstein 1979; Polis 1981, 1984b; Wilson 1986). The benefit is a decrease in resource competition that occurs directly when resources are freed (i.e., not used by competitors) and indirectly as the population size of competing entities is reduced. Individuals with a modified phenotype (and genotype) such that they interfere experience costs (Case and Gilpin 1974; Beddington 1975; Schoener 1976; Buss

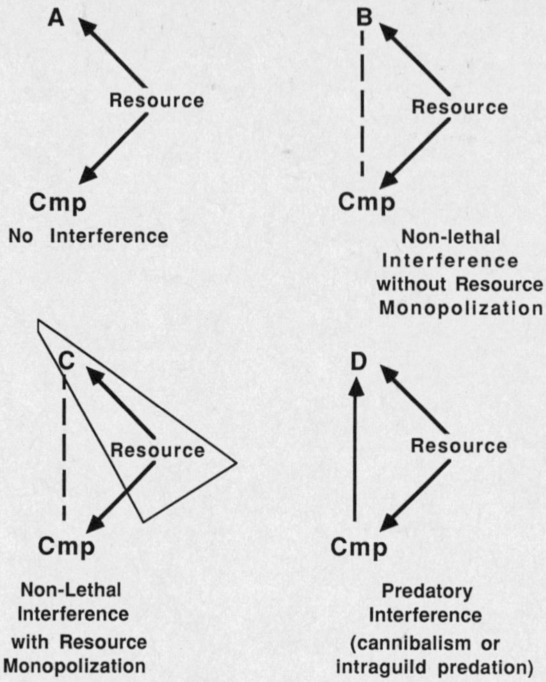

Fig. 5. Restrictions on the evolution of interference against resource competitors (= Cmp). *A, B, C,* and *D* are all members of the same population. *A* does not interfere; *B* interferes nonlethally (e.g., is aggressive) with Cmp in an open population where competing conspecifics and/or resources freely mix; *C* interferes nonlethally with Cmp but is able to monopolize the resources, e.g., by territoriality; *D* interferes with Cmp via cannibalism or intraguild predation. Interference by *B* cannot evolve because *B* incurs costs of interference but *A* and other competing conspecifics receive the same benefits as *B*. Predatory interference by *D* can evolve regardless of population structure and the possibility of resource monopolization

1986) not accrued by other, nonmodified individuals within the same population. Costs are in terms of resource exploitation and therefore the rate of population growth, since interferers spend time and energy in acts of interference that could otherwise be devoted to exploitation. Moreover the morphological, physiological, and behavioral modifications usually necessary for effective interference may reduce further the efficiency of exploitation. Finally, acts of interference can lead to retaliation, injury and death.

In a population with open spatial structure (no territoriality, competitors and/or resources freely mix, e.g., plankton or schooling fish), the cost of interference is borne by the perpetrator yet all benefits are shared (Case and Gilpin 1974; Rothstein 1979; Polis 1981, 1984b; Wilson 1986). This is illustrated (Fig. 5) by contrasting individual A (no interference behavior) with B (a member of the same population exhibiting interference with competitors [Cmp]). When B interferes with Cmp, both A and B benefit. Under these conditions, interference by B is altruistic (or spiteful): B pays the cost of interference, but the decrease in competition is shared equally by all individuals in the population, including freeloaders such as A. This puts B and other interfering individuals at a selective disadvantage within their own population. Under these conditions, resource-related interference is not evolutionarily stable and could only evolve at the level of kin or groups.

Interference will evolve only when the aggressor specifically benefits relative to other individuals within its population. This occurs in two ways. First, in most cases, interfering individuals operate in spatially structured populations so that they retain the greater share of the profit (freed resources – the direct benefit) from their acts of interference. This is the case for organisms that are sessile, relatively immobile or occupy a home range or territory. Interference by C that monopolizes local resources will evolve regardless of A's strategy and is easily explained by individual selection. Both A and C reap the indirect benefits (fewer competitors). Second, acts of interference that supply immediate net energy gains to the perpetrator (individual D, Fig. 5) will evolve regardless of population structure in space. IGP and cannibalism are unique classes of interference whereby the perpetrator benefits in three ways: the direct benefit of freed resources, the direct energetic benefit from feeding, and the indirect benefit of decreasing the population of potential competitors. Thus IGP and/or cannibalism by D can evolve in open populations independent of A's freeloading strategy without the previous restrictions of population structure or resource monopolization.

These restrictions may elucidate why interference is not universally practiced by large entities. The evolution of nonpredatory interference as a response to resource competition depends on the spatial structure of the population and will not be favored if entities cannot monopolize resources (as in freely mixing populations or if resources are not packaged in defendable units). This is one explanation why EC occasionally is the only interaction among differently sized entities (alternately, entities physically may be unable to interfere.) The evolution of cannibalism and IGP is less restrictive; these behaviors can evolve with or without resource monopolization. Consequently, predatory interference should be a frequent coevolutionary responses to EC and the only aggressive response possible in open populations. Although being a predator (as are scorpions) is a preadaptation to cannibalism and intraguild predation, it is not crucial to the evolution of predatory interference. In fact, cannibalism and IGP are common interactions, widespread among predaceous and normally non-predaceous taxa (Fox 1975; Polis 1981, 1984b; Polis et al. 1989). The multiple benefits of cannibalism and IGP further facilitate the evolution of these behaviors (e.g., energy gain, direct reduction in the number of competitors, occasional resource monopolization, and [in wide size range species with mutual IGP such as scorpions] direct reduction in the number of potential future predators on self and offspring).

7 Conclusions

Desert scorpions were used to analyze the interplay between exploitation competition (= EC) and the evolution of interference in size/age structured populations. The analysis focused on the similarities of interactions by differently sized age classes within and among species. Generalizing from these data, I would like to suggest the following:

1. The potential intensity of EC and interference between differently sized entities can vary greatly during growth in "wide size range" species. In general, as the size ratio (large/small) decreases, resource overlap (and, in some cases, potential EC) increases whereas the ease and safety of interference by the larger entity decreases. Interference usually is directed against individuals that are relatively small and/or defenseless.

2. However, given that the demand for resources exceeds the supply, the exact intensity of EC is a function of the relative density, population biomass, and metabolism of the smaller and larger competitors. Small entities are usually more dense and have a higher weight-specific metabolism than their larger counterparts. One or a combination of many small entities often use more energy than larger entities. In at least some of these cases, small entities can cause greater competitive damage than large entities do either to themselves or to the smaller entity. Thus small entities potentially exert considerable exploitative pressure on larger competitors.

3. Deleterious competitive effects from smaller entities establish a selective regime whereby interference may be favored as a response to EC. This is one explanation why size related interference is widespread and common among both conspecifics and heterospecifics.

4. Interference can be a potent force determining temporal, spatial and even trophic patterns of smaller age classes and species. Such patterns reflect avoidance of larger aggressive dominants and are often similar to patterns of "resource partitioning" hypothetically caused by EC. Interference also can decrease the abundance of smaller entities. Thus interference is capable of decreasing resource overlap, density and consequent harmful exploitative effects of smaller competitors.

5. Cannibalism and intraguild predation (= IGP) further decrease EC by directly reducing the density of smaller competitors. When directed against small individuals that eventually grow to an equal or larger size, predatory interference by larger entities will reduce the numbers of future competitors (and in some cases, future predators).

6. The spatial structure of populations is exceedingly important in the evolution of interference. Because interference entails a series of costs, non-lethal interference towards competitors can evolve only if local resources can be monopolized by the aggressor. In open populations without the possibility of resource monopolization, interference behavior cannot evolve at the level of the individual to reduce EC; such resource related interference is altruistic or spiteful and can only evolve via group/kin selection. Cannibalism and IGP are different from all other forms of interference: these behaviors provide direct nutritional benefits that allow them to evolve in spatially open populations. The problem of interference in closed versus open populations suggests one explanation why interference by larger entities is not universal and why EC occurs among some potential competitors.

Acknowledgments. I would like to thank Susan Brawley, Bill Eickmeier, Dave McCauley, Bill Neill, Lennart Persson, Henry Wilbur, and Tsunemi Yamashita for discussion and suggestions that improved this paper. I am especially grateful to Sharon Lee Polis and Tsunemi Yamashita for the help they provided in all aspects of this paper. The University Research Council and the Natural Science Committee of Vanderbilt University provided funds for this research. Finally, I salute the many friends whose efforts throughout the years has facilitated my research and made the field work enjoyable.

8 References

Abrams P (1980) Consumer functional response and competition in consumer-resource systems. Theor Popul Biol 17:80–102

Alstad D (1982) Current speed and filtration rate link caddisfly phylogeny and distributional patterns on a stream gradient. Science 216:533–534

Askew R (1971) Parasitic insects. American Elsevier, New York

Beddington JR (1975) Mutual interference between parasites or predators and the effect on searching efficiency. J Anim Ecol 44:331–340

Bengtsson J (1987) Smaller zooplankton species are not superior in exploitative competition: a comment on Persson. Am Nat 129:928–931

Brown JH, Maurer BA (1986) Body size, ecological dominance and Cope's rule. Nature (London) 324:248–250

Buss L (1986) Competition and community organization on hard surfaces in the sea. In: Diamond J, Case T (eds) Community Ecology. Harper & Row, New York, pp 517–536

Calder WA (1984) Size, function, and life history. Harvard Univ Press, Cambridge, Mass

Carothers J, Jaksic F (1984) Time as a niche difference: the role of interference competition. Oikos 42:403–406

Case T, Gilpin ME (1974) Interference competition and niche theory. Proc Natl Acad Sci (USA) 71:3073–3077

Connell J (1983) On the prevalence and relative importance of interspecific competition: evidence from field experiments. Am Nat 122:661–696

Damuth J (1981) Population density and body size in mammals. Nature (London) 290:699–700

Fox L (1975) Cannibalism in natural populations. Annu Rev Ecol Syst 6:87–106

Gauthroux SA (1978) The ecological significance of behavioural dominance. In: Bateson P, Klopper P (eds) Perspectives in ethology. Plenum, New York, pp 17–54

Gill D (1974) Intrinsic rate of increase, saturation density, and competitive ability. II. The evolution of competitive ability. Am Nat 108:103–116

Hespenheide H (1973) Ecological inferences form morphological data. Annu Rev Ecol Syst 4:213–229

Jeffries M, Lawton J (1984) Enemy free space and the structure of ecological communities. Biol J Linn Soc 23:269–286

May R (1975) Some notes on estimating the competition matrix. Ecology 56:737–741

Morse D (1974) Niche breadth as a function of social dominance. Am Nat 108:818–830

Morse D (1980) Behavioral mechanisms in ecology. Harvard Univ Press, Cambridge, Mass

Nentwig W, Wissel C (1986) A comparison of prey lengths among spiders. Oecologia (Berlin): 595–600

Oksanen L, Fretwell S, Järvinen O (1979) Interspecific aggression and the limiting similarity of close competitors: the problem of size gaps in some community arrays. Am Nat 114:117–129

Pacala S, Roughgarden J (1982) Resource partitioning and interspecific competition in two-species insular *Anolis* lizard communities. Science 217:444–446

Persson L (1985) Asymmetric competition: are larger animals competitively superior? Am Nat 126:261–266

Peters RH (1983) The ecological implications of body size. Univ Press, Cambridge

Peters RH, Wassenberg K (1983) The effect of body size on animal abundance. Oecologia (Berlin) 60:89–96

Polis GA (1979) Diet and prey phenology of the desert scorpion, *Paruroctonus mesaensis* Stahnke. J Zool (London) 188:333–346

Polis GA (1980) Seasonal and age specific variation in the surface activity of a population of desert scorpions in relation to environmental factors. J Anim Ecol 49:1–18

Polis GA (1980) The significance of cannibalism on the populations dynamics and surface activity of a natural population of desert scorpions. Behav Ecol Sociobiol 7:25–35

Polis GA (1981) The evolution and dynamics of intraspecific predation. Ann Rev Ecol Syst 12:225–251

Polis GA (1984a) Age structure component of niche width and intraspecific resource partitioning: can age groups function as ecological species? Am Nat 123:541–564

Polis GA (1984b) Intraspecific predation and "infant killing" among invertebrates. In: Hausfater G, Hrdy S (eds) Infanticide: Comparative and Evolutionary Perspectives. Alsinde, New York, pp 87–104

Polis GA (1988) Trophic and behavioral response of desert scorpions to harsh environmental periods. J Arid Environ 14:123–134

Polis GA, Farley RD (1979) Characteristics and environmental determinants of natality, growth and maturity in a natural population of the desert scorpion *Paruroctonus mesaensis.* J Zool (London) 187:517–542

Polis GA, Farley RD (1980) Population biology of a desert scorpion: survivorship, microhabitat and the evolution of life history strategy. Ecology 61:620–629

Polis GA, McCormick SJ (1982) Arthropods that prey on vertebrates. Biol Rev 57:29–58

Polis GA, McCormick SJ (1986a) Patterns of resource use and age structure among species of desert scorpion. J Anim Ecol 55:59–73

Polis GA, McCormick SJ (1986b) Scorpions, spiders and solpugids: predation and competition among distantly related taxa. Oecologia (Berlin) 71:111–116

Polis GA, McCormick SJ (1987) Intraguild predation and competition among desert scorpions. Ecology 68:332–343

Polis GA, Yamashita T: Population energy budget of the desert scorpion *Paruroctonus mesaensis.* (in preparation)

Polis GA, McReynolds CN, Ford G (1985) Home range geometry of the desert scorpion *Paruroctonus mesaensis*. Oecologia (Berlin) 67:273–277

Polis GA, Myers CA, Holt R (1989) The ecology and evolution of intraguild predation: potential competitors that eat each other. Annu Rev Ecol Syst 20

Rothstein S (1979) Gene frequencies and the selection for inhibitory traits, with special emphasis on the adaptiveness of territoriality. Am Nat 113:317–331

Roughgarden J (1979) Theory of population genetics and evolutionary ecology: an introduction. Macmillan, New York

Schmidt-Nielsen K (1984) Why is animal size so important? Univ Press, Cambridge

Schoener TW (1968) The *Anolis* lizards of Bimini: resource partitioning in a complex fauna. Ecology 49:704–726

Schoener TW (1976) Alternatives to Lotka-Volterra Competition: models of intermediate complexity. Theor Popul Biol 10:309–333

Schoener TW (1983) Field experiments on interspecific competition. Am Nat 122:240–285

Shorthouse D (1971) Studies on the biology and energetics of the scorpion *Urodacus yaschenkoi* (Birula 1904). Thesis, Aust Nat Univ, Canberra

Simberloff D (1982) The status of competition theory in ecology. Ann Zool Fennici 19:241–253

Stiling P, Strong D (1983) Weak competition among *Spartina* stem borers, by means of murder. Ecology 64:770–778

Strong D (1983) Natural variability and the manifold mechanisms of ecological communities. Am Nat 122:636–660

Sutherland W, Parker G (1985) Distribution of unequal competitors. In: Sibley R, Smith R (eds) Behavioural ecology: ecological consequences of adaptive behaviour. Blackwell, Oxford, pp 255–273

Turner M (1979) Diet and feeding phenology of the green lynx spider, *Peucetia viridans* (Araneae: Oxyopidae). J Arachn 7:149–154

Vezina A (1985) Empirical relationships between predator and prey size among terrestrial vertebrate predators. Oecologia (Berlin) 67:555–565

Wilson DS (1975) The adequacy of body size as a niche difference. Am Nat 109:769–784

Wilson DS (1986) Adaptive indirect effects. In: Diamond J, Case T (eds) Community ecology. Harper & Row, New York, pp 437–446

Asymmetries in Competitive and Predatory Interactions in Fish Populations

L. PERSSON

1 Introduction

A fundamental feature of fish and many other taxa is their potential for individuals to grow during most of their life. While increasing in size, fish undergo size-specific changes in niche, generally involving an increase in the mean size of prey eaten. Considering foraging efficiency alone, these ontogenetic niche shifts can be related to morphological changes and increasing energetic costs. The former is a result of an increased capacity to capture larger prey and a loss in the ability to capture and retain smaller prey, while the latter relates to the necessity to switch to increasingly larger prey to compensate for increased energetic demands with increasing body size (Kerr 1971; Mittelbach 1981; Persson 1987a). Since individual growth rates in fish are indeterminant and dependent on resource supply, resource competition will directly affect ontogenetic niche shifts by depressing growth rates. In most cases, however, competition and predation interact and ontogenetic niche shifts in fishes are therefore often a result of both competitive and predatory processes (Gilliam 1982; Werner and Gilliam 1984; Werner 1986; Mittelbach and Chesson 1987; Gilliam and Fraser this Vol.). An illustrative example of how competitive interactions interact with predation is competition-induced reduction in growth rate, which will prolong the time the fish is susceptible to gap-limited predators (Tonn et al. 1986). The strength of competitive and predatory interactions respectively may also change ontogenetically such that while competitive interactions are prevailing between certain size (age) classes, predatory interactions are prevailing between others.

The recognition of complex interactions between predation and competition, related to the presence of size structure in fish populations has a long history in fish ecology. Larkin (1956) suggested that the most important effects of competition at the community level were actually related to the interaction with predation. An often-cited example of this complexity in interactions is the invasion of the redside shiner (*Richardsonius balteatus*) into a lake in North America previously inhabited by rainbow trout (*Salmo gairdneri*). Because adult rainbow trout fed on the redside shiner, their growth rates increased after the invasion. However, the growth rate of small rainbow trout declined due to competition with the redside shiner, and as an overall result the rainbow trout population decreased (Larkin and Smith 1954; Johannes and Larkin 1961).

Although the complex interplay between predation and competition in fish communities has been recognized for a long time, the conceptualization of these

Department of Ecology, Limnology, P.O. Box 65, 22100 Lund, Sweden

Size-Structured Populations
B. Ebenman and L. Persson (Eds.)

complexities into a general and theoretical framework is rather recent (Gilliam 1982; Werner and Gilliam 1984; Werner 1986). This conceptualization has focused on two major aspects: first, mixed competition-predation can be framed in the context of predicting ontogenetic niche shifts given a certain growth rate/mortality rate trade-off (Werner and Gilliam 1984; Werner 1986; Gilliam and Fraser 1987 this Vol.; Werner this Vol.) and second, and perhaps less explored, competitive juvenile bottlenecks. Competitive juvenile bottlenecks refer to the effects of interspecific competition at the juvenile stage on the dynamics of populations (Neill 1975). A common interaction among fish populations is for the larger species to suffer from interspecific competition at the juvenile stage, but to become piscivorous as adults and prey on the smaller species. As a consequence, the interaction between such a species pair is characterized by a high degree of asymmetry.

In this chapter I will provide field experimental evidence for how interspecific competition from a zooplanktivorous species may result in a competitive juvenile bottleneck in the ontogeny of a piscivorous species. The data presented is a review and synthesis of experiments and field studies carried out during the last 7 years (Persson 1983 a, b, c, 1986, 1987 a, b, c; Persson and Greenberg 1989). I will also examine how the interactions between this species pair appear to depend on the productivity of the system such that the relative strength of competition versus predation changes along the productivity gradient. These shifts in competitive intensity along a productivity gradient have implications for processes at the ecosystem level.

2 Competitive Bottlenecks and Mixed Competition/Predation Intraspecifically

The interplay between competitive bottlenecks and predation is usually thought of in the context of interspecific interactions. However, complex interactions of this kind also exist at the intraspecific level. Perhaps the most illustrative example of interactions between competition and predation intraspecifically is Alm's (1946, 1952, 1959) classical studies on slow-growing "stunted" perch (*Perca fluviatilis*) populations. In these populations intense competition between age (size) classes resulted in a very low individual growth rate and a population tending towards small individuals. This can be related to a higher competitive ability of smaller individuals versus larger as a result of the former's lower per-capita metabolic requirements and that the advantage of a larger size with respect to foraging rate was diminished due to prey depression and overall small mean size of the prey (Mittelbach 1983; Persson 1985, 1987 a). On the other hand, larger perch had an advantage in that they could cannibalize young-of-the-year perch, thus counteracting the effects of competition. These perch populations exhibited an interesting dynamics; when adult perch were abundant, cannibalism prevented recruitment for several years. During these years the population was characterized by a more or less constant mean individual size as a result of high resource limitation. When adult mortality (which was low due to absence of other piscivores) had reduced the adult population below a certain level, recruits appeared and this decreased mean individual size. Following this, the mean size again increased over

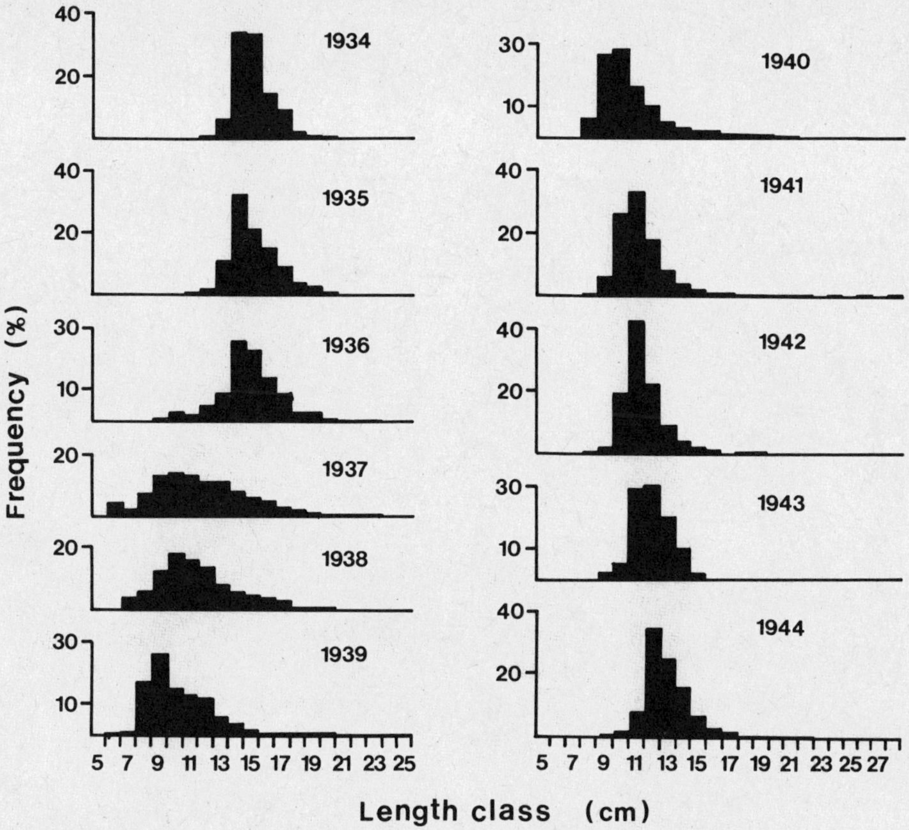

Fig. 1. Change in size structure of a slow-growing perch population over years. Heavy fishing effort was applied during the years 1934–1936 to reduce biomass of adult perch. (After Alm 1952)

time, related to growth of the recruits, and their cannibalism on new recruits again prevented recruitment. However, the increase in mean size was limited, constrained by the low growth rates. To test the prediction that cannibalism prevented recruitment of young perch, Alm (1952) experimentally reduced the number of adult perch in a population that at the time showed limited recruitment. As predicted, a reduction in the adult population size resulted in the appearance of new recruits (Fig. 1). Moreover, after a few years this age class had reached a size where cannibalism totally prevented the appearance of new age classes. The two extreme mean individual sizes exhibited by these perch populations, and the pending between these mean sizes over time, is thus a nice documentation of mixed competition-predation at the intraspecific level.

3 Juvenile Competitive Bottlenecks in Interspecific Competition

A juvenile competitive bottleneck (induced or magnified by interspecific competition) can be defined as the situation where juvenile competition results in pro-

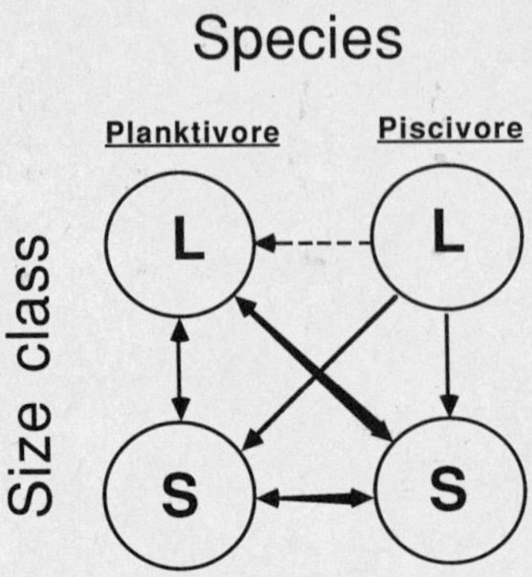

Fig. 2. Suggested size-specific (*S* small, *L* Large) interactions between a planktivorous fish species (*left*) and a fish species undergoing substantial ontogenetic niche shifts leading to piscivory if becoming large (*right*). ←→ indicates competitive and → predatory interaction. Different thicknesses along competition arrows indicate asymmetry in interaction

found effects on population dynamics or population size structure (Johansson and Persson 1986; Persson 1987a). Species that undergo large ontogenetic shifts in prey size may commonly exhibit juvenile competitive bottlenecks, since natural selection operates on morphologies and other traits to cope with a broad spectrum of resources (Werner and Gilliam 1984). As a consequence, trade-offs in foraging efficiency are likely to cause juveniles of this species to be less efficient when feeding on small prey than are the young of other species undergoing less dramatic ontogenetic shifts in resource use (Werner and Hall 1979; Werner 1986; Persson and Greenberg 1989). Experimental demonstrations of such trade-offs have been documented in at least two cases, both of which involved situations where the larger species becomes piscivorous as adults (Werner 1977; Persson 1987c). Thus, as a consequence of its competitive superiority for smaller prey types, the smaller species may limit the recruitment of individuals of the larger species to piscivorous classes. On the other hand, predation from piscivorous classes of the larger species may depress the population size of the smaller species, counteracting the limitations set by the interspecifically induced juvenile bottleneck.

An asymmetry in the competitive and predatory interactions between species may therefore occur with one suggested species becoming piscivorous and mainly affecting the other species through predation, while the effect of the smaller species on the larger is mainly through competition (Fig. 2). This system was in a simplified version (i.e., size-structure excluded) treated by Levins (1979). Interestingly, he predicted that an increase in the abundance of the species being only a competitor could either increase or decrease the abundance of the species being both predator and competitor, depending on the relative strength of competition versus predation. In the following sections I will explore such an asymmetric interaction between two fish species, perch and roach (*Rutilus rutilus*).

4 Resource Use and Population Structure

Perch and roach are quite dissimilar in morphology, are widely distributed over Eurasia and are in many lake systems the two numerically dominant species.

Roach belongs to the family Cyprinidae (order Cypriniformes), a family successful with respect to dispersion and local abundance everywhere in Eurasia. This family is among other things characterized by having their alimentary canal evolutionarily reduced to a long intestine lacking acid and pepsin-secreting cells (Kapoor et al. 1975). Perch, on the other hand, belongs to the family Percidae (order Perciformes) and possess a well-developed stomach with acid and pepsin secreting cells (Fish 1960).

Corresponding to these basic anatomical differences, the diet of roach and perch differ considerably; most importantly the roach is omnivorous, while the perch is strictly carnivorous. Roach, which attains a smaller adult size than perch, feeds on a mixture of microcrustaceans (zooplankton), macroinvertebrates and plant material (Persson 1983b with references), and exhibits an ontogenetic shift in diet from feeding on zooplankton at young stages to omnivory or herbivory/detrivory at older stages (Prejs 1984; Johansson and Persson 1986). Hofer and Wieser (1987) also showed that roach larvae begin life feeding on phytoplankton, then switch to rotifers and microcrustaceans. The capacity of roach to feed on plant material is important because under conditions of high resource limitation, plant material may serve as a competitive refuge from carnivorous species such as perch (Persson 1983b; Johansson and Persson 1986; Persson 1987c). Blue-green algae common in highly productive lakes are most nutritious to roach. Thus, both assimilation efficiency and growth rates are higher when roach are fed bluegreen algae versus other algae and higher plants (Sorokin 1968; Lessmark 1983).

Perch potentially undergo three major ontogenetic shifts in diet. As young they feed on zooplankton, thereafter they switch to macroinvertebrates, and as adults potentially become piscivorous. The shift from zooplanktivory to feeding on macroinvertebrates is associated with a change in habitat use from pelagic to benthic habitat (Persson 1986), suggesting that the shift in food resources used is relatively discrete (Werner and Hall 1979; Mittelbach 1981). The length of the zooplanktivorous stage in perch is negatively correlated with the density of roach. In lakes with no roach, or very few roach, yearlings of perch may feed almost exclusively on zooplankton (Gumáa 1977; Craig 1978; Nyberg 1979) and zooplankton may form a significant portion of the diet even for large perch (Thorpe 1977). In contrast, in highly productive lakes with dense populations of roach, yearlings of perch switch to macroinvertebrates at a small size and young age (Persson 1983a, 1986).

The macroinvertebrate feeding phase has been identified as competitive bottleneck in perch populations; juveniles that switch to this feeding stage early and at small sizes show retarded growth and very few ever turn to piscivory. Perch populations where individuals switch early to macroinvertebrates are thus dominated by small nonpiscivorous individuals. In contrast, in populations where perch shift to feeding on macroinvertebrates at a late stage, this feeding phase is relatively transitory and a shift of piscivory occurs early (Persson 1986; Johansson and Persson 1986). These perch populations are consequently dominated by large piscivorous perch.

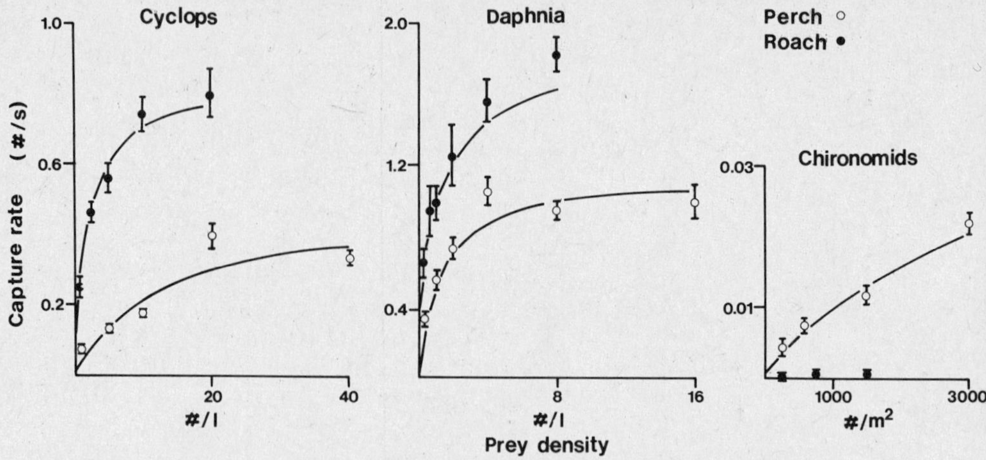

Fig. 3. Foraging efficiency (prey captured/s) of perch and roach on two zooplankton (*Cyclops* 1.5 mm, *Daphnia* 1.3 mm) and one macroinvertebrate prey (chironomids 8.2 mm) in relation to prey density. (Data from Persson 1987a, c, Johansson 1987)

5 Evidence for Asymmetry in Competitive and Predatory Interactions

Competition between roach and perch has been suggested in several field studies (Sumari 1971; Svärdson 1976; Stenson 1979; Persson 1983a, b, c; Lessmark 1983) but has only recently been experimentally demonstrated (Persson 1986; 1987b, c; Persson and Greenberg 1989). That roach and perch compete in highly productive systems, in spite of the dissimilar resource utilization of the species, is due to the fact that the low per-capita effect of roach is offset by their much higher biomass compared to perch (density relationship 10:1) (Persson 1983c, Johansson and Persson 1986). Furthermore, the dissimilar diet observed in natural systems is a result of competition-induced interactive segregation rather than a lack of competition (Johansson 1987; Persson and Greenberg 1989).

Laboratory studies have shown that roach is a more efficient zooplankton feeder than perch (Persson 1987c) (Fig. 3) and even large roarch utilize sizes of zooplankton prey down to 0.4 mm which are used only be perch <10 cm (Lessmark 1983). The superiority of roach when feeding on zooplankton is supported by field experiments showing that roach depress zooplankton densities to much lower levels than perch (Andersson 1984; Persson 1987c). These results suggest that adaptations which allow perch to become piscivorous as adults involve costs in terms of foraging efficiency on zooplankton when young. The inferior foraging efficiency of perch compared to roach when feeding on zooplankton is also apparent in the different search behaviors of the two species. Whereas perch swim slowly (4–5 cm/s) and attack every prey with a burst better suited for attacking large evasible prey, roach swim at speeds ten times that of perch (50 cm/s) and more snap than attack each single prey (Johansson and Persson 1986; Johansson 1987; Persson 1987c).

Laboratory experiments have shown that the perch is a more efficient predator on macroinvertebrates than the roach, suggesting that the main competitive effect of roach occurs at the zooplankton feeding stage (Persson 1987c). The superiority of perch when feeding on macroinvertebrates should have a minor effect on roach as the foraging rate of roach when feeding on chironomids is independent of prey density and overall very low (Fig. 3). This expectation has been supported by field experiments (Persson 1987c). To summarize, both laboratory and field experiments suggest that there exists a strong asymmetry in the competitive relationship between roach and perch and that the impact of roach of perch is much stronger than the reverse. This competitive asymmetry is also reinforced by the fact that roach can switch to a competitive refuge – plant material – not utilized by carnivorous perch.

While the asymmetry in the competitive relationship between perch and roach is a result of several factors, the asymmetry with respect to predator-prey interactions is more simple. Large perch are predators on both roach and perch, whereas roach do not switch to piscivory at all. Perch are gap-limited predators and prey largely on small immature fish (Popova and Sytina 1977; Willemsen 1977), with result that the large roach are not very susceptible to perch predation.

6 Effects of Asymmetry on Competitive Relationship

A mechanistic explanation for how roach affect perch competitively can be advanced based on the above information. Due to its high foraging efficiency on zooplankton, an increase in the density of roach should lead to a reduction in zooplankton availability. This in turn should cause the size and age at which young perch switch from feeding on zooplankton to feeding on the next larger prey, macroinvertebrates, to be advanced. Because microcrustaceans and macroinvertebrates inhabit different habitats (pelagic versus benthic), this feeding shift has also been anticipated to involve a habitat shift (from pelagic to benthic) (Fig. 4). As a consequence of the advanced shift in resource use, intraspecific competition in perch with size (age) classes feeding on macroinvertebrates should increase with a resulting retarded growth rate. This, in turn, has been expected to cause

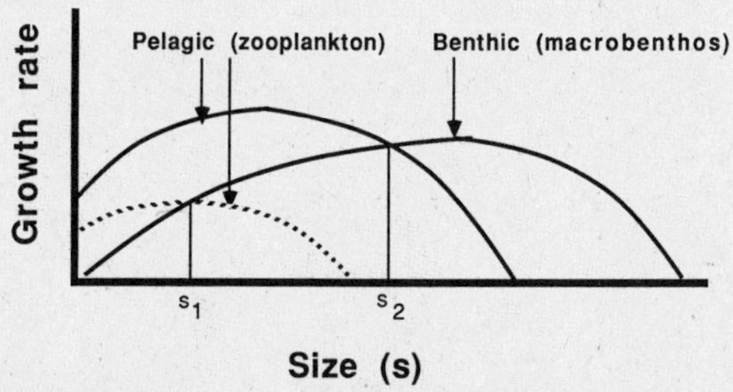

Fig. 4. Expected effect of roach on the ontogenetic niche shift in perch from pelagic (zooplankton feeding) to benthic (macroinvertebrate feeding) habitat. *Solid lines* represent the growth rates of perch in the two habitats in relation to body size in the absence of roach. *Broken line* is the growth rate of perch in the pelagic habitat in relation to body size in the presence of roach. s_1 denotes the optimal size for shift from pelagic to benthic habitat in the presence of roach and s_2 in the absence. (After Werner and Gilliam 1984)

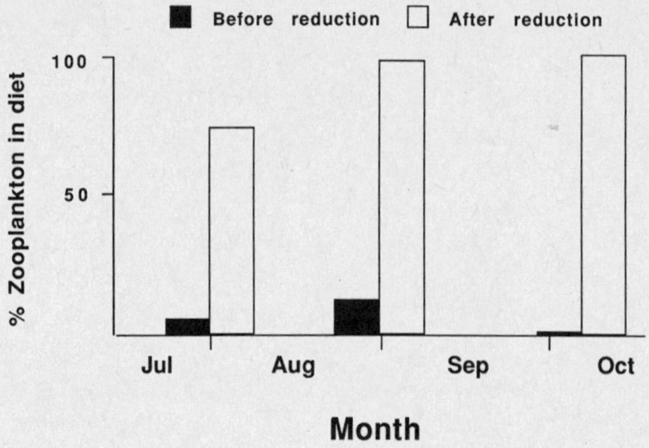

Fig. 5. Percentage zooplankton in the diet of young-of-the-year (0+) perch on different occasions before and after a reduction in the population size of roach in a whole lake experiment. (Data from Persson 1986)

a delay in time and a decrease in magnitude with which perch are recruited to the piscivorous stage. The impact of roach prolonging the time it takes for perch to reach piscivorous size can thus be suggested to be a result of two factors: (1) a direct effect on zooplankton feeding perch and (2) an indirect effect on macroinvertebrate feeding perch by advancing the switch from zooplankton to macroinvertebrate feeding in juvenile perch.

A whole lake experiment supports the interpretation. After a reduction in the population size of roach by more than 70%, zooplankton abundance increased on average four times (Persson 1986). As a result young-of-the-year (0+) perch continued to feed on zooplankton for both a longer time and to a larger size than before the removal experiment (Fig. 5). This change in resource use also involved a change in habitat use, because while 0+ perch moved into the shore (benthic) habitat in late summer before the reduction, they remained in the pelagic habitat after the reduction. The change in resource use of 0+ perch was associated with an increase in individual growth rate and population size. An increase in the growth rate of larger (older) macroinvertebrate feeding perch was also observed and as a consequence they started to shift to piscivory earlier than before the removal experiment. In a subsequent enclosure experiment in the pelagic habitat of the lake, the effect of roach competition on resource use and growth rate of 0+ perch was corroborated (Persson 1987b).

A stronger test of the hypothesis that competition from roach affects the time and size at which 0+ perch switch to macroinvertebrates was carried out in a pond experiment (Persson and Greenberg 1989). Ten pond enclosures were used and 0+ and 1+ (1-year-old) perch were stocked in the same densities in all enclosures. The density of roach was varied between 0 and 35 g/m^2: the higher abundance is similar to that found in highly productive lakes. Major effects of roach density on zooplankton abundance were observed and at the end of the experiment the main difference in zooplankton abundance was found between enclosures with roach present and those absent. As predicted, 0+ perch switched to feeding on macroinvertebrates at higher roach densities and the roach density at which this switch occurred decreased over the experimental period (Fig. 6). The decrease in proportion of zooplankton in the diet was rather abrupt, suggesting

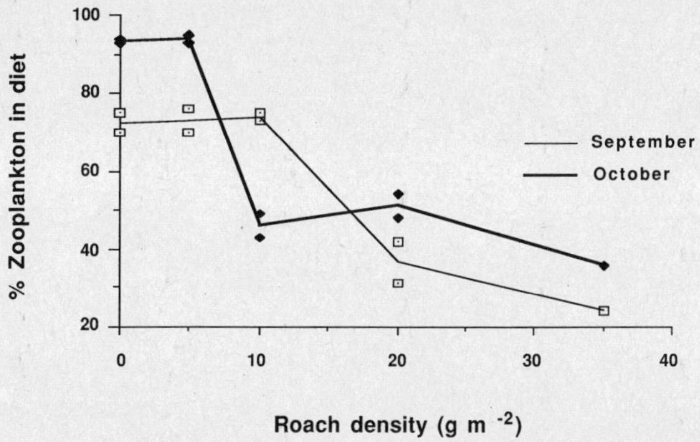

Fig. 6. Percentage zooplankton in the diet of young-of-the-year (0+) perch at the middle and the end of a pond experiment versus roach density. One replicate was present for each roach density except at the highest density. n = 9 on both sampling occasions. (Data from Persson and Greenberg 1989)

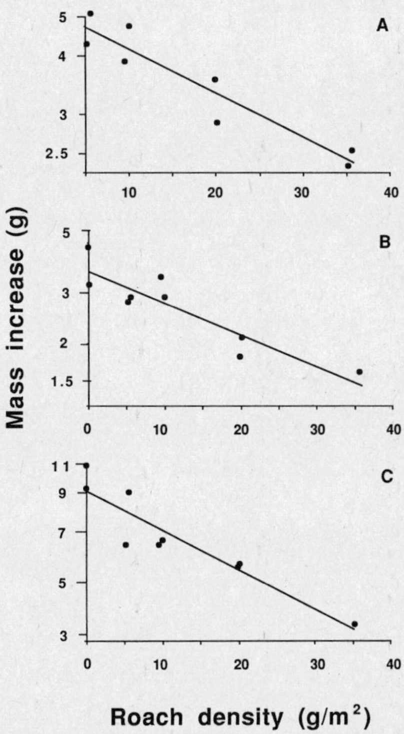

Fig. 7. Growth of roach (**A**), 0+ perch (**B**) and 1+ perch (**C**) as a function of roach density in the pond experiment. Note the different scale for roach. For additional information, see Fig. 6

a discrete switching in habitat even in this shallow system. This interpretation was also supported by the observed time lag between decrease in zooplankton abundance and changes in 0+ perch diet. Strong density-dependent effects of roach on both 0+ and 1+ perch growth rates were observed, supporting the hypothesis that the shift in diet by 0+ perch resulted in an increased competition with

1+perch (Fig. 7). The significance of plant material as a competitive refuge for roach was also supported by this study; the proportion of plant material in the diet of roach increased with increasing resource limitation (reduced growth rate) (Persson and Greenberg 1989).

The above experiments support the hyothesis that roach in highly productive systems reduce zooplankton densities and therefore increase size class competition in perch for macroinvertebrates. The increased size class competition results in a decreased individual growth rate in perch and very few perch are expected to survive to enter the piscivorous stage. These experiments thus give a mechanistic explanation for why perch populations are dominated by small nonpiscivorous individuals in highly productive lakes, where cyprinids are numerically dominant. The rarity by which perch enter the piscivorous stage in these systems is also affected by that decreased individual growth rates will prolong the time period perch is susceptible to predation by coexisting piscivores such as northern pike (*Esox lucius*) and zander (*Stizostedion lucioperca*). Hence, although competition is the cause of the slow individual growth rates, predation may act as a significant mortality agent, which prevents slow-growing perch individuals from living long enough to reach piscivorous size. Behavioral indirect effects of predation are also likely to exert impacts on the habitat distributions of different species/size classes, since these effects may be independent of the densities of the predators (Miller and Kerfoot 1987).

7 Effects of Asymmetry on Predation Relationship

Previous studies have demonstrated that the abundance relationship between perch and roach is dependent on the productivity of the system and in less productive systems perch is more common than roach (Svärdson 1976; Persson 1983c). This can be related to a number of factors which all act to decrease the competitive advantage of roach (Johansson and Persson 1986). With decreasing productivity, the size of the littoral increases relative to the pelagic (Wetzel 1979; Sand-Jensen 1980). The phytoplankton community will shift from a community dominated by bluegreens to a community dominated by other forms which are less nutritious to roach. The increase of the littoral habitat increases the abundance of submerged macrophyte-attached macroinvertebrates on which perch have a higher foraging efficiency than roach. It has also been shown that an increase in the structural complexity per se, as a result of increased abundance of submerged macrophytes, will favor perch relative to cyprinids such as roach (Diehl 1988). These changes, which will reduce the limitations set by the juvenile bottleneck in perch populations, can be classified into two principial factors: (1) an increase in availability of those resources that, when limiting, cause the juvenile bottleneck and (2) decreased impact of interspecific competition from cyprinids affecting the size and age at which juvenile perch switch to macroinvertebrate feeding. These two factors also interact such that an increase in cyprinids will have indirect effects on the availability of littoral resources (Hansson et al. 1987).

The competitive impact of perch on roach is also expected to be low in low productive systems. The available data suggest that the foraging rate of roach on

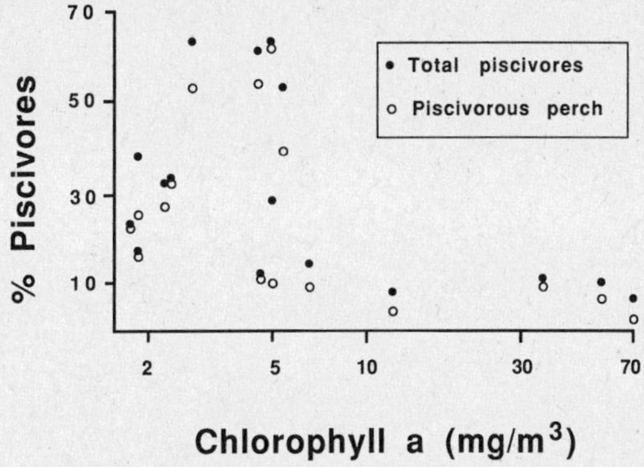

Fig. 8. Percentage piscivorous biomass of total fish biomass in relation to phytoplankton biomass (chlorophyll a) in south Swedish lakes. ● is total piscivorous biomass and ○ is piscivorous perch only. (Data from Persson et al. 1988 and unpublished)

littoral prey is independent of prey density and that, although blue-greens are less common in low productive system, plant material still offers a competitive refuge for roach (Persson 1987c). One simple explanation for the observed abundance relationship between roach and perch in low productive systems is the hypothesis that only weak interactions occur between the species. In this case the abundances of the two species are mere reflections of the availability of different resources. However, the change in the abundance relationship between roach and perch along the productivity gradient also involves a change in the size distribution of the perch populations (Fig. 8) (Persson 1983a; Persson et al. 1988). Thus, an alternative hypothesis based on asymmetric predation may also be advanced. This hypothesis predicts the interaction between roach and perch to change from a mainly competitive interaction in highly productive systems to a mainly predatory interaction (with perch predation limiting roach density) in low productive systems. A support for this hypothesis already exists in the observed change in size distribution of roach populations along the productivity gradient; in highly productive systems the median size of roach is lower than in low productive systems (Persson et al. 1988). The evidence for piscivorous perch affecting roach density in low productive systems is, however, incomplete and under present investigation. In addition, the hypothesis that the competitive advantage to roach in highly productive systems shifts to a predatory advantage to perch in low productive systems must be treated with caution because the shift in abundance relationship may also involve indirect effects. One possible example of this is that along the productivity gradient changes in abundance of alternative forage fish for perch may occur, forage fish that at the same time are competing with roach.

The change in the size structure of perch populations along a productivity gradient (i.e., proportion of piscivorous perch) is expected to have major effects on processes at the ecosystem level. The change in biomass of total piscivores is largely a reflection of changes in biomass of piscivorous perch (Fig. 8). Perch can thus be viewed as a potentially key species, hence understanding th dynamics of perch populations can greatly clarify aquatic ecosystem processes. In very low productive systems (lower than those considered here), the proportion of piscivorous perch is low and increases with productivity to a maximum at a still rather low productivity (Fig. 8). This part of the data suggests that the biomass of piscivores

(4th trophic level) increases with primary productivity, which (although the biomass of piscivores is never zero) partly supports the hypothesis of Oksanen et al. (1981; see also Fretwell 1977). However, the decrease in proportional piscivore biomass in highly productive systems is not predicted by this theory, with the most probable explanation being that the model ignores that the same individual may shift trophic level ontogenetically (see Mittelbach et al. this Vol.). The importance of including size-structure when studying processes at the ecosystem level was considered by Riemann et al. (1986; see also Persson et al. 1988). Riemann et al. (1986) did, for example, show that the change in proportion of piscivorous perch along the productivity gradient had profound effects on the production and biomass of other components of the ecosystem such as zooplankton, phytoplankton and bacteria.

To summarize, the change in the abundance relationship between perch and roach along the productivity gradient appears to be the result of the combined effects of competition and predation. The predation asymmetry is at present less substantially supported than the competitive asymmetry. Still, it can be suggested that the competitive asymmetry is dominating in highly productive systems while the predatory asymmetry is dominating in less productive systems. Because perch is by far the most abundant piscivore in the systems studied, the productivity-related change in the asymmetry of interactions between these species also has implications on processes at the ecosystem level.

8 Asymmetries in Fish Communities in General

I have outlined an asymmetry in the interactions between two species related to the presence of size structure in fish populations. How common and general is this type of asymmetry in fish communities? Because so many studies have focused on either competition or predation, it is to some extent premature to answer this question. Still, Werner (1986) cites a number of studies in which the type of interaction discussed above has been demonstrated or suggested. Besides these studies, the classical interaction between charr (*Salvelinus alpinus*) and trout (*Salmo trutta*), (which mainly has been discussed as a competitive interaction, Nilsson 1960, 1965, 1967; Schoener 1982; Werner 1986), is actually an example of a mixed competition/predation and juvenile competitive bottleneck situation (Filipson and Svärdson 1976; Svärdson 1976). Whereas charr affect trout mainly competitively, large trout may prey on charr fingerlings. This asymmetry in interaction became evident due to unintended manipulations of trout-charr communities. High fishing pressure on trout in some lakes has caused a drastic decrease in the numbers of large predatory trout. As a consequence, the survival of small charr increased, leading to an intensified competition both between charr individuals and between charr and trout. This has resulted not only in an increase in the abundance of charr and a decrease in trout, but also in a reduced mean size of charr and in many cases stunted charr populations (Svärdson 1976).

The charr-trout example suggests that many mixed competition/predation interactions may be hidden in studies that have attributed the patterns observed to either competition or predation. Support for these mixed interactions is contained in Svärdson's (1976) review of species interactions in Scandinavian lakes. He showed that the most abundant species in lakes is generally the most zooplanktivorous species, which may be related to a high productivity of pelagic prey compared to benthic and littoral prey (Mittelbach and Chesson 1987; Persson 1987a). Due to its numerical dominance, this species is likely to have profound effects on piscivorous species undergoing ontogenetic niche shifts. A piscivorous species which utilizes zooplankton when young will thus be strongly susceptible to competition from planktivores. At the same time, it may be speculated that, due do the high abundance of pelagic prey in lakes, such a species may, under favourable conditions, be able to build up a higher biomass and hence have a more significant impact on ecosystem processes than piscivores relying to a smaller extent on this resource.

All the above studies suggesting asymmetries in interactions with respect to competition and predation have three things in common. Firstly, the species having a smaller mean size is usually a zooplanktivor, hence the asymmetric competitive interaction usually affects the juvenile size class of the larger species. Secondly, adults and juveniles of the larger species are affected differently by the small species in that growth rates of adults are increased while the opposite is the case for juveniles. Finally, the general effect of the smaller zooplanktivorous species is a reduction in abundance and mean size of the larger species related to increased intraspecific competition and reduced recruitment to piscivorous stages. These examples of mixed predation/competition interactions from fish communities have focused mainly on the competitive asymmetry. Besides the perch-roach interaction, there is to my knowledge no study that has suggested that the asymmetry in competitive and predatory interactions changes in a predictable way along an environmental gradient. Production-related and size-structure-dependent interactions have been demonstrated for invertebrate planktivorous organisms, showing other configurations than those presented in Fig. 2 (Neill 1981; Neill this Vol.; Neill and Peacock 1980).

I have outlined above how the strength of different components in an asymmetry may change between systems along an environmental gradient. A relevant question to ask is whether the asymmetry may vary in direction in the same lake in different years, for example due to recruitment variation. The data from highly productive systems suggests that the competitive advantage to roach is so predominant, that this is unlikely to occur. It can be hypothesized that such between year shifts in asymmetry are most likely in systems that are on the border between dominance of the predatory and competitive asymmetry respectively (see Fig. 8). The presence of such a shift in asymmetry within a system remains, however, to be demonstrated. Finally, the strong coupling between lake productivity and structure of the fish community presented here also lends some warning against drawing the consequences of nonequilibrium conditions in natural fish communities to far.

Acknowledgments. Over the years of studies I have benefited from collaboration and discussions with many permanent or temporary members of the Fish Ecology Group, University of Lund. Of these I would especially like to thank Gunnar Andersson, Lars Collvin, Larry Greenberg, Stellan Hamrin, Lars Johansson, Olof Lessmark and Ian Winfield. Special thanks to Gary Polis and Gary Mittelbach whose comments substantially improved the quality of this paper. The research has been supported by the Swedish Natural Science Research Council and the Swedish National Environmental Protection Board.

References

Alm G (1946) Reasons fo the occurrence of stunted fish populations with special reference to perch. Rep Inst Freshwater Res Drottningholm 25

Alm G (1952) Year class fluctuations and span of life of perch. Rep Inst Freshwater Res Drottningholm 33:17–38

Alm G (1959) Connection between maturity, size and age in fishes. Rep Inst Freshwater Res Drottningholm 40:5–145

Andersson G (1984) The role of fish in lake ecosystems – and in limnology. In: Bosheim S, Nicholls M (eds) Interaksjoner mellan trofiske nivåer i ferskvann, Nordisk limnologisymposium, Oslo, 1984, pp 189–197

Craig JF (1978) A study of the food and feeding of perch, *Perca fluviatilis* L. in Windermere. Freshwater Biol 8:59–68

Diehl S (1988) Foraging efficiency of three freshwater fish: Effects of structural complexity and light. Oikos 53:207–214

Filipsson O, Svärdson G (1976) Principer för fiskevården i rödingsjöar. Inform Inst Freshwat Res Drottningholm 2 (Mimeographed in Swedish with English summary)

Fish GR (1960) The comparative activity of some digestive enzymes in the alimentary canal of Tilapia and Perch. Hydrobiology 15:161–178

Fretwell SD (1977) The regulation of plant communities by the food chains exploiting them. Perspect Biol Med 20:169–185

Gilliam JF (1982) Habitat use and competitive bottlenecks in size-structured fish populations. Thesis, Michigan State Univ, East Lansing

Gilliam JF, Fraser DF (1987) Habitat selection when foraging under predation hazard: A model and a test with stream-dwelling minnows. Ecology 68:1856–1862

Gumaá SA (1978) The food and feeding habits of young perch, *Perca fluviatilis*, in Windermere. Freshwater Biol 8:177–187

Hansson L-A, Johansson L, Persson L (1987) Effects of fish grazing on nutrient release and succession of primary producers. Limnol Oceanogr 32:723–729

Hofer WMR, Wieser W (1987) Diet spectra and resource partitioning in the larvae and juveniles of three species and six cohorts of cyprinids from a subalpine lake. Oecologia (Berlin) 71:388–396

Johannes RD, Larkin PA (1961) Competition for food between redside shiners and rainbow trout in two British Columbia lakes. J Fish Res Board Can 18:203–220

Johansson L (1987) Experimental evidence for interactive habitat segregation between roach (*Rutilus rutilus*) and rudd (*Scardinius erythrophthalmus*). Oecologia (Berlin) 73:21–27

Johansson L, Persson L (1986) The fish community of temperate, eutrophic lakes. In: Riemann B, Søndergaard M (eds) Carbon dynamics of eutrophic, temperate lakes: The structure and functions of the pelagic environment. Elsevier, Amsterdam, pp 237–266

Kapoor BG, Smit H, Verighina IA (1975) The alimentary canal and digestion in teleosts. In: Russell F, Younge M (eds) Advances in marine biology, vol 13. Academic Press, New York London, pp 109–239

Kerr SR (1971) Analysis of laboratory experiments on growth efficiency of fishes. J Fish Res Board Can 28:801–808

Larkin PA (1956) Interspecific competition and population control in freshwater fish. J Fish Res Board Can 13:327–340

Larkin PA, Smith SB (1954) Some effects of introduction of the redside shiner on the Kamloops trout in Paul Lake, British Columbia. Trans Am Fish Soc 83:161–175

Lessmark O (1983) Competition between perch (*Perca fluviatilis*) and roach (*Rutilus rutilus*) in South Swedish lakes. Thesis, Univ Lund

Levins R (1979) Asymmetric competition among distant taxa. Am Zool 19:1097–1104

Miller TE, Kerfoot WC (1987) Redefining indirect effects. In: Kerfoot WC, Sih A (eds) Predation: direct and indirect impacts on aquatic communities. Univ Press New Engl, Hanover London, pp 33–37

Mittelbach GG (1981) Foraging efficiency and body size: a study of optimal diet and habitat use by bluegills. Ecology 62:1370–1386

Mittelbach GG (1983) Optimal foraging and growth in bluegills. Oecologia (Berlin) 59:157–162

Mittelbach GG, Chesson PL (1987) Predation risk: Indirect effects on fish populations. In: Kerfoot WC, Sih A (eds) Predation: direct and indirect impacts on aquatic communities. Univ Press New Engl, Hanover London, pp 315–332

Neill WE (1975) Experimental studies of microcrustacean competition, community composition and resource utilization. Ecology 56:809–826

Neill WE (1981) Impact of *Chaborous* predation on the structure and dynamics of crustacean zooplankton community. Oecologia (Berlin) 48:164–177

Neill WE, Peacock A (1980) Breaking the bottleneck: Interactions of invertebrate predators and nutrients in oligotrophic lakes. In: Kerfoot WC (ed) Evolution and ecology of zooplankton communities. Univ Press New Engl, Hanover, NH, pp 715–724

Nilsson N-A (1960) Seasonal fluctuations in the food segregation of trout, char and white fish in 14 North-Swedish lakes. Rep Inst Freshwater Res Drottningholm 41:185–205

Nilsson N-A (1965) Food segregation between salmonid species in North Sweden. Rep Inst Freshwater Res Drottningholm 46:58–73

Nilsson N-A (1965) Interactive segregation between fish species. In: Gerking SD (ed) The biological basis of freshwater fish production. Blackwell, Oxford, pp 296–313

Nyberg P (1976) Production and food consumption of perch in two Swedish forest lakes. Thesis, Univ Uppsala

Nyberg P (1979) Production and food consumption of perch, *Perca fluviatilis* L., in two Swedish forest lakes. Rep Inst Freshwater Res Drottningholm 58:140–157

Oksanen L, Fretwell SD, Arruda J, Niemelä P (1981) Exploitation ecosystems in gradients of primary productivity. Am Nat 118:240–261

Persson L (1983 a) Food consumption and competition between age classes in a perch *Perca fluviatilis* population in a shallow eutrophic lake. Oikos 40:197–207

Persson L (1983 b) Food consumption and the significance of detritus and algae to intraspecific competition in roach *Rutilus rutilus* in a shallow eutrophic lake. Oikos 41:118–125

Persson L (1983 c) Effects on intra- and interspecific competition on dynamics and size structure of a perch *Perca fluviatilis* and a roach *Rutilus rutilus* population. Oikos 41:126–132

Persson L (1985) Asymmetrical competition: Are larger animals competitively superior? Am Nat 126:261–266

Persson L (1986) Effects of reduced interspecific competition on resource utilization in perch (*Perca fluviatilis*). Ecology 67:355–364

Persson L (1987 a) The effects of resource availability and distribution on size class interactions in perch *Perca fluviatilis*. Oikos 48:148–160

Persson L (1987 b) Competition-induced diet shift in young-of-the-year perch (*Perca fluviatilis*): An experimental test of resource limitation. Environ Biol Fish 19:235–239

Persson L (1987 c) Effects of habitat and season on competitive interactions between roach (*Rutilus rutilus*) and perch (*Perca fluviatilis*). Oecologia (Berlin) 73:170–177

Persson L, Greenberg LA (1989) Competitive juvenile bottlenecks: The perch (*Perca fluviatilis*)-roach (*Rutilus rutilus*) interaction. Ecology (in press)

Persson L, Andersson G, Hamrin SF, Johansson L (1988) Predator regulation and primary productivity along the productivity gradient of temperate lake ecosystems. In: Carpenter SC (ed) Complex interactions in lake communities. Springer Verlag, New York Berlin Heidelberg

Popova OA, Sytina LA (1977) Food and feeding relations of Eurasian perch (*Perca fluviatilis*) and pikeperch (*Stizostedion lucioperca*) in various waters of the USSR. J Fish Res Board Can 34:1559–1570

Prejs A (1984) Herbivory by temperate freshwater fishes and its consequences. Environ Biol Fish 10:281–296

Rieman B, Søndergaard M, Persson L, Johansson L (1986) Carbon metabolism and community regulation in eutrophic, temperate lakes. In: Riemann B, Søndergaard M (eds) Carbon dynamics of eutrophic, temperate lakes: the structure and functions of the pelagic environment. Elsevier, Amsterdam, pp 267–280

Sand-Jensen K (1979) Balancen mellan autotrofe komponenter i tempererade søer med forskellig næringsbelastning. Vatten 2/80:104–115 (In Danish with English summary)
Schoener TW (1982) The controversy over interspecific competition. Am Sci 70:586–595
Sorokin JI (1968) The use of ^{14}C in the study of nutrition of aquatic animals. Mitt Int Verein Theor Angew Limnol 16
Sumari O (1971) Structure of perch populations in some ponds in Finland. Ann Zool Fennici 8:406–421
Svärdson G (1976) Interspecific population dominance in fish communities. Rep Inst Freshwater Inst Drottningholm 56:144–171
Thorpe JE (1977) Daily ration of adult perch, *Perca fluviatilis* L., in Loch Leven, Scotland. J Fish Biol 11:55–68
Tonn WM, Paszkowski CA, Moermond TC (1986) Competition in *Umbra-Perca* fish assemblages: experimental and field evidence. Oecologia (Berlin) 69:126–133
Werner EE (1977) Species packing and niche complementarity in three sunfishes. Am Nat 11:553–578
Werner EE (1986) Species interactions in freshwater fish communities. In: Diamond J, Case T (eds) Community ecology. Harper & Row, New York, pp 344–358
Werner EE, Hall DJ (1979) Foraging efficiency and habitat switching in competing sunfishes. Ecology 60:256–264
Werner EE, Gilliam JF (1984) The ontogenetic niche and species interactions in size-structured populations. Annu Rev Ecol Syst 15:393–425
Wetzel RG (1979) The role of the littoral zone and detritus in lake metabolism. Arch Hydrobiol Beih Ergebn Limnol 13:145–161
Willemsen J (1977) Population dynamics of percids in Lake Ijssel and some smaller lakes in the Netherlands. J Fish Res Board Can 34:1710–1719

Trophic Relations and Ontogenetic Niche Shifts in Aquatic Ecosystems

G. G. MITTELBACH, C. W. OSENBERG and M. A. LEIBOLD

1 Introduction

One of the most pronounced and consistent patterns in freshwater lakes is that the biomasses of producers and consumers increase along a gradient of increased nutrient (phosphorous) loading. For example, phytoplankton biomass is positively correlated with total phosphorus concentration (Stockner and Shortreed 1985 and references therein), as is the biomass of zooplankton (Hanson and Peters 1984; Pace 1984), and fish (Hanson and Leggett 1982). Given that the biomass of aquatic producers, herbivores, and carnivores each positively covaries with phosphorus loading, the abundances of adjacent trophic levels should also be positively correlated across productivity gradients. Indeed, such positive correlations between consumers and their resources are common; i.e., zooplankton biomass is positively correlated with phytoplankton biomass (references in McQueen et al. 1986), plantivore density is positively related to zooplankton biomass (Mills and Shiavone 1982), and fish density positively covaries with macrobenthos biomass (Hanson and Leggett 1982; Nakashima and Leggett 1975).

Aquatic ecologists are well aware of the above patterns, indeed, they are often cited as support for the view that aquatic communities are regulated from the "bottom-up" by nutrient input (see McQueen et al. 1986 for discussion). In this paper, we suggest that most models of exploitative trophic-level interactions in fact do not predict the above patterns. We than discuss some ways in which simple equilibrium models may be modified to account for these patterns. In particular, we discuss how the inclusion of population size (or age) structure and a separation in resource use between size (or age) classes, provides a simple mechanism to generate positive correlations between consumer and resource densities across a productivity gradient. We illustrate this result using a simple population model containing two life stages. In addition, we provide data for two species of sunfishes (Centrarchidae) in a series of Michigan lakes, supporting predictions of the model.

2 A Theoretical Perspective

The equilibrium abundance of organisms at each trophic level represents a balance between mortality rates (in part, a function of predator densities) and

---

Kellogg Biological Station and Department of Zoology, Michigan State University, Hickory Corners, MI 49060, USA

Size-Structured Populations
B. Ebenman and L. Persson (Eds.)

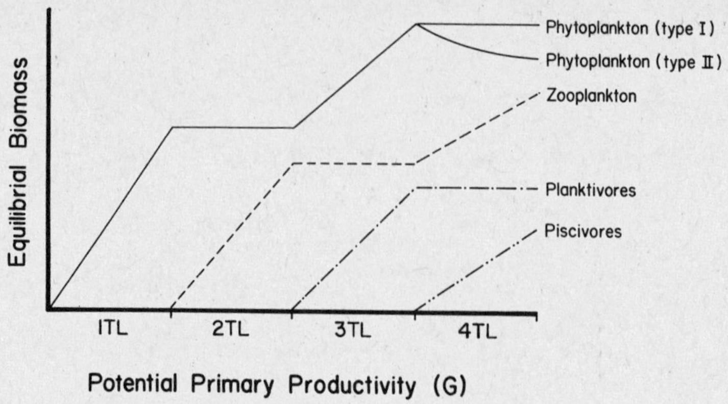

Fig. 1. Relationships between equilibrium biomass (standing stock) at various trophic levels (e.g., phytoplankton, zooplankton, planktivores, piscivores) and potential primary productivity (*G*), e.g., phosphorus loading. Relationships are based on the model in Oksanen et al. (1981), which assumes that increases in potential primary productivity permit the addition of higher trophic levels (denoted by *TL* in the figure). Potential discontinuities and instabilities in the transition regions between trophic states along the productivity axis are ignored in this simplified presentation (see Oksanen et al. 1981 for details). Two potential responses are shown for phytoplankton, depending on the functional response of the planktivores (see text for details)

birth rates (in part, a function of resource densities). Determining how birth and mortality rates equilibrate to set the abundances of organisms at various trophic levels, therefore, requires a theory of biological exploitation among a number of trophic levels. There have been a few notable attempts to build such a theory (Rosenzweig and MacArthur 1963; Smith 1969; Rosenzweig 1973, 1977; Wollkind 1976; Wiegert 1977; Oksanen et al. 1981), although many of these studies have concentrated on the question of system stability rather than species abundance. The model of Oksanen et al. (1981) provides the most thorough look at how the equilibrium biomasses of several different trophic levels should change across a gradient in potential primary productivity (i.e., nutrient loading). One interesting result of Oksanen et al.'s model is that it predicts no correlation between the equilibrium biomasses of consumers and resources within a given trophic structure.

In Fig. 1, we have drawn the predicted relationships between equilibrium biomass (standing stock) at various trophic levels (e.g., phytoplankton, zooplankton, planktivores, and piscivores) and potential primary productivity (G), e.g., phosphorus loading. The relationships are based on the model in Oksanen et al. (1981), which assumes that increases in potential primary productivity permit the addition of higher trophic levels. In their model, mortality on the highest existing trophic level is assumed constant and density-independent. For a one-trophic level system (containing only phytoplankton), any increase in G results in an increase in phytoplankton biomass. In a two-trophic level system (containing only phytoplankton and zooplankton), increases in G translate into increases in zooplankton biomass and no change in phytoplankton biomass. This constancy in phytoplankton biomass is due to the fact that per-capita mortality on zooplankton (the top trophic level in this case) is assumed constant and density-independent. A constant per-capita mortality rate implies a constant per-capita birth rate at equilibrium. Thus, the biomass of the zooplankton's resource (phytoplankton) should remain constant across the productivity gradient in a two-trophic level system (see Oksanen et al. 1981 for details). By the same argument, for three trophic levels (phytoplankton, zooplankton, and planktivores), increases in potential productivity (G) lead to an increase in the biomass of planktivores and phytoplankton, but no change in zooplankton biomass. Finally, in a four-trophic level system (containing phytoplankton, zooplankton, planktivores and piscivores), increases in G result in an increase in the biomass of piscivores and zooplankton, no change in planktivore biomass, and no change or a decrease in the biomass of phytoplankton. A decrease in phytoplankton biomass in the four-trophic level system is expected if the planktivore has a Holling type II functional response (as in Oksanen et al. 1981); i.e., for a given density of planktivores the per-capita mortality rate of the zooplankton decreases as the density of zooplankton increases. In this case, although planktivore density is constant, the increase in zooplankton abundance leads to a decreasing per-capita mortality rate of the zooplankton; therefore, the abundance of phytoplankton should decrease. If a type I functional response is assumed (per capita mortality is independent of prey density), then the phytoplankton density remains constant.

From Fig. 1, we see that within a given trophic structure (i.e., holding the number of trophic levels constant), there should be *no correlation* between the biomass of a consumer and its resource. However, the nearly ubiquitous pattern in aquatic systems is that the biomasses of consumers and resources *positively covary* across a gradient in nutrient input. Therefore, we are left with something of a paradox. An inspection of Fig. 1 suggests one simple resolution to this problem. If lake surveys documenting a positive correlation between trophic level biomass and nutrient loading include lakes with different numbers of trophic levels, then we would expect to see a general increase in the biomass at each trophic level with an increase in nutrient loading. However, in the one study reporting the number of trophic levels contained in each lake surveyed (Mills and Schiavone 1982), positive correlations between consumers and resources were observed across lakes that each contained four trophic levels. Furthermore, empirical analyses of food webs by Pimm (1982) and by Briand and Cohen (1987) show no relationship between productivity and the number of trophic levels in either terrestrial or aquatic ecosystems. Thus, we are skeptical that the general pattern of increased trophic level biomass with increased nutrient loading is simply a consequence of surveying a series of lakes with varying numbers of trophic levels.

Using conventional predator-prey isocline analyses (Rosenzweig and MacArthur 1963; Rosenzweig 1971), we can examine the general conditions necessary for a positive correlation between consumer and resource densities over productivity gradients, when the number of trophic levels remains constant. In standard consumer-resource models, the consumer isocline is perpendicular to the resource axis (Fig. 2a). As resource productivity increases (which elevates the resource isocline but does not affect the consumer isocline), the equilibrium abundance of the consumer increases (C_L^* to C_H^*), but the equilibrium abundance of the resource remains constant ($R_L^* = R_H^*$). In order to produce a positive response of both trophic levels to the increased productivity, the consumer isocline must bend

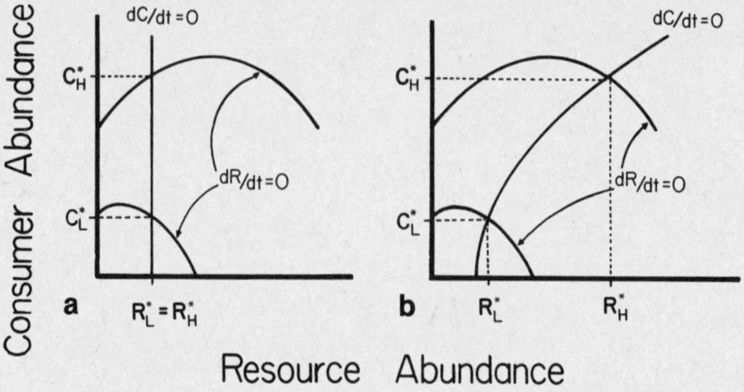

Fig. 2 a, b. Graphical depiction of standard and modified models of consumer-resource interactions under conditions of low and high resource productivity. In both **a** and **b**, the *lower curve* represents the zero-growth isocline for the resource under conditions of low productivity, while the *upper curve* represents the resource isocline at high productivities. **a** shows the prediction from standard models, where increasing productivity produces no change in R* (the equilibrium density of resources) and tends to destabilize the interaction (Rosenzweig 1971). In **b**, increasing productivity produces an increase in R* and C* (the equilibrium densities of the resource and consumer, respectively) and the model is stabilized by the curvature in the consumer isocline

away from the consumer axis as consumer density increases (Fig. 2b). A bent consumer isoline occurs when there is density dependence in the mortality or the energetic requirements of the consumers. These density-dependent effects are absent in Oksanen et al.'s model and their model does not predict positive correlations between consumers and resources. Furthermore, while Fig. 2 shows only two trophic levels, the density-dependence experienced by the consumer population in Fig. 2 cascades down the food web. If all predators exhibit type I functional responses (i.e., per-capita prey mortality is independent of prey density), then the equilibrium abundances of all trophic levels will be positively correlated. If, however, the functional responses are nonlinear (e.g., type II), then the signs of the resulting correlations at lower trophic levels will depend on the relative effects of the predator's numerical and functional responses on the preys' per-capita mortality rates.

We can now ask: what are some of the mechanisms that can cause density-dependent increases in consumer mortality or energetic requirements? One factor that can increase consumer mortality is an increase in the equilibrium density of the consumer's predator. However, postulating an increase in the equilibrium density of a consumer's predator with an increase in the equilibrium density of the consumer simply shifts the problem up a trophic level without really resolving it. Again, we have to ask what allows the densities of the consumer and its predator to positively covary with increased productivity. Another factor that can produce density-dependent consumer mortality or energetic requirements is direct interference among the consumers (Wollkind 1976). Thus, in aquatic systems, direct interference among piscivores could explain the general pattern of positive covariance of biomass at all trophic levels with increasing productivity. Below, we show how an alternate mechanism involving population structure and associated ontogenetic niche shifts can also produce positive correlations between the density of fish and their resources.

3 Population Size Structure and Ontogenetic Niche Shifts

Most aquatic organisms grow substantially in size from birth to death, resulting in pronounced size structure within populations. Associated with this change in body size, there is often an abrupt shift in diet or habitat use with ontogeny (Werner and Gilliam 1984). These ontogenetic niche shifts can be a consequence of changes in an organism's foraging ability, such that it feeds more efficiently on different sizes or types of prey, or in different habitats, as it grows. Size-dependent changes in predation risk can also cause organisms to shift from feeding in protected habitats when vulnerable to feeding in more open areas at relatively invulnerable sizes (examples in Werner and Gilliam 1984; Mittelbach and Chesson 1987; see also Werner this Vol.).

Shifts in an organism's diet or habitat use during ontogeny can have important consequences for the expected patterns between consumer and resource abundance. Ontogenetic niche shifts act to partially decouple the various life stages. Thus, density-dependent performance (e.g., survival or fecundity) at one life stage can result in the other life stage exhibiting a positive correlation between its equilibrium density and the equilibrium density of its resource. Below we use a very simple two life-stage model to illustrate this effect.

A Two Life-Stage Model

The model presented below was developed by Mittelbach and Chesson (1987) and is similar in form to the model of Tschumy (1982). Here we consider a single species consisting of two life stages (juveniles and adults). Juveniles and adults are assumed to occupy separate habitats and/or feed on separate resources. That is, there is a complete niche shift during ontogeny. We also assume that density dependence occurs at each life stage. At the adult stage, per-capita birth rate declines with an increase in adult density; at the juvenile stage, per-capita juvenile survival declines with an increase in juvenile density. Density-dependent fecundity and juvenile survivorship have been demonstrated in many groups of organisms and are likely to be common phenomena in size-structured species. The densities of juveniles and adults in the population can be expressed as:

$$J(t) = A(t)F(A(t)) \tag{1}$$

$$A(t+1) = J(t)\ell(J(t)) \ , \tag{2}$$

where J(t) is the number of juveniles at time t, A equals the number of adults, F is the per-capita fecundity and ℓ is the probability of surviving through the juvenile stage. These equations can be written as a single difference equation describing the change in adult numbers from one time to the next:

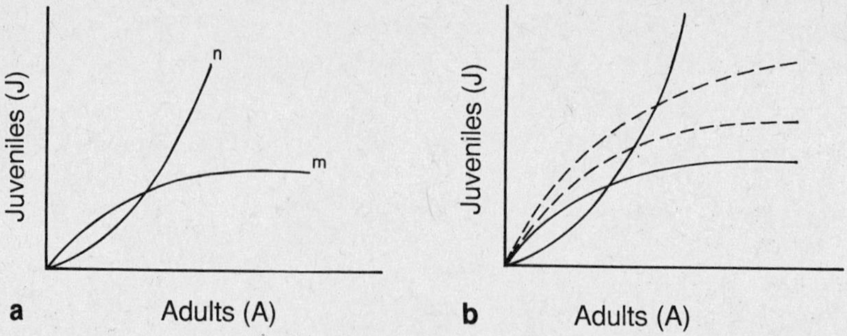

Fig. 3 a, b. Production curves for adults and juveniles. In **a**, curve *m* describes the number of juveniles produced at a given adult density and is the product of per capita adult fecundity [F(A)] and adult density (*A*); curve *n* describes the number of adults produced from a given density of juveniles and is the product of per capita juvenile survival [ℓ(J)] and juvenile density (*J*). The intersection of the adult and juvenile production curves defines the equilibrium densities of the two life stages and is the point where $F(A^*)\ell(J^*) = 1$. **b** shows the effects of increasing per capita fecundity [f(A)] on the equilibrium densities of adults and juveniles. *Dashed lines* represent increased fecundities. Equilibrium densities determined as in **a**

$$A(t+1) = A(t)F[A(t)]\ell\{A(t)F[A(t)]\} \tag{3}$$

In this simple model, we assume no overlap in generations. Mittelbach and Chesson (1987) show how the above model may be generalized (without affecting the major conclusions) to allow for overlapping generations, and for juvenile density to affect time to maturity as well as survival.

We can examine the behavior of the model and determine the equilibrium densities of juveniles and adults by constructing juvenile and adult production curves as shown in Fig. 3a. Curve (m) describes the number of juveniles produced at a given adult density and it is the product of per-capita adult fecundity [F(A)] and adult density (A); curve (n) describes the number of adults produced from a given density of juveniles and is the product of per-capita juvenile survival [ℓ(J)] and juvenile density (J). The intersection of the adult and juvenile production curves defines the equilibrium densities of the two life stages and is the point where $F(A^*)\ell(J^*) = 1$. In Fig. 3a, the production of juveniles is shown as a monotonically increasing function of adult density and recruitment into the adult class is shown as a monotonically increasing function of juvenile density. Humped-shaped production curves are possible if density dependence at a life stage is sufficiently strong (Tschumy 1982; Mittelbach and Chesson 1987; Wilbur this Vol.). However, in the following analysis, we limit consideration to those cases where both the juvenile and adult production curves increase monotonically; this simply requires that density dependence at either life stage be not too severe.

We can now ask what happens to the equilibrium densities of juveniles and adults when resource productivity for a life stage changes. For example, an increase in adult resources should result in an increase in per-capita fecundity at a given adult density. Figure 3b illustrates how such in change in the fecundity curve affects the equilibrium densities of adults and juveniles in the system, as well as the per-capita performances of adults and juveniles (i.e., fecundity and survival respectively). From Fg. 3b we see that an increase in adult resource productivity leads to:

1. an increase in the equilibrium density of adults,
2. an increase in the equilibrium density of juveniles,
3. an increase in per-capita frecundity (the ratio of juveniles/adult at equilibrium), and
4. a decrease in per-capita juvenile survival probability (the ratio of adults/juvenile at equilibrium).

Changes in juvenile resources, given that they affect juvenile survival, can be similarly modeled. An increase in juvenile resources and a corresponding increase in the juvenile survival curve will also generate predictions (1) and (2) above, however, predictions (3) and (4) will be changed, namely, per-capita fecundity will decrease and per-capita juvenile survival will increase.

Therefore, the above model predicts a positive covariance between equilibrium consumer density and the equilibrium density of the resource whose productivity was altered. This positive covariance occurs because density dependence at one life stage prevents the other stage from responding completely to a change in its own resource productivity. Thus, the response between consumers and resources in this simple stage-structured model is quite different from that shown by the models of Oksanen et al. (1981) and others. In aquatic systems, the vast majority of taxa are size-structured and many species exhibit distinct shifts in diet and/or habitat between life stages. Therefore, we suggest that the recognition of these ontogenetic niche shifts and the fact that life stages often show density-dependent responses to different resources, provides a likely mechanism whereby the densities of consumers and their resources will positively covary within a life stage. In the following section, we present field data for the bluegill sunfish (*Lepomis macrochirus*) illustrating how a species exhibiting a distinct ontogenetic niche shift may respond to variation in its resources in a manner consistent with the model presented above.

4 Ontogenetic Niche Shifts in the Bluegill

In small, Michigan lakes the bluegill shows a pronounced habitat shift with ontogeny; small bluegills ($\lesssim$75 mm standard length) feed on benthic invertebrates in or near littoral-zone vegetation, whereas larger bluegills shift to feeding on open-water zooplankton (*Daphnia* sp.) (Mittelbach 1981, 1984; Werner and Hall 1988). We have shown that the use of vegetation by small bluegills is primarily a response to high predation risk in the open-water and that when predators (i.e., largemouth bass, *Micropterus salmoides*) are removed, small bluegills will feed on zooplankton (Werner et al. 1983). Bluegills >75 mm are relatively immune to predation by bass and consequently can feed in the open water under little risk. Thus, in these lakes the bluegill exhibits essentially a two-stage life history with regard to resource use; juvenile bluegills occupy the littoral zone and feed on benthic invertebrates while adult bluegills feed extensively on open-water zooplankton. For simplicity, we ignore a brief (2–3 wks) early life history period, when bluegill fry feed in large schools in the limnetic zone (Werner and Hall 1988). The life history to the bluegill therefore roughly corresponds to the simple stage-structured model presented above. While we do not know the degree of which juvenile and adult bluegill exhibit density-dependent survival and fecundity, we have shown that juvenile bluegill growth (and thus perhaps survival) is density-dependent in our study lakes (Osenberg et al. 1988, Mittelbach 1988) and Beyerle and Williams (1972) have observed density dependence in bluegill fecundities. Adult bluegill growth is also food-limited in our lakes (Osenberg et al. 1988). Thus, the bluegill life history fits the model reasonably well, although we cannot as yet make quantitative tests of the model's predictions. We can, however, use a large data set on bluegill growth rates and densities from seven Michigan lakes to compare a number of important qualitative predictions from the model regarding how variation in adult and juvenile resource productivity should affect the performances and densities of the two life stages.

In the model, adult and juvenile performances are measured in terms of per-capita fecundity and per-capita survival, respectively. Both of these parameters are extremely difficult to measure accurately for natural fish populations. Fortunately, we can use the rate of individual body growth as a reasonable surrogate for both parameters. Body size (the result of growth history) and fecundity are highly correlated in bluegills and other fishes (Carlander 1977; Bagenal 1978). Further, if some fraction of the net energy gained by a foraging adult is allocated to growth and reproduction, then growth and fecundity will be positively correlated. Therefore, we expect adult growth rates to be positively correlated with fecundities. Juvenile survival to adulthood should also positively covary with juvenile growth rates because, in part, vulnerability to predators generally decreases with size in juvenile fishes (Ware 1975; Craig 1982; Werner 1986). Larger individuals are also less susceptible to starvation and physicochemical factors (Oliver et al. 1979). Thus, an increase in juvenile growth should decrease the time spent in vulnerable size classes and therefore decrease the overall probability of death. In the following analyses, we measure adult and juvenile performances in terms of individual growth rates, and assume that increases in growth rate reflect increases in fecundity or juvenile survivorship. Furthermore, our field data for bluegills demonstrate that adult bluegill growth rates within a lake are positively correlated with predicted energetic returns available from the zooplankton (Fig. 4). Thus, adult bluegill growth rates increase with an increase in their resource availability, and we can use increased growth rates of adults as an index of increased adult resources as well as increased fecundity. Similarly, juvenile growth is positively correlated with predicted energetic returns from the littoral habitat (Mittelbach 1988), so we can also use increased growth of juveniles as an index of increased juvenile resources.

Table 1 presents the correlations between bluegill growth rates and densities from seven Michigan lakes. These correlations can be compared with the predicted correlations based upon variation in either adult resources or juvenile resources. There are three significant ($p < 0.05$) correlations in Table 1, and each of these is consistent with what the model would predict if the lakes varied primarily in adult (zooplankton) resources (Fig. 3b). First, adult bluegill density is positively correlated with adult growth rates. That is, adult bluegill density is

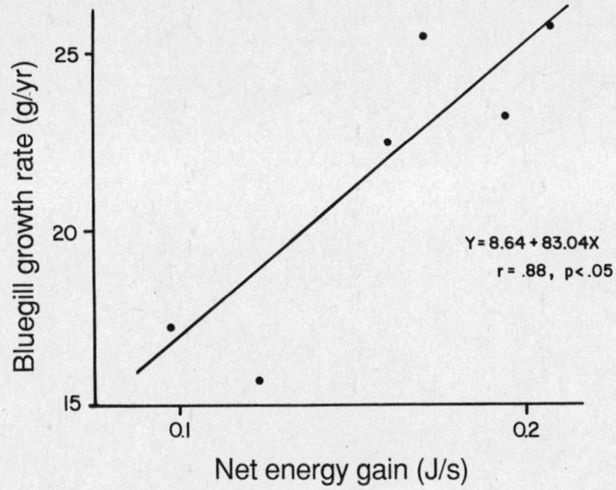

Fig. 4. Relationship between predicted net energy gain averged over the growth season (Joules/second foraging time) and average annual growth rates of large bluegills in four Michigan lakes (two lakes represented by 2 years of data). Energy gains were calculated using the optimal foraging model in Mittelbach (1981) and are the average energy gains derived from the size-density distributions of zooplankton sampled in each lake (6 dates/summer/lake). (See Osenberg et al. 1988 for calculation of bluegill growth rates)

Table 1. Correlations among bluegill growth rates and densities from seven Michigan Lakes. Small bluegills are ≤ 75 mm SL by mid-summer; large bluegill > 75 mm SL. (Data from Osenberg et al. (1988) and Osenberg et al. unpubl.)

	Small Bg growth	Large Bg growth	Small Bg density	Large Bg density
Small Bg growth	–	−0.30	−0.88*	−0.48
Large Bg growth	–	–	+0.52	+0.76*
Small Bg density	–	–	–	+0.92*
Large Bg density	–	–	–	–

* $p<0.05$

greater in lakes with greater adult resource densities. (A significant positive relationship between adult bluegill density and foraging return from the zooplankton (E/T) is also found in the four lakes for which we have zooplankton data (Osenberg et al. 1988); $r = 0.96$, $p<0.05$.) Second, adult bluegill density is positively correlated with small bluegill density. Third, small bluegill growth is negatively correlated with small bluegill density. The above correlations agree with the model's predictions regarding the effects of changes in adult resources on adult and juvenile densities and performances. In addition, the other three correlations in the matrix (while not significant at $p<0.05$), are all in the direction predicted by variation in adult resources.

Thus, variation in adult resources correctly predicts all six of the observed correlations for the bluegill. In particular, note that the key model prediction of a positive correlation between adult performance and adult density, in conjunction with a negative correlation between juvenile performance and juvenile density, is supported. Model predictions based upon variation in juvenile resources across lakes are consistent with only one of the three significant correlations and only two of the six total correlations (Table 1). Thus, the bluegill data provide an example where adult consumer and resource densities positively covary across a series of lakes with the same trophic structure, and the entire entire data set is consistent with predictions of the two life-stage model presented in the text.

5 Ontogenetic Niche Shifts and Species Interactions

The above analyses have concentrated on how the inclusion of size (age) structure can affect the relations between consumers and resources within a single species. However, it is clear that a given trophic level rarely contains a single species and that consumer species will often interact with each other as well as with their resources. In the case of species undergoing ontogenetic niche shifts, it is common for species to share habitats or food resources at some life stages, but not others (Werner and Gilliam 1984). When competition occurs for these shared resources, the population dynamics of the species become linked due to interactions at specific life stages. This linkage of populations can greatly modify the predicted patterns of covariation in density, growth and resources derived from the single-species model presented earlier. Below we examine the particular case where consumer species compete as juveniles but not as adults. We expect this to be a common interaction among-size structured species where small (young) individuals often share a common protective habitat (refuge) due to their vulnerability to size-limited predators (Mittelbach 1986; Mittelbach and Chesson 1987). First, we show how the two life-stage model for a single species can be modified to include a second species that is a juvenile stage competitor. We than look at model predictions for how juvenile and adult performances and densities for two species should vary as resource production for juveniles and adults of each species vary. Then, we return to the field data for sunfish growth and density and examine the patterns of covariation in density and growth for the bluegill and its congener, the pumpkinseed (*L. gibbosus*), in light of model predictions.

A Two Life-Stage Model for Competing Species

We now consider a system of two interacting species, where juveniles compete interspecifically but adults do not. An empirical example of this situation is illustrated by the interaction between the bluegill and pumpkinseed sunfish in small Michigan lakes. Small bluegills and pumpkinseeds (ca. <75 mm) commonly co-occur in the vegetated littoral zone, where they compete for invertebrate resources (Mittelbach 1984, 1988). At large sizes, each species shifts to feeding on a separate resource, with the bluegill concentrating on zooplankton and the pumpkinseed feeding extensively on snails (Mittelbach 1984). Thus, each spieces undergoes an ontogenetic niche shift, with the juvenile stages occupying the same habitat and competing for resources, while the adults show little direct interspecific interactions.

We can examine this type of interaction using a two species, two life-stage model developed in Mittelbach and Chesson (1987). In this model, we leave the total reproduction functions the same as in the single species model [Eq. (1)], except for the addition of a subscript to indicate the species and to allow the possibility that these functions differ between species. Thus, we have:

$$J_i(t) = A_i(t)F_i[A_i(t)] \ , \tag{4}$$

i = 1, 2. The juvenile survival equations become:

$$A_i(t+1) = J_i \ell[J(t)] \ , \tag{5}$$

where $J(t) = J_1(t) + J_2(t)$ is now the total number of juveniles in the system. Here per capita survival of juveniles of species i declines as a function of total juvenile density, not just the density of species i. For simplicity, we assume that juvenile survival probabilities for both species are equal. For the bluegill and pumpkinseed this appears to be a reasonable approximation (Mittelbach 1986). Mittelbach and Chesson (1987) show how Eqs. (4) and (5) can be used to specify the equilibrium densities of juveniles and adults of both species. Further, these equilibrium densities will be locally stable under the conditions previously specified for the single-species case (see Appendix A in Mittelbach and Chesson 1987).

As in the single-species case, the two-species, two-life stage model makes a series of predictions concerning how variation in adult and juvenile resource productivities will affect the densities and performances of juveniles and adults of each species. These predictions can then be compared to the observed patterns of variation in bluegill and pumpkinseed performances across lakes. For example, if the productivity of the bluegill's adult resource (i.e., zooplankton) varies across lakes, the model predicts that an increase in zooplankton productivity will result in:

1. an increase in the equilibrium density of juvenile and adult bluegills,
2. an increase in adult bluegill and adult pumpkinseed performance (fecundity or growth),
3. a decrease in juvenile bluegill and juvenile pumpkinseed performance (survival or growth),
4. a decrease in the equilibrium density of juvenile and adult pumpkinseeds,
5. an increase in the total density of juveniles (bluegills and pumpkinseeds combined).

These increases and decreases predict a particular set of correlations among the growth and density responses of the two species. Different sets of predictions will be generated if there is variation in adult pumpkinseed resources (snails) or juvenile sunfish resources (littoral invertebrates). These predictions can then be compared to the observed patterns of growth and density variation exhibited by the two species (Osenberg et al. 1988).

Table 2 presents the correlations among growth rates and fish densities of bluegills and pumpkinseeds from seven Michigan lakes. Five of the 21 correlations listed in this matrix are significant at $p < 0.05$, with a sixth correlation being marginally significant at $p < 0.07$. We compare the signs of these six correlations with the predicted signs of the correlations based upon the two-species, two-life stage model, assuming variation between lakes occurs primarily in bluegill adult resources (zooplankton), pumpkinseed adult resources (snails), or juvenile resources (littoral-zone invertebrates). Table 3 shows that variation in productivity of adult bluegill resources correctly predicts the sign of all six correlations, while variation in the productivities of snails or littoral-zone invertebrates predicts only

Table 2. Correlations among growth rates and fish densities from seven Michigan Lakes. Large and small fish defined as in Table 1. Correlations > 0.75 or < −0.75 are significant at $p < 0.05$. (Data from Osenberg et al. 1988, Table 4)

	Small bluegill growth	Small pumpkinseed growth	Large bluegill growth	Large pumpkinseed growth	Small fish density	Large pumpkinseed density	Large bluegill density
Small bluegill growth	–	+0.86	−0.30	−0.59	−0.88	+0.18	−0.48
Small pumpkinseed growth	–	–	−0.55	−0.40	−0.92	+0.36	−0.73
Large bluegill growth	–	–	–	+0.20	+0.52	−0.64	+0.76
Large pumpkinseed growth	–	–	–	–	+0.38	+0.39	+0.21
Small fish density	–	–	–	–	–	−0.30	+0.81
Large pumpkinseed density	–	–	–	–	–	–	−0.38
Large bluegill density	–	–	–	–	–	–	–

Table 3. Comparisons between model predictions and observed correlations in Table 2. Signs of the six strongest correlations in Table 2 are compared to their predicted signs if variation between lakes is primarily in bluegill adult resources (zooplankton), pumpkinseed adult resources (snails), or juvenile resources (littoral-zone invertebrates). Listed at the bottom of the prediction columns are the number of observed correlations whose sign is correctly predicted by variation in each of the three resources. Also listed are the number of cases where variation in each of the three resources correctly predicts the sign of the 21 total correlations presented in Table 2

Comparison	Observed correlations		Predicted correlations		
			zooplankton	snails	Juv. resources
Small Bg growth versus small Ps growth	+		+	+	+
Small Ps growth versus small fish density	–		–	–	+
Small Bg growth versus small fish density	–		–	–	+
Large Bg density versus large Bg growth	+		+	–	–
Large Bg density versus small Ps growth	–		–	+	+
Large Bg density versus small fish density	+		+	–	+
Total	6	no. correct	6	3	2
All correlations	21		20	12	10

about half of the correlations. We can also ask how well variation in the productivity of zooplankton, snails, or juvenile resources predicts the sign of all 21 of the correlations presented in Table 2, irrespective of the magnitude of the individual correlation. In this case, variation in zooplankton correctly predicts over

95% (20/21) of the correlations, while variation in the other two resources predicts only about 50% of the observed correlations correctly.

While comparisons between the model predictions and the field data for the bluegill and pumpkinseed are not as direct as we would like (e.g., they measure individual performance in different units), the correpondence between predictions based upon variation in adult bluegill resources and the observed data is compelling. The overall response by bluegills and pumpkinseeds suggests that variability in zooplankton production between lakes, in conjunction with juvenile competition linking that dynamics of two species, may be a primary factor determining the species growth and abundance patterns we observe in the fish community.

6 Discussion

Size structure is nearly ubiquitous among aquatic animals and often leads to dramatic niche shifts during ontogeny. We have shown that these niche shifts, which lead to segregation among members of the same population, can lead to positive correlations among consumer and resource abundances (for one stage of the life history) when measured along a gradient in resource productivity. This positive correlation arises despite competition among consumers. These results are quite different than those predicted by standard intraspecific competition models, and are consistent with the patterns we see for the adult bluegill and pumpkinseed. Of course, if we were to measure the resource abundances for the juvenile stages, they would be negatively correlated with consumer densities. Often, however, resources for the entire life history are not assessed, and in general it is the adult stage of the population and its resources that are monitored. Resolution of the compete relationship between consumer and resource abundances can only be made by recognizing the effect of ontogeny of the consumer's ecology.

Although niche shifts can produce positive correlations among consumers and resources, niche shifts can in some cases obscure these patterns. For example, during the early stages of their ontogenies, many piscivorous fish feed on zooplankton (Werner 1986). Therefore, these piscivores, which are predators of planktivorous fish species, may also compete with their prey early in their life histories. Bottlenecks in the recruitment of piscivores can arise through these competitive interactions (Werner and Gilliam 1984; Persson, this Vol.), and these bottlenecks can prevent piscivore biomass from increasing with increased production of its prey, therefore leading to diminished (or even negative) correlations among the abundances of the piscivore species and their prey (Persson et al. 1988). These types of mixed interactions where a population occupies two (or more) trophic levels seem to be more common at higher levels in the food chain (e.g., piscivorous fish often feed on zooplankton, planktivores, and juvenile piscivores, and can be thought of as occupying up to four trophic levels). This mixing (or blurring) of trophic relationships at higher trophic levels may be partially responsible for the reduction in the correlation between the abundances of consumers and resources as the focal trophic level is increased (a pattern suggested by McQueen et al. 1986).

As discussed in the introduction, the positive correlation between the biomass of consumers and their resources requires a bend in the consumer isocline. The niche shift explanation we have developed in this paper (in which resources refer only to those for one of the life history stages) causes the consumer isocline to bend due to the increased mortality experienced by the other life stage. This buffering of the consumer population's response to resources requires that, as consumer densities increase, more resources be present to offset the loss of individuals (or energy) incurred during the other life history stage. Interference among consumers can also produce positive correlations between consumer and resource abundances (e.g., Wollkind 1976; Wiegert 1977). In this case, an increased consumer density requires a greater resource density due to increased energy losses and/or decreased time available for foraging.

Differential edibility among components of the resource category can also lead to positive correlations between consumer and "resource" abundance. It has commonly been observed that increased consumer densities lead to shifts in the resource community toward less desirable resource types (e.g., McCauley and Briand 1979; Zaret 1980). Increased consumer densities should favor lower quality resources; therefore, in reference to the general relationship shown in Fig. 2b, higher total resource densities would be needed to maintain the consumer population (thus bending the consumer isocline). Walters et al. (1987) have recently suggested differential edibility as a possible cause of the positive correlation between zooplankton and phytoplankton abundances, though their discussion of the process is not consistent with the earlier work of Phillips (1974). Leibold (1988), in a general synthesis of the models of Oksanen et al. (1981) and Phillips (1974), demonstrates more clearly the conditions under which differential edibility can affect consumer and resource abundances across gradients in resource productivity and consumer mortality. He shows that the equilibrium resource availabilities, measured across productivity gradients, can be positively correlated with consumer densities, but that the positive correlation arises primarily through changes in the densities of the relatively inedible components of the resources: the total abundance of resources, when weighted by their contribution to the consumer's birth rate (e.g., a measure of their edibilities), does not change across the productivity gradient (unless the death rate also changes).

The distinction between edible and relatively inedible resources raises a general point with regard to measuring resources. Clearly, not all resources contribute equally to the population growth of the consumer. Further, measures of resource density in the environment are not necessarily good estimates of resource availability. In our studies, we have been able to resolve this problem by using foraging models that incorporate differences in prey availability and value (Mittelbach 1981; Osenberg 1988). These models provide good descriptions of foraging rates and growth rates (Fig. 3; Mittelbach 1981, 1983, 1988; Osenberg 1988; Osenberg et al. 1988), and can therefore be used to estimate resource availability based on the abundances and size distributions of potential prey organisms in the environment. Discrepancies between resource abundance (e.g., biomass/m^2) and resource availabilities (e.g., measured as a feeding rate) can be large if the resources change much in terms of their relative abundances and/or size structure. For example, estimates of the foraging returns for bluegills feeding on zooplankton are not correlated with zooplankton biomasses from these same samples ($n = 38$ for the data summarized in Fig. 3, $r = 0.05$, $p > 0.75$). Obviously, the patterns that we have documented using foraging return and growth rates would not have emerged if we had used a less accurate estimator of resource availability (e.g., biomass of all "prey").

In addressing what factors create and maintain positive correlations between consumers and resources along productivity gradients, we kept the analysis simple. For example, we assumed that increased resources would lead to increased performance (e.g., birth and growth rates) of a consumer. In the real world, where trophic levels are not clearly defined and many species comprise a trophic level, this might not be the case, as the previous discussion demonstrates. In these cases a thorough understanding of the processes that control resource and consumer abundances must include information about the way consumers utilize resources (and therefore influence the resources' mortality rates). Foraging theory is a powerful tool that can be used to approach this problem. This study illustrates one useful application of these mechanistic models to the study of ecosystem patterns.

Finally, the distinction between resource quality and quantity helps focus our attention on the beneficial effects of the environment to individuals (e.g., as reflected in birth rates or growth rates), and allows us to explore how these effects are balanced by detrimental effects of the environment (e.g., as reflected in mortality rates or energetic losses). Ultimately, population densities are the outcome of the summation of many rate functions that measure the gain and loss to a specified population. If population densities remain constant or fluctuate within defined limits, average gains and losses (defined over an appropriate time scale) must balance: e.g., birth rates equal mortality rates. However, ecologists too often tend to dichotomize the processes that regulate populations, which is evident in the recurring debates over whether communities are structured by competition versus predation, or top-down versus bottom-up processes. Indeed, control is mediated in the ways population densities and the environment interact to equilibrate gain and loss rates; neither top-down nor bottom-up processes control population densities, instead densities are set by the equilibration to both. Advances in our understanding of the structure and function of communities and ecosystems must advance through the recognition of the important interrelation of these processes.

Acknowledgments. We thank Bill Neill, Jim Gilliam, and Lennart Persson for helpful comments on the manuscript, and Beth Kohler and Jim Zablotney for drafting the figures. This work was partially supported by a grant from the National Science Foundation (USA) to GGM (#BSR-8796261). This is contribution No. 621 from the W. K. Kellog Biological Station.

7 References

Bagenal TB (1978) Aspects of fish fecundity. In: Gerking SD (ed) Ecology of freshwater fish production. Blackwell, Oxford

Beyerle GB, Williams J (1972) Survival, growth, and production by bluegills subjected to population reduction in ponds. Michigan Dep Nat Resour Rep 273

Briand F, Cohen JE (1987) Environmental correlates of food chain length. Science 238:956–959
Carlander KD (1977) Handbook of freshwater fishery biology. Iowa State Univ Press, Ames
Craig JF (1982) Population dynamics of Windermere perch. Freshwater Biol Assoc Annu Rep 50:49–59
Hanson JM, Leggett WC (1982) Empirical prediction of fish biomass and yield. Can J Fish Aquat Sci 39:257–263
Hanson JM, Peters RH (1984) Empirical prediction of crustacean zooplankton biomass and profundal macrobenthos biomass in lakes. Can J Fish Aquat Sci 41:439–445
Leibold MA (1988) Predation, competition, habitat structure and the structure of freshwater plankton communities. Thesis, Michigan State Univ, East Lansing
McCauley E, Briand F (1979) Zooplankton grazing and phytoplankton species richness: field tests of the predation hypothesis. Limnol Oceanogr 24:243–252
McQueen DJ, Post JR, Mills EL (1986) Trophic relationships in freshwater pelagic ecosystems. Can J Fish Aquat Sci 43:1571–1581
Mills EL, Shiavone A (1982) Evaluation of fish communities through assessment of zooplankton populations and measures of lake productivity. N Am J Fish Manag 2:14–27
Mittelbach GG (1981) Foraging efficiency and body size: a study of optimal diet and habitat use by bluegills. Ecology 62:1370–1380
Mittelbach GG (1983) Optimal foraging and growth in bluegills. Oecologia (Berlin) 59:157–162
Mittelbach GG (1984) Predation and resource partitioning in two sunfishes (Centrarchidae). Ecology 65:499–513
Mittelbach GG (1986) Predator-mediated habitat use: some consequences for species interactions. Environ Biol Fish 16:159–169
Mittelbach GG (1988) Competition among refuging sunfishes and effects of fish density on littoral zone invertebrates. Ecology 69:614–623
Mittelbach GG, Chesson PL (1987) Predation risk: indirect effects on fish populations. In: Kerfoot WC, Sih A (eds) Predation. Univ Press New Engl, Hanover, NH
Nakashima BS, Leggett WC (1975) Yellow perch (*Perca flavescens*) biomass responses to different levels of phytoplankton and benthic biomass in Lake Memphremagog, Quebec-Vermont. J Fish Res Board Can 32 (10):1785–1797
Oksanen L, Fretwell SD, Arruda J, Niemela P (1981) Exploitation ecosystems in gradients of primary productivity. Am Nat 118:240–261
Oliver JD, Holeton GF, KE Chua (1979) Overwintering mortality of fingerling smallmouth bass in relation to size, energy stores, and environmental temperature. Trans Am Fish Soc 108:130–136
Osenberg CW (1988) Body size and the interaction of fish predation and food limitation in a freshwater snail community. Thesis, Michigan State Univ, East Lansing
Osenberg CW, Werner EE, Mittelbach GG, Hall DJ (1988) Growth patterns in bluegill and pumpkinseed sunfish: environmental variation and the importance of ontogenetic niche shifts. Can J Fish Aquat Sci 45:17–26
Pace ML (1984) Zooplankton community structure, but not biomass, influences the phosphorus-chlorophyll a relationship. Can J Fish Aquat Sci 41:1089–1096
Persson L, Andersson G, Hamrin SF, Johansson L (1988) Predator regulation and primary productivity gradient of temperate lake ecosystems. In: Carpenter SR (ed) Complex interactions in lake communities. Springer, Berlin Heidelberg New York Tokyo
Phillips OM (1974) The equilibrium and stability of simple marine biological systems II. herbivores. Arch Hydrobiol 73(3):310–333
Pimm SL (1982) Food webs. Chapman & Hall, London
Rosenzweig ML (1971) Paradox of enrichment: destabilization of exploitation ecosystems in ecological time. Science 171:385–387
Rosenzweig ML (1973) Exploitation in three trophic level. Am Nat 107:275–294
Rosenzweig ML (1977) Aspects of biological exploitation. Q Rev Biol 52:371–380
Rosenzweig ML, MacArthur RH (1963) Graphical representation and stability conditions of predator-prey interactions. Am Nat 97:209–223
Smith FE (1969) Effects of enrichment in mathematical models. In: Eutrophication: causes, consequences, correctives. Nat Acad Sci, Washington, DC
Stockner JG, Shortreed KS (1985) Whole lake fertilization experiments in coastal British Columbia lakes: empirical relationships between nutrient inputs and phytoplankton biomass and production. Can J Fish Aquat Sci 42:649–658

Tschumy WO (1982) Competition between juveniles and adults in age-structured populations. Theor Popul Biol 21:255–268

Walter CJ, Krause E, Neill WE, Northcote TG (1987) Equilibrium models for seasonal dynamics of plankton biomass in four oligotrophic lakes. Can J Fish Aquat Sci 44:1002–1017

Ware DM (1975) Relation between egg size, growth, and natural mortality of larval fish. J Fish Res Board Can 32:2503–2512

Werner EE (1986) Species interactions in freshwater fish communities. In: Diamond J, Case T (eds) Ecological communities. Harper & Row, New York

Werner EE, Gilliam JF (1984) The ontogenetic niche and species interactions in size structured populations. Annu Rev Ecol Syst 15:393–426

Werner EE, Hall DJ (1988) Ontogenetic habitat shifts in the bluegill sunfish (*Lepomis macrochirus*): the foraging rate-predation risk tradeoff. Ecology 69:1352–1366

Werner EE, Mittelbach GG, Hall DJ, Gilliam JF (1983) Experimental tests of optimal habitat use in fish: the role of relative habitat profitability. Ecology 64:1525–1539

Wiegert RG (1977) Population models: experimental tools for analysis of ecosystems. In: Horn DJ, Stairs GR, Mitchell RD (eds) Analysis of ecological systems. Ohio State Univ Press, Columbus

Wollkind DJ (1976) Exploitation in three trophic levels: an extension allowing intraspecies carnivore interaction. Am Nat 110:431–447

Zaret TM (1980) Predation and freshwater communities. Yale Univ Press, New Haven

Community Responses to Experimental Nutrient Perturbations in Oligotrophic Lakes: The Importance of Bottlenecks in Size-Structured Populations

W. E. NEILL

1 Introduction

Most aquatic metazoans exhibit major changes in size as they mature. Accordingly, they often use different foods and habitats, and are differentially vulnerable to various predators, competitors, and abiotic conditions as they grow. In short, small/young individuals may have very different ecologies than larger/older individuals in the same population. In zooplankton, for example, larger individuals may be strongly limited by size-selective predation by fish planktivores (Brooks and Dodson 1965). Small organisms, on the other hand, are more susceptible to invertebrate predators (Zaret 1980) and possibly to resource limitation (Romanovsky and Feniova 1985; McCauley and Murdoch 1987). In consequence, the dynamics and evolution of various sizes/ages may be set by a complex array of ontogenetically changing environmental factors. Similarly, among fish, large individuals are often nearly invulnerable to piscivore attack, but may interfere with each other's access to resources. On the other hand, small individuals may suffer heavy piscivore predation (Mittelbach 1981), resource shortages during "critical stages" of growth (Hjort 1926), and both exploitative and interference competition (Werner et al. 1983). At population and community levels, then, fitness limitation through various size-dependent processes may strongly influence patterns of relative success, the form and strength of ecological interactions, and potentially the flow of mass through the ecosystem.

If major players in aquatic community structure are affected by size-dependent limitation, their influence over community organization may change markedly with environmental circumstance. For example, Mills and Forney (1988) indicate that variation in survival of small/young yellow perch (*Perca flavescens*) in Oneida Lake, New York, can strongly affect patterns of planktivory on zooplankton and consequently patterns of grazing on phytoplankton. Zooplankton and phytoplankton dynamics and community structure in this lake appear to be influenced by temporal variations in the constraints on survival of small/young perch, viz., piscivore predation. Thus, the sensitivity of the control processes limiting juvenile planktivore recruitment seems to be crucial to the persistence of this community.

My research is concerned largely with identifying the "sensitive" links in community organization. I use a variety of perturbations to communities to test the accommodating capabilities of particular direct and indirect interactions to

The Ecology Group, Department of Zoology, The University of British Columbia, Vancouver, B.C., Canada V6T 2A9

Size-Structured Populations
B. Ebenman and L. Persson (Eds.)

define their limits. In this chapter, I discuss responses of a plankton community to artificial nutrient perturbations that have been applied to evaluate the importance of specific size-dependent constraints on component populations. Roughly mimicking natural nutrient variations (e.g., storm run-off), these "bottom-up" disturbances must be accommodated by existing physiological, behavioral, and demographic mechanisms to maintain current community organization. The magnitude of disturbance that can be accommodated measures the sensitivity of particular links within the community. For this symposium, I will focus on one particular size-structured interaction that seems significant in explaining observed community differences among fishless oligotrophic ponds and lakes in coastal western Canada. The key players are age/size classes of a predatory planktonic insect, the phantom midge *Chaoborus trivittatus*, several species of solitary rotifers, and the cladoceran *Daphnia rosea*.

2 Background Natural History

Phantom midges, so called because of their transparent larval stages, are non-biting flies of the family Chaoboridae. The aquatic larval stages comprise over 95% of the duration of the life cycle and the larvae prey upon a great variety of zooplanktonic (and benthic) organisms in ponds and lakes. Their broad died range implies that, given sufficient densities, they may affect abundances of many components of the herbivore community. Large (13–18 mm) third and fourth instar larvae prey upon nearly all metazoan plankton up to 3 mm length, though with differing success rates among prey taxa and sizes (Fedorenko 1975; Pastorok 1981). Small (2–3 mm) first instar larvae are also plankton predators, relying on large protozoans, solitary rotifers, and very small crustaceans.

Most species produce one to several nonoverlapping larval cohorts that pass through the population over the course of the year. At any one time, most individuals will be at the same developmental stage. In oligotrophic ponds and lakes that I have surveyed, *Chaoborus trivittatus* usually have a single episode of breeding by the flying adults in June or July, develop quickly to reach fourth larval instar by late August and then gradually accumulate reproductive reserves over the next 22 months before pupating. In fishless farm ponds or when provided with extra food, development can be completed in a year or less (Fedorenko and Swift 1972). Plankton ecologists are most interested in this organism during its long fourth larval instar because it has its largest per capita predation rates on nearly all crustacean taxa at this stage (Giguere 1981; Riessen et al. 1984). In eutrophic lakes and farm ponds, fourth instar *Chaoborus* are often sufficiently abundant to have major impacts on the prey community (e.g., Kajak and Ranke-Rybicka 1970; Hillbricht-Ilkowska et al. 1975).

3 *Chaoborus* Dynamics in Oligotrophic Lakes

Sampling of over 100 oligotrophic lakes in the coastal mountains of British Columbia, Canada, reveals that nearly all *Chaoborus trivittatus* populations have

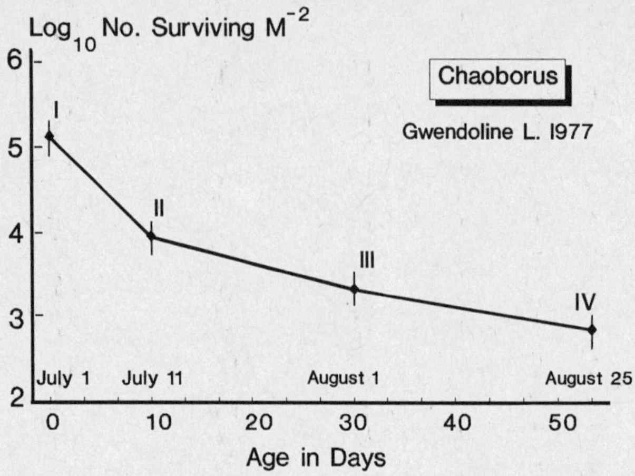

Fig. 1. *Chaoborus* survivorship in Gwendoline Lake. *Error bars* represent 95% confidence intervals around mean of five density estimates on each date

2-year life cycles (Neill unpubl.). Figure 1 summarizes the survival of this species in Gwendoline Lake in the University of British Columbia Research Forest in 1977. Annual and lake to lake variation in egg deposition is quite large, varying from about 70000/m^2 to more than 250000/m^2. Mortality of the resulting first instar larvae is high (89–98%) and density dependent ($k_{1-2} = 0.91$ log Density, $r^2 = 0.69$, $P<0.05$). The force of mortality declines during subsequent instars (means: 72%, 59%, and 20% for instars II through IV, respectively) and shows little or no density dependence for these instars. Less than 1% of the initital cohort is still alive 2 months after oviposition. These individuals persist as fourth instar predators upon crustaceans until they pupate 2 years later.

The net effect of very heavy mortality in first and second instar is low densities of third and fourth instar larvae in these lakes. Further, when fourth instar larvae distribute themselves over their full depth range in summer (5–15 m depth strata), they decline to only 0.05–0.15/l. These densities are 3 to 15 times lower than those reported elsewhere. At such low densities, *Chaoborus* are ineffectual at limiting the densities of any species in the summer crustacean community (Neill 1981). Even when I tripled fourth instar larval density in enclosures, most crustaceans compensated for higher predation mortality by elevated reproduction, thereby minimizing predation effects. Poor recruitment by early instars of *Chaoborus* larvae continued without detectable improvement under these conditions. However, following one exceptionally cold May and June, predatory impacts of *Chaoborus* on the crustacean community were detected when I tripled predator densities. The resulting lower crustacean densities were correlated with higher than average recruitment of first and second instar *Chaoborus* (Neill 1981). Temperature interacts with predation and *Chaoborus* recruitment.

In short, except under unusually cold summer temperature conditions, experimental variations within the natural density range of instar IV *Chaoborus* had minimal effects on any of its prey or on its own recruitment. The survival bottleneck at first and second instars seems a nearly unavoidable constraint on *Chaoborus* population ecology under oligotrophic conditions in these fishless lakes. This bottleneck seems not to be so severe under more productive limnological conditions, judging from high *Chaoborus* densities and growth rates reported from such sites (Hillbricht-Ilkowska et al. 1975).

What causes such an enormous bottleneck in the survival of first and second instar *Chaoborus* such that prey suppression by third and fourth instars is negligible? Observations suggested the importance of food limitation among the smallest larvae. Of more than 2600 instar I and II larvae examined from three 24-hour sampling series collected in July 1976–1978 in Gwendoline Lake, only 14 contained any prey at all, 1 copepod nauplius and 13 rotifers. In later laboratory tests, 95% of 200 starved first instar larvae died within 8 days, all were dead within 2 weeks, and none molted to second instar without a meal (Neill unpubl.). A scarcity of solitary rotifers in these fishless oligotrophic lakes may be the cause of *Chaoborus* recruitment failures.

4 Nutrient Perturbations

4.1 Big Pulses and Big Effects

Study Design. Like other such lakes in the region, oligotrophic Gwendoline Lake supports few rotifers (<1/l) after grazable algal biomass declines in May and June and crustaceans become abundant (Neill 1984 and unpubl.; Walters et al. 1987). In more productive systems, however, rotifers persist at modest densities and often track pulses of algal productivity that irregularly occur in summer. I hypothesized that rotifer biomass in summer was directly or indirectly limited in oligotrophic fishless lakes by shortages of appropriate algal foods. Pulses of algal productivity might be pre-empted by other grazers or be too small to stimulate rotifer increases. The latter might be particularly important if predators play a substantial role in limiting rotifer survival. To test this food limitation hypothesis and, subsequently, to evaluate the sensitivity of the process(es) inhibiting rotifer population growth, I artificially produced pulses of algal productivity of several magnitudes via pulsed nutrient manipulations in enclosures.

To generate a strong primary productivity signal roughly analogous to eutrophic conditions, I added large amounts of inorganic fertilizers (ca. 35× the mean annual loading of PO_4 to Gwendoline Lake) in weekly doses to triplicate 27 m^3 enclosures (1.5 m diam.×15 m depth) containing Gwendoline lakewater and plankton. This manipulation was designed to grossly stimulate algal production, saturate crustacean functional and numerical responses to increased food, and stimulate rotifer production. Heavy fertilization was required because many cladocerans are good trackers of resource increases (Hall et al. 1976). There was no intention to reproduce true eutrophic lake conditions, only to massively stimulate rotifer production to test the hypothesis of a food constraint on the early life stages of the predator. Flying adult *Chaoborus* were permitted to oviposit naturally in half the enclosures in July, 6 weeks after nutrient additions began.

Results: Community Responses to Heavy Fertilization. Figures 2, 3, and 3 summarize the major results. A large nutrient manipulation to this oligotrophic system strongly stimulated algal growth, crustacean and rotifer biomass and

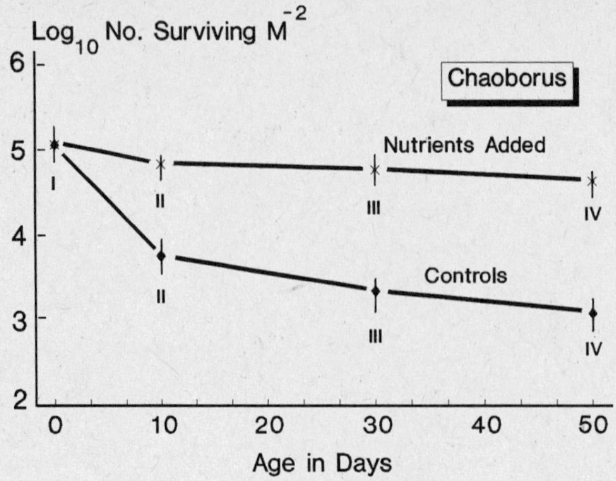

Fig. 2. *Chaoborus* survivorship in control and high nutrient enhancement enclosures. *Error bars* represent 95% confidence intervals around mean density estimate from three replicate enclosures under each treatment

Chaoborus survival. Figure 2 shows that mortality of *Chaoborus* during the first two instars declined from about 98% under oligotrophy to only about 40% after heavy nutrient enrichment, which resulted in third instar larveae being nearly 20 times more abundant than under oligotrophic conditions. Crops of 1st and 2nd instar larvae in enriched enclosures contained 5.2 ± 4.9 solitary rotifers, but only 0.02 ± 0.08 in controls, suggesting that food conditions were very much improved for the small larval predators. In addition, abundant crustaceans under heavy fertilization apparently provided a major survival boost to older larvae as well. Mortality of third instars declined from about 65% (2500/m^2 to 900/m^2) to about 15% (47600/m^2 to 41200/m^2). By the end of August, when larvae molted to fourth instar, *Chaoborus* were about 45 times denser under eutrophic conditions than under oligotrophic conditions. Large doses of nutrients provided strong primary productivity signals that increased the grazers required by small (and medium sized) *Chaoborus* larvae. These foods allowed realization of *Chaoborus*'s enormous recruitment potential in a single generation. The predatory consequences to the prey were extreme, as indicated below.

Figure 3 shows mean grazer biomasses of three replicates/treatment, pooled in four categories, at two periods during the experiment – 7 weeks and 13 weeks after initiation of the manipulations. At these times, *Chaoborus* recruits were in second or fourth instars, respectively. Unfertilized treatments in late July showed that about half the zooplankton biomass was composed of large cladoceran species (mostly *Daphnia*), with large calanoid copepods making up most of the balance of the community. Small cladoceran species (*Bosmina, Ceriodaphnia, Diaphanosoma*) were relatively rare, making up only about 5% of the total biomass. Rotifers were too scarce to count. No differences between unfertilized enclosures with and without *Chaoborus* were detected. *Chaoborus* apparently had no observable impact on their prey under oligotrophy, as had been found in earlier experiments (Neill 1981).

In contrast, enhanced primary productivity under heavy fertilization produced over six times more zooplankton biomass by late July in both *Chaoborus*

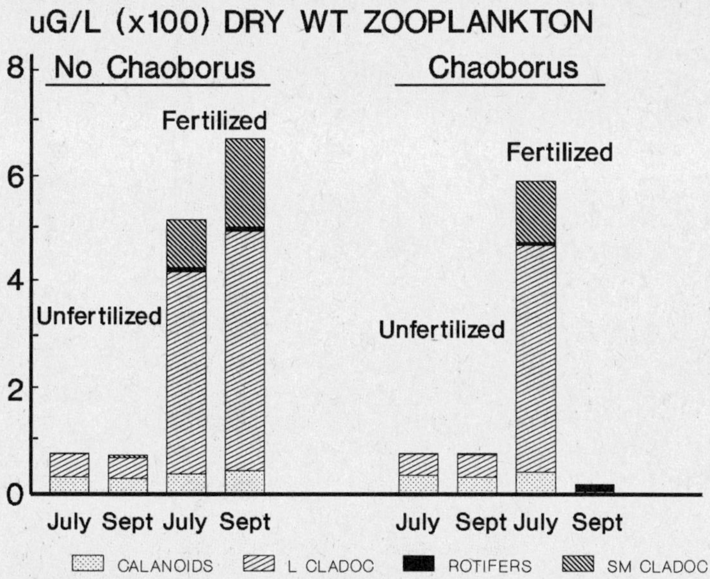

Fig. 3. Mean zooplankton biomass densities in enclosures under high and low nutrient conditions when *Chaoborus* had been exluded or allowed to recruit naturally. Data for four major groups of zooplankton (calanoid copepods, large cladocerans, solitary rotifers and small cladocerans) are shown when *Chaoborus* were mainly in second instar (late July) or in fourth instar (September). n = 3 enclosures for each treatment combination

and *Chaoborus*-free enclosures, compared to unfertilized enclosures. Large cladocerans, particularly *Daphnia*, sequestered about 75% of this increase, while smaller cladocerans, especially *Bosmina*, and solitary rotifers made up most of the balance. Rotifers were very successful recruiters and densities were about 220 times higher in fertilized/no *Chaoborus* vs. unfertilized enclosures in late July. Heavy predation by recruiting *Chaoborus* in the fertilized/*Chaoborus* enclosures reduced rotifer densities to about half that in the fertilized/no *Chaoborus* enclosures. This enormous rotifer and small cladoceran productivity fueled *Chaoborus*' survival during the first three larval instars. As anticipated in the original hypothesis, rotifers were apparently directly or indirectly food limited under oligotrophy and were released from limitation by large productivity pulses.

The data for early September reflect conditions approximately 7 days after *Chaoborus* molted into its highly predatory fourth instar. Densities of *Chaoborus* in unfertilized enclosures were so low by this time that no differences in herbivore biomass between predator and predator-free enclosures could be detected. This pattern echoes earlier results from *Chaoborus* manipulation studies (Neill 1981). In fertilized enclosures where *Chaoborus* had been excluded, zooplankton biomass bloomed to about nine times their biomass in unfertilized enclosures. *Daphnia* was the major benefactor of the enhanced productivity, but nearly all taxa were much more abundant, including some contaminating nonplanktonic chydorid cladocerans. In enclosures where *Chaoborus* had been permitted to oviposit, their extremely high survival under fertilization produced immediate and catastrophic mortality among vulnerable prey populations after the predators molted to fourth instar. All small planktonic cladoceran species, all copepods and all large cladoceran species except for a few of the very largest individual *Daphnia rosea* were eliminated within a week of the molt to fourth instar. Only rotifers, both solitary and colonial forms, were abundant, with stragglers of the other groups occasionally observed. Protozoan taxa, including many large ciliates and flagellates never previously observed in the lake were also abundant.

In short, enhanced *Chaoborus* survival in one generation had more than compensated for enhanced crustacean productivity under high nutrient enrichment. They overwhelmed the crustaceans, produced massive local extinctions, and opened up the herbivore community to forms previously unable to invade the assemblage. Rosenzweig's (1971) paradox of enrichment was dramatically illustrated by the community response to a large increase in primary production. The release of predators from food limitation destabilized the whole prey community.

Changes in nutrient availability and grazing pressures generated major changes in phytoplankton composition that were noticeable as early as July (Fig. 4). Without nutrient stimulation, phytoplankton was relatively sparse (about 850 cells/ml of all taxa) and included mostly "grazable" forms (thin-walled, <35 μm, lacking spines or a gelatinous covering). Filamentous, large colonial or thick-walled taxa were not well represented, while flagellates and two to eight cell colonies were common. Under the heavy fertilization perturbation, however, total

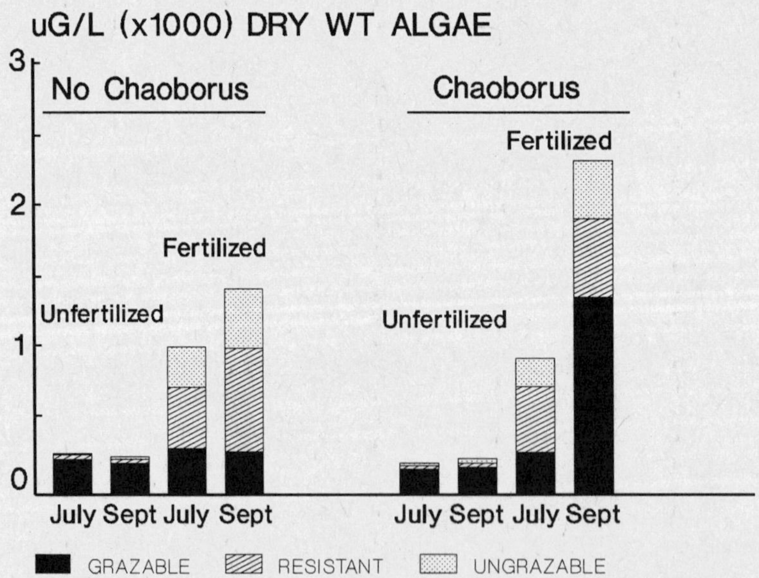

Fig. 4. Mean phytoplankton biomass in enclosures under high and low nutrient conditions when *Chaoborus* had been excluded or allowed to recruit naturally. Data for three major classes of algae are shown for same times as zooplankton biomasses in Fig. 3. n = 3 for each treatment combination. *Grazable* = thin-walled, <35 μm diameter; *resistant* = thick-walled, spiny or gelatinous and <35 μm in diameter; *ungrazable* = filamentous or colonial and >35 μm in diameter

phytoplankton biomass increased about eight times by late July compared to unfertilized controls, with filamentous, gelatinous, thick-walled or spiny forms predominating. Grazable taxa were only about 1.5 times more abundant than in the unfertilized enclosures while resistant (thick-walled or gelatinous) and ungrazable (filamentous of colonial) forms made up the balance. Despite their probable competitive advantage for nutrients and consequent high productivity, thin-walled flagellates and otherwise grazable forms were likely strongly regulated by heavy grazing from the elevated zooplankton biomass (Walters et al. 1987). The nutrient signal thus accumulated in the ungrazed fraction, as well as in the grazing herbivore and predator biomasses.

By September, total algal biomass was about 15 times larger in the fertilized enclosures than in the unfertilized enclosures lacking *Chaoborus*. Much of the algae was resistant or ungrazable, but small, thin-walled forms made up about 10% of the biomass. These small, rapidly dividing cells presumably formed the food base for most grazers. Bacterial biomass (and presumably productivity) was also very high compared to unfertilized enclosures (7.6×10^7 cells/ml vs. 5×10^5 cells/ml) and probably contributed to protozoan and possibly metazoan nutrition. After *Chaoborus* eliminated most crustacean grazing in September, total algal biomass doubled, with over 90% of the increase deriving from the grazable forms (flagellates, desmids, small dinoflagellates, small colonial greens). These taxa apparently were released from grazing pressure with the death by predation of the crustacean fauna. Only protozoans, rotifers, and some chydorid cladocerans were present to exert any grazing pressure on these normally vulnerable algal forms.

Conclusions. Clearly, providing the community with large pulses of primary productivity via weekly inputs of nutrient had a major destabilizing effect on the system. The top predator, normally severely limited under oligotrophic conditions by food shortages in early instars, was able in a single generation to become an overwhelming suppressor of almost the entire grazer community after its survival bottleneck had been broken. Community composition and dynamics were set, not by algal-herbivore interactions as under oligotrophy, but from the top of the food web by predation, freeing up resources for other taxa less vulnerable to *Chaoborus* predation. Of course, the magnitude of the *Chaoborus* response, viz, increasing from ca. 900/m^2 to about 41 000/m^2 under strong algal stimulation, could not possibly have been sustained even if the enclosures were to have remained in the lake over winter. Virtually all usable prey had been eliminated during the first 2 weeks of September, with no prospect of colonization by other taxa from external sources. *Chaoborus* at this density would have suffered major density-dependent overwinter mortality and any survivors would undoubtedly have had minimal energy reserves for pupation or reproduction. The community resulting from this nutrient perturbation thus does not represent a viable configuration. Nevertheless, the speed and magnitude of the *Chaoborus* response to the large nutrient signal implies a hidden potential for major community impact from the normally highly limited, size-structured *Chaoborus* population. Severe ecological constraints on survival of early larval instars, and secondarily on later juvenile instars, prevent the animal from exercising this potential. Release of those ecological constraints could overwhelm virtually any prey productivity response to out-run predatory mortality.

4.2 Low-Level Nutrient Perturbations

Introduction. Productivity signals of the magnitude just discussed are very unlikely in oligotrophic systems, except perhaps as a result of human intervention. However, lower pulses of primary production are well known in natural systems (Wetzel 1983). Failure to accommodate these signals within the existing community framework can, of course, generate the kinds of instabilities seen above, given that young/small *Chaoborus* are very sensitive to increases in food supplies. I was interested in assessing how small a nutrient perturbation was sufficient to destabilize the then unknown control process that limited rotifers and generated *Chaoborus*' survival bottleneck. How little nutrient was required to provide early *Chaoborus* instars with sufficient rotifers that their survival could have major mortality consequences for the crustacean prey community? Would a rainy spring with its attendant higher nutrient inputs into Gwendoline Lake allow *Chaoborus* to break its survival bottleneck and subsequently restructure the crustacean community via predation?

In 1984 and 1985, the probable culprit in suppressing rotifer densities under oligotrophic conditions was identified: *Daphnia*. I showed that removal of *Daphnia rosea* from enclosures greatly stimulated algal growth and rotifer productivity (Neill 1984) while Gilbert and Stemberger (1985) showed in the laboratory *Daphnia*'s interference effects on many rotifer taxa. To free rotifers of exploitative and interference competition exerted by *Daphnia*, it is necessary to stimulate algal biomass to the point where *Daphnia*'s functional and numerical responses to food are saturated. Food becomes available to rotifers on a sustained basis, while mechanical damage to rotifers by *Daphnia* is reduced. To detect a threshold level of primary productivity capable of releasing rotifers from this competition, I used a graded series of nutrient enrichment levels that would not substantially alter phytoplankton composition.

1984 Experiment. A 32:1 N:P fertilizer mix was chosen because it approximated natural input ratios to Gwendoline Lake at that time and had produced few relative changes among phytoplankton in trial runs. Nutrient levels were chosen to span a range typical of quite oligotrophic lakes to near mesotrophic lakes, viz., 1×, 2×, 3.5×, and 5× mean June loadings estimated for Gwendoline Lake (data courtesy of M. Feller). Nutrients were added into 27 m^3 enclosures at 4-day intervals beginning in early June. *Chaoborus* were permitted to recruit via aerial deposition of eggs by adults and by supplemented first instar larvae from nearby Shirley Lake (due to low oviposition rates in Gwendoline Lake). Three additional enclosures at the 5× nutrient loading treatment were screened to prevent egg deposition and allow assessment of *Chaoborus* and *Chaoborus*-free zooplankton communities. Figure 5 summarizes the major results for phytoplankton and zooplankton.

Phytoplankton composition was relatively well-behaved along the nutrient loading gradient, hopefully minimizing confounding effects of altered food selec-

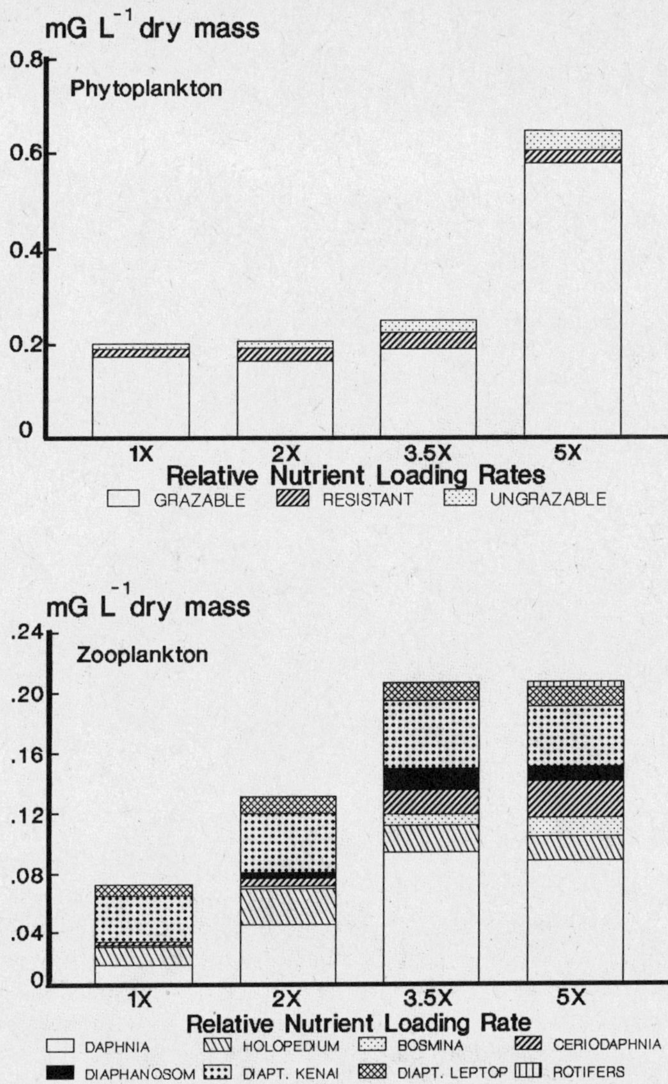

Fig. 5. Mean phytoplankton and zooplankton biomass in enclosures receiving nutrient additions during June at loading rates 1, 2, 3.5, or 5 times the mean N and P loading rates (mg/m^2/month) for Gwendoline Lake. First panel shows phytoplankton densities in early July, 1984 classified as in Fig. 4. Second panel shows zooplankton classified by species. n = 3 enclosures for each nutrient treatment

tivity by zooplankton. Further, most of the algae appeared to be "grazable" (<35 μm, thin-walled), though at very low concentrations (<800 cells/ml). Various flagellates predominated. Algal productivity 1 day after nutrient additions was strongly positively correlated ($r = 0.89$, $P < 0.01$) with nutrient loading rate (data not presented here), but algal biomass in early July was only weakly correlated with nutrients ($r = 0.52$, $P < 0.05$). Most of this latter correlation was actually due to biomass at extremes of the nutrient series, while no relation existed at all at intermediate loadings.

The reason that algal production was not realized as increased biomass during July of course was a strong response in total grazing rates. Grazing estimates indicate that *Daphnia* removed 75–90% of the increased production and that small *Daphnia* (because of their numbers and spatial distribution) actually were far more important in this process than adults >1.5 mm. Nutrient additions within the "oligotrophic lake" range (1×, 2×, 3.5× natural Gwendoline Lake loadings, Wetzel 1983) stimulated algal productivity that was apparently harvested quickly by crustacean grazers. They, of course, converted it to additional grazer biomass. The crustaceans' functional and numerical responses to enhanced production seem to have regulated algal abundance so no algal biomass differences were detected between 1×, 2×, and 3.5× treatments.

Figure 5 shows the biomass responses of the grazer community to enhanced primary production along the nutrient gradient. Total community biomass increased nearly three times before apparently saturating at the 3.5× treatment. Cladocerans, especially *Daphnia*, responded strongly while longer lived copepods responded little. *Daphnia* increased most across the series, comprising nearly 50% of the grazer biomass in the 3.5× treatment. However, its growth apparently was saturated at this nutrient loading rate as its biomass was no greater under 5× loading than at 3.5× loading. Correlatively, in the 5× treatment, there was a substantial increase in algal biomass during June and early July as grazing rates apparently could not keep up with high algal productivity at this time. Rotifer reproduction leaped more than tenfold in this treatment and rotifer abundance increased very substantially over the 6 weeks of nutrient enhancement. Peak rotifer densities exceeded 100/l in the 5× treatment, while 1–3.5× treatments had rotifers less than 1% of this value.

Rapid adjustments in total crustacean grazing (largely *Daphnia*) appear capable of quickly suppressing algal biomass to low concentrations throughout the oligotrophic nutrient input range, so that little is available for rotifer recruitment. In the presence of *Daphnia*, only relatively large inputs of nutrient seem to have the potential for releasing rotifers during summer. Crustacean control of algal biomass is a key feature of this system and it is seemingly quite insensitive to enhancement disturbances within the oligotrophic productivity range.

The responses of *Chaoborus* recruitment to this range of nutrient enhancements were consistent with the algal biomass responses. Figure 6 shows the densities of third instar *Chaoborus* in these enclosures in late July. These larvae represent the survivors of any bottleneck in the life history in first and second instar. As expected from low rotifer abundance, there was no statistically significant difference in the (low) level of predators produced among the three lower nutrient loadings (1–3.5×). At 5× Gwendoline Lake's loading rate, *Chaoborus* recruitment success was high, eventually resulting in about 5700 fourth instar larvae/m^2 – six times historical *Chaoborus* abundance estimates for the lake. Thus, a perturbation to June nutrient inputs equal to about 5× Gwendoline Lake's average input would be required to release *Chaoborus* from its juvenile bottleneck. This is a very large increase in nutrient supply and I suspect is unlikely to occur without human intervention.

Fig. 6. Densities of third instar *Chaoborus* larvae in August, 1984 in enclosures receiving nutrient additions during June at loading rates 1, 2, 3.5, or 5 times the mean N and P loading rates for Gwendoline Lake. Values shown are estimates for each replicate

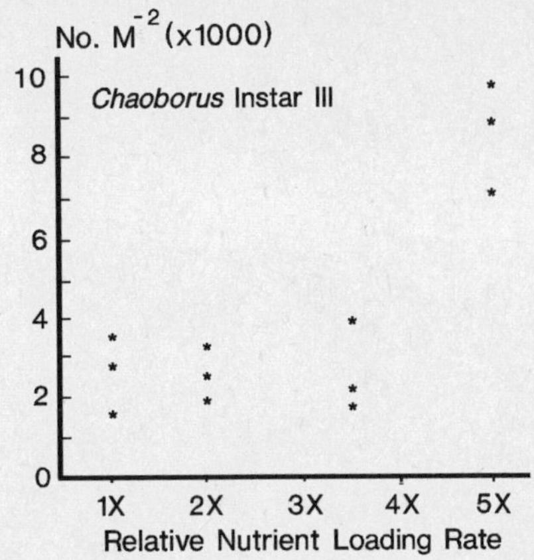

Temperature-Nutrient Interactions. Algal and zooplankton production are of course temperature-mediated processes. Temperature dependence of the rate functions may differ among taxa or age classes at both trophic levels, such that community response patterns to a productivity signal may vary with temperature. It is thus unclear whether phytoplankton-grazer-predator interactions would process a nutrient-generated pulse of primary production through the food web in similar ways in warm vs. cold years. An opportunity to examine this question occurred in the year following the low level nutrient perturbation study reported above. 1984 was about 4.5 °C warmer on average in the epilimnion during June than was 1985 (16.0° vs. 11.5 °C). I repeated the 1984 experiment of adding nutrients to enclosures for 6 weeks in June and July, and the results for 1984 and 1985 are summarized in Fig. 7.

The main response of the algal community in cooler 1985 to the same nutrient inputs as in 1984 was substantially greater biomass and somewhat different taxonomic composition. Algal biomass by early July 1985 was about 50% larger at 1× mean June loading rates than it had been in July 1984. Large flagellates (mainly chrysophytes and cryptomonads) and some diatoms that are usually gone from the epilimnion by May characterized most of the phytoplankton biomass in 1985. Smaller flagellates, small colonial greens and some blue-greens made up the assemblage in warmer 1984. All taxa in both years, however, appeared quite grazable. Further, in contrast to 1984, where algal biomass did not respond to increased nutrient inputs until 5× the mean June loading to the lake, algal biomass in cooler 1985 increased by more than 60% between 1× and 2× loadings. At 3.5× and 5× loadings it was more than 400% larger than the 1× treatment, with blooms of 15–30 μm flagellates and diatoms within two weeks of starting the enrichment. In 1984, only in the 5× treatment was algal biomass substantially greater than in the 1× treatment, yet its increase there was still only half of the increase seen in cooler 1985 at that same loading rate. Clearly, the nutrient signals were proceeding into the food web in quite different ways in the 2 years.

The reason for the different algal responses in the 2 years seems to lie in grazer responses. In warmer 1984, cladoceran and copepod growth and reproduction tracked nutrient input rates across treatments very closely until saturated in the

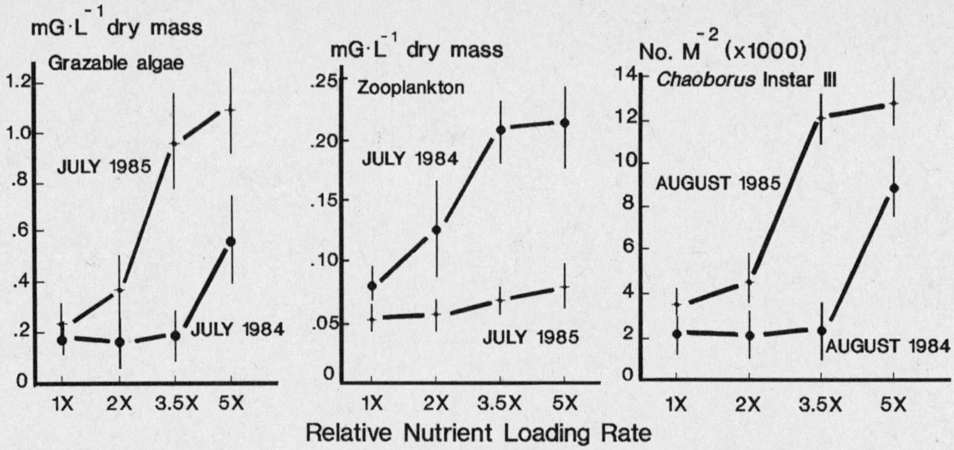

Fig. 7. Mean phytoplankton (*first panel*), zooplankton (*second panel*), and *Chaoborus* (*third panel*) concentrations in warm 1984 and cool 1985 in enclosures receiving nutrients as in Figs. 5 and 6. *Error bars* represent 95% confidence intervals. n = 3 enclosures per nutrient level per year

5× treatment. Crustacean biomass increased almost linearly with nutrient input rate from 1× to 3.5× while rotifers were nearly excluded. In cooler 1985, crustacean biomass was only about two-thirds the 1984 value in the 1× treatment and responded only weakly to differences in algal productivity across the nutrient treatments. Despite higher numbers of eggs in the clutches of most species in 1985 vs. 1984 treatments, pre- and postembryonic development rates were slowed in 1985, reducing crustacean biomass accumulation and rates of harvesting algae. Apparently, low temperature in late spring and early summer effectively created a recruitment bottleneck for the crustacean grazers, reducing their abilities to use enhanced primary productivity.

On the other hand, solitary rotifers, many of which are spring taxa in these lakes, bloomed in 1985 in all enclosures receiving nutrients above background levels (1×), achieving densities between 200 and 400/l in some samples from 3.5× and 5× treatments. Most significantly, juvenile *Chaoborus* survival leaped to very high levels at 3.5× and 5× natural loadings as a consequence of abundant rotifers (Fig. 7). More than 9200 fourth instar larvae survived the juvenile bottleneck in these two nutrient treatments by late August. Even the 2× treatment (i.e., twice the average nutrient loading rate to Gwendoline Lake and presumably not an unlikely natural variation) produced significantly denser fourth instar larvae than the 1× treatment at the lower temperature. Because there is always some uniqueness in each lake's nutrient budget, temperature-nutrient interactions of the sort seen here may be one way that higher-than-average *Chaoborus* densities are occasionally generated in some oligotrophic lakes.

In summary, given the absence of warm and cold year replication, the above results are highly suggestive that environmental conditions such as temperature may strongly affect the responses of a plankton community to accommodate a nutrient perturbation signal from the bottom of the food web. Temperature and possibly the size and taxonomic composition of the primary producers, can apparently affect the physiological and behavioral responses of grazers such that the efficiency of resource use by the community differed in the two years. Under cold conditions, a small productivity perturbation was sufficient to saturate the crustacean grazer's demographic capacities and allow inferior competitors to proliferate. Under warmer conditions, much larger perturbations were required to release the inferior rotifer competitors from competitive limitation. The survival consequences for rotifer-limited predators were then virtually determined.

Chaoborus Impacts and Crustaceans, 1984 and 1985. Whether the unusually large numbers of *Chaoborus* reaching fourth instar in the 5× treatment in 1984 and the 3.5× and 5× treatments in 1985 were sufficient to seriously impact the crustacean community is the most important issue from a community perspective. Because *Chaoborus* were prevented from ovipositing in several 5× enclosures in both years, a comparison of high *Chaoborus* and predator-free conditions can be done to examine this question. Figure 8 summarizes the impact of the high densities of predators on the late summer communities in these enclosures. In 1984, approximately 5700 larvae/m^2 had substantial negative effects in September upon some cladoceran and copepod populations compared to 5× predator-free controls, but not upon other populations. Small to medium sized cladocerans (*Ceriodaphnia, Bosmina, Diaphanosoma*) experienced 25 to 60% reductions (MANOVA, $P<0.05$, $n = 3$), while the small copepod *Diaptomus leptopus* was about 30% less abundant (MANOVA, $P<0.05$). Predation upon *Daphnia rosea* and *Diaptomus kenai* occurred mainly on the smaller/younger individuals (according to crop analyses) but had no measurable negative effect on these species' population densities. In fact compensatory increases in *Daphnia* reproduction more than made up for predatory losses to *Chaoborus*, resulting in higher

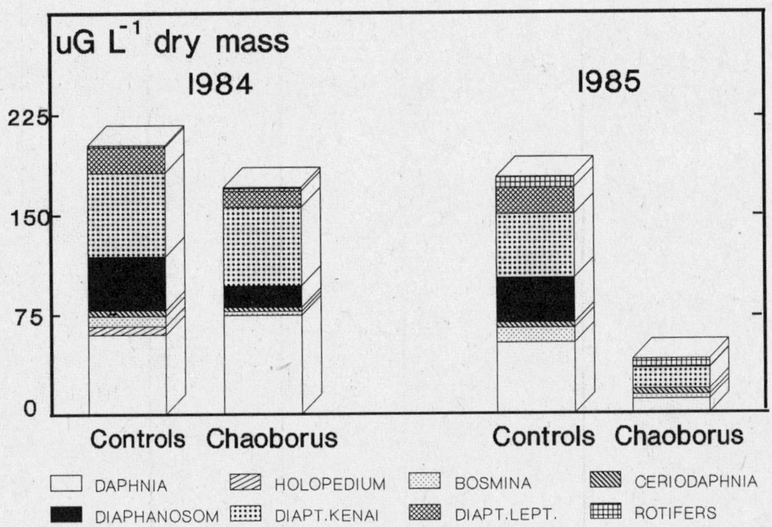

Fig. 8. Mean zooplankton biomass by species in September 1984 and 1985 under 5× nutrient treatments in the absence (controls) or presence of naturally recruiting *Chaoborus* fourth instar larvae. Shown are mean biomasses for three replicates per treatment

Daphnia densities than in the controls ($P<0.05$). At the end of the experiment in October 1984, there were nearly four times the number of *Daphnia* resting eggs (ephippia) counted in the high nutrient/high *Chaoborus* enclosures compared to the high nutrient controls. Numbers of resting eggs of the copepod *Diaptomus kenai* and the cladoceran *Holopedium gibberum* were also substantially higher than in the controls (140% and 85%, respectively). Apparently, elevated predation upon the small to intermediate sized crustaceans in August and September freed up resources for the somewhat less vulnerable survivors, resulting in their higher than normal resting egg production. It appears that *Chaoborus* predation at this intensity acts mainly to further entrench *Daphnia* as a community dominant by reducing the overwintering capabilities of small to intermediate-sized species already suffering competition difficulties (Kerfoot 1987; Neill unpublished).

Very much larger *Chaoborus* densities in 1985 ($>9000/m^2$) in 3.5× and 5× enclosures had very major negative effects on all crustaceans (except *Holopedium*) during August and September (Fig. 8). *Bosmina, Ceriodaphnia*, and *Diaphanosoma* were nearly eliminated from the enclosures, while the copepods and *Daphnia* experienced 65–95% losses compared to *Chaoborus*-free controls that had received 5× nutrients in June and July. *Holopedium*, which often does poorly in enclosures with high algal abundance, unfortunately disappeared from both predator and predator-free enclosures in 1985. Consequently, no crustacean species was common in the 3.5× and 5× *Chaoborus* enclosures. Were this condition present in the lake rather than in enclosures, some currently rare species might have been able to exploit the unused resources and perhaps invade the unoccupied grazer community.

In any event, it seems apparent that a 3.5× or 5× nutrient flush in the early summer of a colder than average year could have drastic effects on the late summer and fall community. Unless over winter mortality of *Chaoborus* at these densities were very high, some 6000 to 7000 larvae/m^2 would be present in the spring following the cold year. These abundant predators would likely have a major effect on survival of young cladocerans and copepods emerging from the less-than-usual number of resting eggs in the sediments from the previous year. Unless there were a large bank of resting eggs in the sediments from previous years, abundant *Chaoborus* could potentially limit crustacean recruitment, thereby maintaining grazing opportunities for weak competitors like rotifers and so food for first and second instar *Chaoborus*. The domination of the grazer community by *Chaoborus* predation may thus be sustainable once *Chaoborus* reach high densities in a cold, rainy year. It is somehow fitting that heavy fourth instar *Chaoborus* predation on recruiting crustaceans in spring may be able to produce a juvenile bottleneck in crustacean demography, since their competitive exclusion of rotifers is the apparent cause of *Chaoborus*' juvenile bottleneck under normal circumstances.

5 Discussion

The life history strategy of ecologically decoupling different life stages via specializations in their use of habitats and resources is phylogenetically widespread (Wilbur 1980). A major advantage of this strategy is that, in the absence of parental care, small/young stages should, on the average, be better able to avoid predators or disease and gain access to limited resources than undifferentiated forms which interact with all life stages of their species. However, on ecological time scales, temporal and spatial variation in the distribution of resources, predators, etc. may actually limit growth and survival of particular age/size classes within some local populations of such differentiated species. Thus, some populations may occur in marginal habitats where their specializations at one or more stages may actually be a liability to fitness. For example, a bottleneck in cohort maturation or survival could occur in a habitat where (specialized) young/small individuals require particular resources that vary unpredictably, that are occasionally (or consistently) outcompeted by other organisms, or that are themselves sequestered by competitively superior consumers.

The particular processes generating the condition may differ among habitats or species, but the consequences are the same. That is, mortality during a specific stage rarefies the cohort at that stage or, if resource shortages are implicated, may retard ontogenetic development such that densities at subsequent stages depend more on conditions during the bottleneck than on their own environmental conditions. Bottlenecks functionally recouple stages of the life history by limiting the maximum number of survivors at a stage to the number of survivors of the previous stage. Thus, overall commonness or rarity of a species with strongly size-structured ecology may depend largely on survival during one or two early life stages.

Examples of this kind of demography are common enough among invertebrates and lower vertebrates when life table data are available. In a crustacean community that I studied in the laboratory some years ago (Neill 1975), large-bodied *Daphnia magna* was prevented from colonizing the community because small/young *Daphnia* were competitively inferior to small bodied adult *Ceriodaphnia quadrangula*. Survival and reproduction of large, adult *Daphnia* were unaffected as their food resources remained in excess. But juvenile *Daphnia* were unable to survive at low densities of their own required foods, so adult *Daphnia* senesced without being replaced. In the *Chaoborus*-rotifer-*Daphnia* situation described in this paper, the bottleneck in small *Chaoborus* was caused by the competitive superiority of (small) *Daphnia* over rotifers, which are the principal food of small *Chaoborus*. Large *Chaoborus* were the indirect casualties of an interaction they were unable to influence because so few survived the bottleneck. External changes in circumstances (substantially elevated primary productivity, cold spring temperatures) were required to release rotifers from a very resilient strong competitor, to allow *Chaoborus* to recruit in large numbers and suppress the competitively superior *Daphnia*. *Chaoborus trivittatus* in deep oligotrophic lakes thus appears to be surviving in a marginal habitat for its evolved characteristics. One lifestage requires high densities of particular prey that can be plentiful (in the absence of fish) only when competition from large crustaceans is minimal. They are best adapted to shallow, productive ponds and lakes where their prey can compete effectively.

One final example comes from Stockner (Stockner and Porter 1988) who interprets a bottleneck in the growth of sockeye salmon fry (*Oncorhynchus nerka*) as being caused by a competitive interaction among food resources used by fry in an analogous manner to the above rotifer-*Daphnia* interaction. Like most young planktivorous fish, salmon fry are well adapted to prey on large zooplankton such as *Daphnia*. Though normally widespread and able to withstand salmon predation in most localities, *Daphnia* seems to be outcompeted in certain types of lakes by rotifers and small crustaceans that are best adapted to some unusual foods present there. Thus, only small, relatively unprofitable prey are available for these salmon fry and growth is severely retarded, limiting juvenile salmon survival after migration to the marine environment.

Curiously, most general theory in population biology and community ecology fails to actively incorporate age- and, especially, size-structure, despite its dynamical and evolutionary significance to large numbers of organisms (Neill 1975; Werner 1984). Fisheries biologists and economic entomologists, for example, have long appreciated the potential of population limitation and release by stage-specific events, especially on young/small individuals (e.g., Hjort 1926; Morris 1957). On the other hand, many terrestrial vertebrate ecologists, especially bird ecologists, whose views have influenced much of general animal population theory, usually deemphasize age/size dependencies in population dynamics. Recruitment bottlenecks from juvenile stages are not commonly seen to limit adult numbers. Instead, adult-adult interactions dominate the dynamics of birds and other resource limited, near-equilibrium terrestrial vertebrates. Opportunities for young/small individuals to influence adult numbers occur more commonly in populations displaced (temporarily?) far from resource equilibrium.

The mathematics required to model these latter dynamics can sometimes be complex and nongeneral, especially if size and age are partially decoupled by indeterminant growth and phenotypic plasticity. Indeed, relatively little general theory has been developed that is explicitly size-dependent, not age-dependent. Nevertheless, the dynamics that often occur in size-structured populations, especially among fish, many planktonic invertebrates, some insects, most higher plants and others, require theoretical models and empirical studies that have good resolution at the life stage level. Stage-specific constraints on survival and growth readily generate thresholds and bifurcations leading to startling dynamics, whose etiology is perplexing without an appreciation of the finer scales of interaction.

Whether the constraining processes on stage-specific growth and survival, especially the density-dependent ones like parasitism, disease, competition and (some) predation, are resilient to variation in environmental conditions is especially important to maintenance of the limitation. Environmental conditions obviously vary and regulatory processes must respond accordingly. In the plankton system that I have been manipulating, the rapid functional and numerical responses of the dominant herbivore *Daphnia* are essential in suppression of micrograzers (rotifers). These small grazers are widely regarded as excellent opportunists that take advantage of short pulses of food production (Hutchinson 1967). Within insects, opportunistic species such as these are regarded as among the most likely to outbreak because of the vulnerability of the regulating processes to disturbance (Stenseth 1987). However, because *Daphnia*'s functional responses are far from saturated in these lakes (Walters et al. 1987) and substantial numbers of them persist at low food levels, most of the enhanced productivity disappears before rotifers increase in density. Unusually high dose rates of nutrients, such as those from logging, agricultural run-off, waste water disposal and other human activities, are required to saturate *Daphnia*'s functional and numerical responses to elevated primary productivity in these lakes. On the other hand, if some other factor such as low temperatures or possibly fish predation limits *Daphnia*'s response to food, rotifers bloom on the unused phytoplankton, and the bottleneck on *Chaoborus*'s survival is broken. Thus, within the normal temperature range and in the absence of fish, the control process indirectly regulating *Chaoborus*' survival is relatively robust to natural productivity fluctuations. Such weather-related releases from endemic levels to epidemic levels are commonly invoked by entomologists.

Finally, I would like to make a few observations about the curious nature of size-structuring as perceived by many aquatic ecologists. With the major exception of microbial forms, age/size-structured populations are the norm in aquatic communities. Yet constraints on early/small life history stages of species are not integrated into aquatic community theory. Plankton ecologists over 20 years ago (Brooks and Dodson 1965) recognized the conceptual importance of body size differences among *species* to the ecology of planktonic food webs. But almost no one approaches the ecology and evolution of a large species as if the organism were influenced by the ecology of its small offspring. Romanovsky (1985) and Romanovsky and Feniova (1985) are important exceptions. I am convinced that aquatic ecologists do not avoid population substructure for fear of the work involved. Difficulties like that merely attract aquatic biologist's attention. I suspect that a principal culprit in this adult-biased species-centered view of zooplankton community ecology is the insidious notion that the "Purpose" of zooplankton (and benthic invertebrates) is to feed fish. Because many differences in the abundances of the most obvious and "important" species (i.e., the big ones) can be explained by fish predation upon larger individuals, there is little reward in examining their substructure (size/age) dynamics. The dynamics of juveniles of these large species can be defined as unimportant, or at least non-limiting, with little or no loss of predictive power. Whether this is also true of the majority of taxa that are small and less affected by fish or of larger taxa in the absence of fish predators (McCauley and Murdoch 1987) is doubtful (Neill 1975).

In many communities, at least 75% of the biomass, grazing, or nutrient regenerating capacities is found below the threshold size at which fish planktivores show interest (Le Cren and Lowe-McConnell 1980). Thus, the potential for ecosystem, community and food-web dynamics to be affected by processes occurring among or generated by the immature and small individuals in an assemblage cannot be ignored in our theory. In my opinion, much greater empirical and theoretical attention needs to be focussed on the ecology of these organisms. That the responses of systems to perturbations like nutrient pulses can be mediated by the success and failure of juvenile stages implies that our horizons may be too closely bounded.

Acknowledgments. The experimental work reported here could not have been accomplished without the excellent technical assistance of Danuta Dolecki. Discussions with students and faculty in the Ecology Group at the University of British Columbia helped with many aspects. I also thank Gary

Mittelbach and Lennart Persson for helpful comments on an earlier draft of this paper. Ann Tautz and Linda Blaine helped prepare the manuscript, for which I am grateful. Support for this research was provided an operating grant from the Natural Science and Engineering Research Council of Canada.

6 References

Brooks JL, Dodson SI (1965) Predation, body size, and composition of plankton. Science 150:28–35

Fedorenko AY (1975) Feeding characteristics and predation impact of *Chaoborus* (Diptera, Chaoboridae) larvae in a small lake. Limnol Oceanogr 20:250–258

Fedorenko AY, Swift MC (1972) Comparative biology of *Chaoborus americanus* and *Chaoborus trivittatus* in Eunice Lake, British Columbia. Limnol Oceanogr 17:721–730

Giguere LA (1981) The energetics of predation: a components study of *Chaoborus trivittatus* larvae. Thesis, Simon Fraser Univ

Hall DJ, Threkeld ST, Burns CW, Crowley PH (1976) The size-efficiency hypothesis and the size structure of zooplankton communities. Annu Rev Ecol Syst 7:177–208

Hillbricht-Ilkowska A, Kajak Z, Ejsmont-Karabin J, Karabin A, Rybak J (1975) Ecosystem of Mikolajskie Lake: the utilization of the consumers' production by invertebrate predators in pelagic and profoundal zones. Pol Arch Hydrobiol 22:53–64

Hjort J (1926) Fluctuations in the year classes of important food fishes. J Conserv Int Explor Mer 1:1–38

Hutchinson GE (1967) A treatise on limnology. Vol. 2. John Wiley, New York

Kajak Z, Ranke-Rybicka A (1970) Feeding and production efficiency of *Chaoborus flavicans* Meijen (Diptera, Culicidae) larvae in eutrophic and dystrophic lakes. Pol Arch Hydrobiol 17:225–232

Kerfoot WC (1987) Cascading effects and indirect pathways. In: Kerfoot WC, Sih A, Predation: direct and indirect impacts on aquatic communities. Univ Press New Engl, Hanover, NH, pp 57–70

Kerfoot WC, Sih A (eds) (1987) Predation: direct and indirect impacts on aquatic communities. Univ Press New Engl, Hanover NH

Le Cren ED, Lowe-McConnell RH (1980) The functioning of freshwater ecosystems. Univ Press, Cambridge

McCauley E, Murdoch WW (1987) Cyclic and stable populations: plankton as paradigm. Am Nat 129:97–121

Mills EL, Forney JL (1988) Trophic dynamics and development of pelagic food webs. In: Carpenter SR (ed) Complex interactions in lake communities. Springer, Berlin Heidelberg New York London Tokyo

Mittelbach GG (1981) Foraging efficiency and body size: a study of optimal diet and habitat use by bluegills. Ecology 62:1370–1386

Morris RF (1957) The interpretation of mortality data in studies of population dynamics. Can Entomol 89:49–69

Neill WE (1975) Experimental studies of microcrustacean competition, community composition and efficiency of resource utilization. Ecology 56:809–826

Neill WE (1981) Impact of *Chaoborus* predation upon the structure and dynamics of a crustacean community. Oecologia (Berlin) 48:164–177

Neill WE (1984) Regulation of rotifer densities by crustacean zooplankton in an oligotrophic montane lake in British Columbia. Oecologia (Berlin) 61:175–181

Pastorok RA (1981) Prey vulnerability and size selection by *Chaoborus* larvae. Ecology 62:1311–1324

Riessen HP, O'Brien WJ, Loveless B (1984) An analysis of the components of *Chaoborus* predation on zooplankton and the calculation of relative prey vulnerabilities. Ecology 65:514–522

Romanovsky YE (1985) Food limitation and life-history strategies in cladoceran crustaceans. Arch Hydrobiol Beih 21:363–372

Romanovsky YE, Feniova IY (1985) Competition among Cladocera: effect of different levels of food supply. Oikos 44:243–252

Rosenzweig ML (1971) Paradox of enrichment: destabilization of exploitation ecosystems in ecological time. Science 171:385–387

Stenseth NC (1987) Evolutionary process and insect outbreaks. In: Barbarosa P, Schultz JC (eds) Insect outbreaks. Academic Press, San Diego New York London

Stockner JG, Porter KG (1988) Microbial food webs in freshwater planktonic ecosystems. In: Carpenter SR (ed) Complex interactions in lake communities. Springer, Berlin Heidelberg New York London Tokyo

Walters CJ, Krause E, Neill WE, Northcote TG (1987) Equilibrium models for seasonal dynamics of plankton biomass in four oligotrophic lakes. Can J Fish Aquat Sci 44:1002–1017

Wetzel R (1983) Limnology. Saunders, Philadelphia

Werner EE (1984) The mechanisms of species interactions and community organization in fish. In: Strong DR, Simberloff D, Abele LG, Thistle AB (eds) Ecological communities: conceptual issues and the evidence. Univ Press, Princeton, NJ

Werner EE, Mittelbach GG, Hall DJ, Gilliam JF (1983) Experimental tests of optimal habitat use in fish: the role of relative habitat profitability. Ecology 64:1525–1539

Wilbur HM (1980) Complex life cycles. Annu Rev Ecol Syst 11:67–94

Zaret TM (1980) Predation and freshwater communities. Yale Univ Press, Hanover, NH

Size-Abundance Relations in Plants and Animals

Self-Thinning of Plant Populations Dictated by Packing Density and Individual Growth Geometry and Relationships Between Animal Population Density and Body Mass Governed by Metabolic Rate

R. Å. NORBERG

1 Introduction

When individual plants grow in an uncrowded population there is little or no mortality from competition. In a diagram displaying the logarithm of the volume of an average plant against the logarithm of population density, growth trajectories therefore are nearly vertical. But as plants begin to fill up the available space, crowding commences, and the ensuing mortality reduces population density and causes the growth trajectories to bend off towards the ordinate (Figs. 1, 2). As plants grow, the data points representing consecutive growth stages of survivors in a log-log diagram lie along this thinning line, from a high population density and small average volume per plant at lower right, upward and leftward toward the ordinate (Figs. 1, 2).

The growth trajectories of populations with different starting densities eventually approach a common asymptote, which constitutes the so-called limiting thinning line (Figs. 1, 2; White and Harper 1970; Kays and Harper 1974; Westoby 1976; White 1981). It is a line of constraints in that no data-points occur above it. For an even-aged plant population in dynamic equilibrium at some crowding density, the power function

$$V = k_1 N^a , \quad (1)$$

relates the maximum, average plant volume V to the crowding population density N. Empirically obtained exponents come close to -1.5 for terrestrial plants as different as mosses, ferns, grasses, herbs, and some trees (Fig. 1; Yoda et al. 1963; White and Harper 1970; Gorham 1979; White 1980). Depending on the kind of plant, the line's elevation (the y-intercept k_1) in a log-log diagram may differ by a factor of eight on a linear scale (White 1981, p. 494), but the exponent usually remains close to $-3/2$. Provided that the specific gravity of the plants does not change systematically with growth, the volume of plant biomass and the biomass itself can be used interchangeably, albeit after adjustment of the coefficient k_1 (the y-intercept) as required.

This is the well-known-3/2 power-law in plant ecology,

$$V = k_1 N^{-3/2} . \quad (2)$$

The relationship described by Eq. (2) between the average plant volume and the population density often applies to natural populations, provided that the popu-

Department of Zoology, University of Gothenburg, Box 250 59, 40031 Gothenburg, Sweden

Size-Structured Populations
B. Ebenman and L. Persson (Eds.)

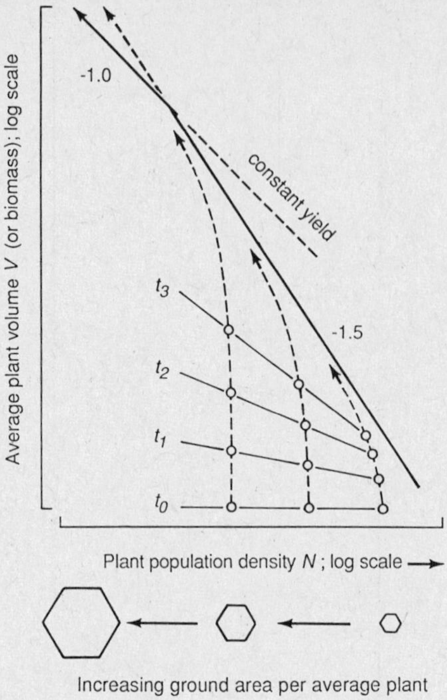

Fig. 1

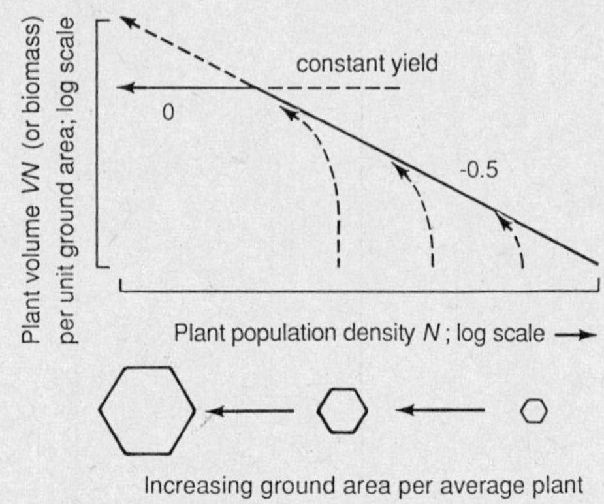

Fig. 2

Fig. 1. Generalized diagram showing two limiting thinning lines from the power function $V = kN^{\alpha}$, relating average plant volume V to plant population density N, with α taking the values -1.5 and -1, which become gradients of the lines in a log-log plot. The -1.5 slope is consistent with geometric similarity of growth, but may be mimicked by other growth modes. The *three steep, broken arrows* are growth trajectories for different populations, from various initial population densities, upward to the constraining line of slope -1.5. The "competition-density" lines labeled $t_0 - t_3$ connect data-points representing the same age of the three populations. The -1 line represents constant yield (see legend to Fig. 2). (Modified from Fig. 1 in Norberg 1988; based on White and Harper 1970; Kays and Harper 1974; Westoby 1976; White 1981)

Fig. 2. The same generalized diagram as Fig. 1, but with plant volume per unit of ground area plotted against population density. The slopes change from -1.5 to -0.5 and from -1 to 0. *The 0 line* represents constant yield, that is, the growth in biomass of individual plants is exactly compensated for by the concomitant population thinning, resulting in a constant plant biomass per unit of ground area. (Modified from Fig. 2 in Norberg 1988)

lation is subjected to such severe crowding that competition leads to mortality and self-thinning of the population (see also Ebenman this Vol. and Wilbur this Vol.).

This power law applies to the following three kinds of data from terrestrial plants of profoundly different form and structure, spanning the entire size range from mosses to trees: (1) to successive growth stages of plants in single-species populations, that is, to time trajectories for single-species populations, and probably also for mixed-species populations (White and Harper 1970); (2) to different growth stages of plants in different populations of the same species and, again, probably also to mixed-species populations; and, most notably, (3) it applies also to regressions across single-species populations but with the different populations composed of different plant species of different sizes. The same rule, or very similar rules, therefore seem to govern the size-versus-density relationships observed among plants of profoundly different size, form and structure. The $-3/2$ self-thinning rule is supposed to apply to populations of even-aged plants, provided that any one population is crowded (Yoda et al. 1963; Gorham 1979; White 1980); the possibility that it also applies to size-structured populations is discussed in Section 10 below.

The self-thinning rule is sometimes written in an alternative form,

$$NV = k_2 N^{\beta} , \tag{3}$$

with volume of plant material NV per unit of ground area versus population density N (Fig. 2). Equation (3) is obtained by multiplying Eq. (1) by N, and when α is -1.5, β becomes $-1/2$. Westoby (1981, 1984) and Weller (1987) advocated the version of the thinning law given in Eq. (3) and gave several reasons why it is better for fitting curves to empirical data. But the original version [Eq. (1)], relating volume of an average plant to population density, is the obvious choice for theoretical models depending upon plant shape, structure, and growth mode (Norberg 1988).

The dynamic equilibrium between individual growth and mortality, represented by the limiting thinning line, is identified by a lack of further flattening of growth trajectories of individual populations and a lack of further steepening of the "competition-density" lines, like line $t_0 - t_3$ in my Fig. 1 (term from White 1981, Fig. 1), each one of which connects equal-age points for different populations.

The $-3/2$ thinning rule is one of the most important principles in plant ecology (White 1980), considered by John Harper to be "the only generalization worthy of the name of a law in plant ecology" (cited in Hutchings 1983, p. 765). It was derived empirically, and has subsequently been corroborated by new empirical data, even though many empirical studies have produced important deviations from the $-3/2$ exponent (Weller 1987). The theoretical explanation of the $-3/2$ thinning rule is disputed (Westoby 1977, 1984; White 1977, 1980, 1981; Miyanishi et al. 1979), and it was recently described as an ecological law in search of a theory (Hutchings 1983). There has been some confusion about the reasons why populations of plants of drastically different size, form, and structure should obey the $-3/2$ thinning rule, or one or other of a few notable deviations from it, such as exponents near -1.8 for many trees, and exponents of about -1 for stands of big trees.

The causal basis of the thinning rules obviously does not depend on details of plant structure and shape, but rather on general, geometric principles of growth and packing. In this chapter I attempt to clear up significant misunderstandings in previous discussions about the $-3/2$ thinning law. In order to understand the thinning exponents it is necessary to consider the growth mode of individual plants, that is, the way in which plant form and proportions change with increasing size. For between-species comparisons of plants of different sizes the size-related design principle of plants has the same central role for the exponent as has the growth mode for intrapopulation comparisons at different growth stages. I treat mathematical relationships between design principles of plants for structural strength, the growth mode of individual plants, packing geometry of plants, and population thinning resulting from individual growth and ensuing competition. I also show that for functional reasons, different plant parts must sometimes exhibit different growth modes, so that the thinning function for whole plants is the result of a combination of different thinning functions for different plant parts (while in earlier models one growth mode has, sometimes erroneously, been applied to the whole plant). Thus, for instance, when the root has a different growth mode than above-ground structures, and when root competition drives population thinning, the geometry of aerial structures may easily be at variance with the thinning function for above-ground plant parts (Norberg 1988). This might explain perplexing discrepancies between population thinning and allometry of above-ground plant parts.

One purpose of this chapter is to explore to what extent the self-thinning law is applicable to plant populations with a mixed age, size, and species structure. But before this can be done, the self-thinning law must be genuinely understood. Therefore the main part of this chapter is devoted to explaining the self-thinning law. I will focus on explanations of the gradient of the thinning curve in a log-log diagram [α in Eq. (1)], using packing theory and various growth modes, but will also consider in detail the nature of the coefficient k (the y-intercept). The following presentation builds upon my recent theoretical treatment of plant population thinning (Norberg 1988), which is developed further here in several respects. New developments presented here include: (1) a new derivation of the self-thinning law based upon geometric similarity; (2) explanations of differences between stands and species in the local elevation of self-thinning curves and of their y-intercepts; (3) a new, simple, final expression for the thinning slope following from the elastic-similarity principle; (4) a brief discussion of self-thinning of plant populations with different age, size, and species structures; and, finally, (5) a comparison with self-thinning functions for animal populations.

2 Geometric Similarity in Individual Plant Growth; Derivation of the −3/2 Self-Thinning Law of Plant Populations

The simplest growth mode is when plants grow isometrically so that individual plants remain geometrically similar throughout individual growth. Geometric similarity of growth results when all lengths increase in proportion to a representative length l, whence it follows that areas increase as l^2 and volumes as l^3. Before going into more complex growth modes and the associated thinning functions, I first give an account of close-packing of geometrically similar plants, based on the original idea of Yoda et al. (1963). They derived the −1.5 thinning exponent from empirical data and also gave a geometric explanation for it – later elaborated by White and Harper (1970). The following derivation is similar in principle to those given by Yoda et al. (1963), White and Harper (1970), Whittington (1984), and Norberg (1988), but differ in detail. In particular, the following derivation for the first time clearly identifies the factors that dictate the local elevation of the self-thinning line and its y-intercept k.

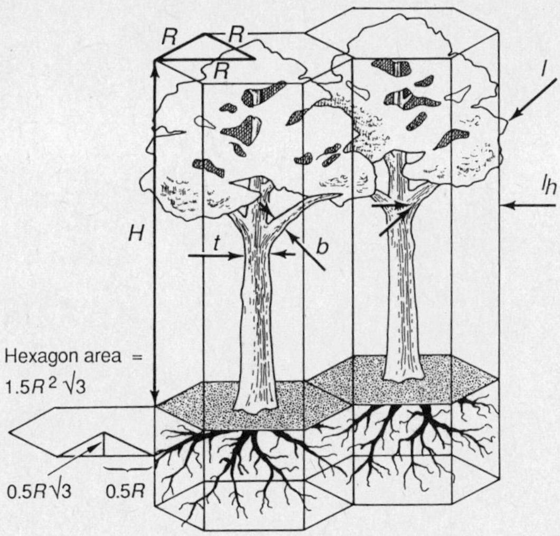

Fig. 3. Scheme of the geometry of closely packed plants. R is the radius of the average, hexagonal, ground area occupied per plant in a crowded population; H is height; t, trunk diameter; b, branch diameter; l, branch length, and l_h, the horizontal projection of branch length. The ground area of the average spatial volume per plant may be the projected area of the crown or root, or any larger or smaller area from which the plant draws resources (light, water, and nutrients) to the exclusion of others. (Modified from Fig. 3 in Norberg 1988)

When plants are close-packed, the average ground area occupied per plant equals the area of the plot divided by the number of plants, and similarly for spatial volumes. It is convenient to think of the spatial volume of an average plant as a hexagonal prism (for close-packing), as shown in Fig. 3. When plants retain isometry throughout growth, the height-to-breadth ratio of a plant's exclusive space also remains constant, that is, the height H remains a fixed multiple h of the diameter (or strictly, the largest diagonal) 2R of the ground area occupied per plant;

$$H = h2R \quad \text{or} \quad h = H/2R \ . \tag{4}$$

The hexagonal ground area A is

$$A = 1.5 \cdot 3^{1/2} R^2 \tag{5}$$

and the spatial volume occupied per average plant, v, is

$$v = H \cdot A = h2R \cdot 1.5 \cdot 3^{1/2} R^2 = 3 \cdot 3^{1/2} h R^3 \ . \tag{6}$$

When plants are densely packed, their number per unit of ground area, N, is inversely related to the average ground area occupied per plant, A [Eq. (5)], so

$$N = 1/(1.5 \cdot 3^{1/2} R^2) \ . \tag{7}$$

Rewriting Eq. (7) for our purposes,

$$R = (1.5 \cdot 3^{1/2} N)^{-1/2} \tag{8}$$

and inserting into Eq. (6), the volume of the hexagonal prism, v, becomes

$$v = 3 \cdot 3^{1/2} h (1.5 \cdot 3^{1/2} N)^{-3/2} = 1.24 h N^{-3/2} \ . \tag{9}$$

Let

$$k_2 = 1.24 h \ , \tag{10}$$

yielding

$$v = k_2 N^{-3/2} \ . \tag{11}$$

When the measuring units match, k_2 does not contain any matching factor for units.

Now consider the plant that claimed the prism-shaped spatial volume in the first place. Let packing density, p be the proportion of the plant's spatial volume (the hexagonal prism) that is filled up with plant material, or biomass, including also those parts of the plant that may lie outside of the hexagonal prism when there is mutual spatial overlap between neighbours. When different-sized plants remain geometrically similar, the packing density p is the same in comparisons across individuals. This follows from the geometric similarity criterion, because the volume of plant material and the volume of the exclusive space occupied by the plant both scale as the cube of any representative linear dimension [Eq. (6)]. From Eq. (9) the volume of plant material, V, then becomes

$$V = 1.24 h p N^{-3/2} \ . \tag{12}$$

Let

$$k_1 = 1.24 h p \ , \tag{13}$$

and the famous $-3/2$ thinning law results,

$$V = k_1 N^{-3/2} \ . \tag{2}$$

When the self-thinning law is derived in this way, the nature of k_1 is fully identified [Eq. (13)]. The factor 1.24 is derived from the hexagonal shape of the exclusive ground area occupied per average plant (Fig. 3). If the ground area were a square, this factor would take the value 1. The factor h is the height-to-width ratio of the exclusive space, that is, the hexagonal prism, occupied by an average plant [Eq. (4)], and p is the packing density of plant material within the exclusive spatial volume of the hexagonal prism.

The two necessary and sufficient assumptions behind the above derivation are that (1) plants retain geometric similarity throughout growth and that (2) crowding prevails so that self-thinning occurs. This is an extremely simple interpretation of the self-thinning law. But the close fit of a large amount of empirical data to this power function (Gorham 1979; White 1980) suggests that it is an adequate, albeit crude, explanation in many cases, even though important deviations from this rule occur (Weller 1987). Factors that dictate the values of a and k in Eq. (1) are considered below, followed by discussions of the confounding effects of branches and of the foliage of trees.

3 The Exponent a

The Growth Mode of Individual Plants Dictates the Exponent a

The geometric similarity criterion implies that the volume of plant material is proportional to the cube of any characteristic linear dimension of the plant itself and of the hexagonal prism-shaped space that it occupies. Thus, for instance,

$V \propto R^3 \propto H^3$. It is important to note that this convenient relationship applies only when different-sized plants maintain geometric similarity. With most other growth modes (or design principles) there is no single linear dimension, either of the plant itself or of its exclusive spatial volume, whose cube is proportional to the volume of plant material. Likewise, and as emphasized by White (1981), except for the growth mode which results in geometric similarity, there is no single plant measure whose square is proportional to the exclusive ground area of the plant.

The coefficient a in Eq. (1) is strictly determined by the growth mode of the plants in the self-thinning population. The a is absolutely insensitive to differences in shape and structure between plants. It only reflects the way plants *change* shape and structure throughout growth (Norberg 1988). The failure to realize this has caused much confusion and mystification as to why populations of plants with drastically different shape and structure often conform with the $-3/2$ thinning rule.

The value $-3/2$ for a may signify no change at all of shape and structure, i.e., the maintenance of geometric similarity throughout individual growth and population thinning. But other growth modes may mimic the $-3/2$ power. It is important to note that in an across-species regression with a given k, and with $a = -1.5$, it is still eminently possible for the plants to be geometrically dissimilar, as indeed they often are. Data-points for different plant species will fall along the same line despite differences in plant size, structure, and form, if only the height-to-breadth ratio is the same (throughout growth and for all species), and the same proportion of the spatial volume occupied per plant is filled with plant material (throughout growth and for all species) – or if these two variables change compensatorily. This is how the geometric-similarity-effect on a (turning it into $-3/2$) may be mimicked despite geometric dissimilarities among plants. Conversely, with given k and height-to-breadth ratio, constancy of packing density (within the space volume occupied per plant) obtains only when the exponent is -1.5, because only then do empty spaces between plant parts scale up to the same extent as do plant parts themselves, i.e., only then will plant parts fill the same proportion of the space volume occupied per plant, despite differences in plant size.

With certain growth modes, such as for prostrate plants (Norberg unpubl.), or when elastic similarity prevails (Norberg 1988), the thinning function does not give a straight line in a log-log diagram of plant volume versus population density. Self-thinning is instead described by a curve, as explained below.

4 The y-Intercept k

4.1 Increasing Packing Density of Plant Material Within a Plant's Exclusive Space Increases the y-Intercept k

In a log-log diagram of average plant volume versus population density, the coefficient k determines the elevation of regression lines with the same slope a. A convenient comparison between different k values may be done at the lines' y-intercepts at population density 1 [since log (1) = 0]. As long as the slope is the

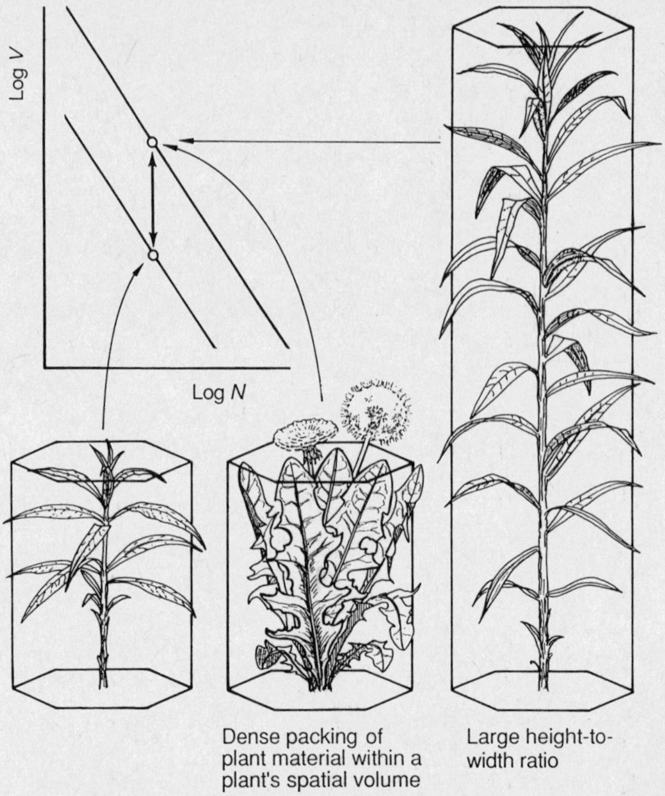

Fig. 4. Increased packing density of plant material within the spatial volume occupied by a plant, and increased height-to-width ratio of the plant, both result in more plant material being stacked onto the average ground area per plant. This gives a higher elevation of the self-thinning line because of an upward, vertical, displacement of data points for given population densities, that is, for given average ground areas per plant (x-coordinates)

same for the lines being compared, it need cause no concern that the y-intercept is often outside the observation range, therefore representing hypothetical plant volumes at population densities which may never occur; with identical slope the ratios between the elevations of different lines remain identical regardless of the x-coordinate chosen for a comparison. Even when slopes differ, y-intercepts are useful for descriptive purposes, but biological comparisons and interpretations must then be restricted to some specified x-coordinate, preferably within the range of observation.

The coefficient k equals 1.24 h p [Eqs. (1), (13)]. Therefore, the packing density p of plant material within the spatial volume occupied per plant directly affects the y-intercept k. Packing is determined by, for instance, the girth and taper of stem and branches as well as by the size and number of leaves. The more densely that plant parts are packed within the plant's exclusive space, and the more solid they are, the larger p, and hence k, become. Increasing packing density within the spatial volume occupied by a plant results in a higher elevation (or y-intercept) of the thinning line because of an upward, vertical, displacement of the data-points in the log-log thinning diagram (Fig. 4).

4.2 Increasing Height-to-Width Ratio Increases the y-Intercept k

Because the coefficient k equals 1.24 h p [Eqs. (1), (13)], the characteristic height-to-width ratio h of the exclusive spatial volumes occupied by individual plants in a population directly affects the y-intercept k. The taller and narrower that plants are, the higher the y-intercept k should be (Fig. 4). Therefore, grasses and tall but narrow coniferous trees, like *Picea*, should have higher y-intercepts than most other plants. This has also been observed (Lonsdale and Watkinson 1983).

Increasing the height-to-width ratio of the spatial volume occupied by a plant results in a higher elevation (or y-intercept) of the thinning line because of an upward, vertical, displacement of the data-points in the log-log thinning diagram (Fig. 4). The net effect therefore is as with increasing packing density of plant material within the plant's spatial volume (Sect. 4.1). Indeed, what matters in determining the elevation of the thinning line is of course only how much plant material that is stacked onto the ground area occupied by an average plant; it is immaterial whether a large volume of plant material is achieved by dense packing of plant material within the plant's spatial volume, or by growth to large heights.

4.3 Increasing Overlap Between Neighboring Plants Increases the y-Intercept k

Consider plants of a particular form and structure, and growing at a particular population density. Then increase the population density but maintain everything else constant. Because the spacing between individual plants is now smaller, but the plants have the same size and form, there is more mutual overlap between neighbors and more shared space. The average ground area occupied per plant is smaller, but the volume of plant material per plant is the same. The resulting increase in elevation of the thinning line (and of its y-intercept) is brought about by a horizontal displacement of the data-points to the right (Fig. 5). So even though the plants maintain the same form, structure, and volume, increased crowding results in higher local elevations and a higher y-intercept k.

Packing density of plant material (Sect. 4.1) and a plant's height-to-width ratio (Sect. 4.2) are species-characteristics features. But to some extent both may be affected also by the environment. For instance, the characteristic light levels may restrict the packing density of plant material within an individual plant's spatial volume. The degree of overlap between neighboring plants that plants can tolerate may also be species-characteristic (this section), but it should be particularly sensitive to environmental variables, such as light, water, and nutrients. The elevation k of the thinning line may thus be influenced in various ways by the interaction of species-characteristic features and environmental variables.

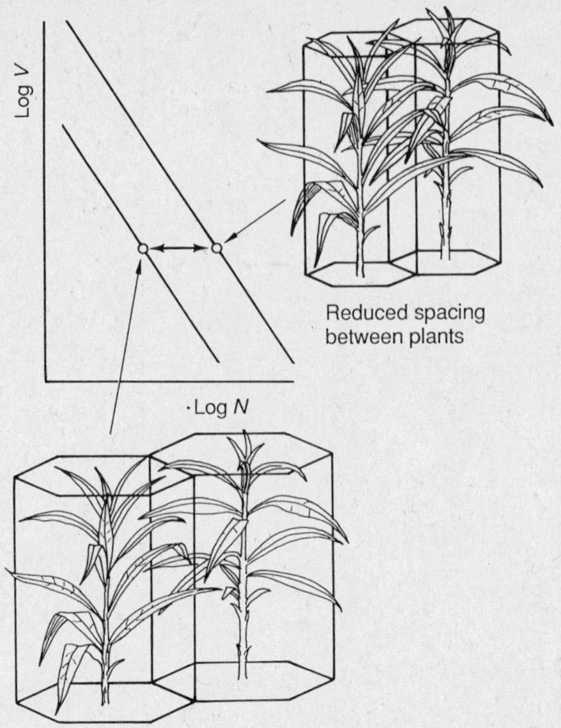

Fig. 5. Increased overlap between neighboring plants results in a smaller average ground area per plant and a higher elevation of the self-thinning line because of a horizontal, rightward, displacement of data points for given plant volumes (y-coordinates)

5 Geometric Similarity and Branches

Strict maintenance of geometric similarity throughout growth requires a constant number of branches and the subjection of branches, and their individual separation, to only isometric enlargement while their proportions remain constant. But new branches obviously develop as plants grow taller, and their spacing may remain more or less constant, thereby violating the geometric-similarity criteria. Norberg (1988, p. 228) showed that under such circumstances, the total branch volume per plant $\sum V_b$ should be related to the radius R of the average ground area per plant, and to plant population density N, as

$$\sum V_b \propto R^4 \propto N^{-2} . \tag{14}$$

On the other hand, many plants shed low branches as new ones grow at the top, and self-thinning usually occurs among branches in a growing crown. This process is governed by light shortage near the ground and inside crowns. By these means, geometric similarity might sometimes be approximated throughout plant growth, so that the thinning exponents remain close to $-3/2$ (Norberg 1988).

In those cases, however, when branches increase in number with plant growth, branches will tend to reduce the thinning gradient of the whole plant below $-3/2$. This might help explain some empirically observed exponents between -1.5 and -2. But this explanation cannot apply to trees, because measured tree volumes are usually only the stemwood volume, exclusive of branches. The elastic-similarity model seems better able to explain exponents near −2 for trees, as shown below.

6 The Growth Modes of Tree Leaves and Prostrate Plants May Mimic Geometric Similarity

Tree Leaves. If geometric similarity is maintained in above-ground plant parts, so that leaf dimensions, including leaf thickness, also increase isometrically with respect to stem and branches, then the −3/2 thinning exponent [Eq. (2)] applies also to the leaves taken separately. But in most trees the leaves remain about the same size throughout tree growth, be they needles of conifers or any size and shape of leaf in most broad-leaved trees.

The maintenance of a constant leaf size throughout tree growth is a means of making the combined foliage-surface area per tree, A_f, increase at the same rate as the combined foliage-volume per tree, V_f (Norberg 1988, p. 242). That is, a given increment in leaf-surface area requires less leaf-biomass if more small leaves are grown,

$$V_f \propto A_f \ , \tag{15}$$

than if existing, small, leaves grow larger, in which case the geometric-similarity principle dictates

$$V_f \propto A_f^{3/2} \ . \tag{16}$$

The maintenance of a constant leaf size throughout tree growth is an economical use of materials to achieve a certain goal, here light interception.

Most tree leaves thus depart drastically from the growth mode adopted by trunk and branches. This constancy of leaf size might be expected to result in radically different thinning slopes for leaves than for other plant structures, like stem and branches, that maintain geometric or elastic similarity. When total plant volume is related to the population density, different thinning slopes for different plant parts combine into a compromise slope for the whole plant. Because leaves may thus affect the overall thinning slope for the entire plant, the thinning function for the total volume of constant-sized leaves per average tree versus tree population density was explored by Norberg (1988).

While the foliage layer still grows in depth, the thinning exponent should be −3/2 regardless of whether leaves remain constant-sized or grow isometrically with respect to trunk and branches. The explanation why the −3/2 exponent applies despite this drastic difference in leaf growth-mode is as follows (Norberg 1988). With geometric similarity extending to the leaves, leaf number should remain constant, and empty spaces between leaves would scale up isometrically with the leaves and other plant structures. But with constant-sized leaves, the spacing between leaves may remain constant as new leaves grow, and crown volume increases isometrically with stem and branches, as usual. In effect, this mimics geometric similarity; since packing density of leaf biomass per unit of crown volume remains constant in both cases, total leaf volume scales isometrically with crown volume, which in turn grows isometrically with stem and branches.

But after the foliage layer has attained its maximum thickness, the slope for leaf volume and area should shift to -1 (Norberg 1988). From then on, tree leaves thus tend to increase the compromise thinning exponent for the entire tree to values somewhat above -1.5 (but, still, much lower than -1). Recall from Section 5 that if the number of branches per plant increases rather than remains constant along with plant growth, the thinning exponents tend to be lower than -1.5. Tree leaves thus tend to balance this effect. But neither branches, nor leaves are usually included in empirical measures of tree mass or weight, which are usually for stemwood only.

Prostrate Plants. Norberg (MS) considered prostrate plants among which different plant parts (runners, petioles, and leaves) exhibit different growth modes, resulting in theoretically derived thinning exponents between -2.00 and -1.00, with the most realistic range being from -1.25 to -1.75. This is another case where the geometric similarity exponent -1.5 may be mimicked by other growth modes, here a combination of two growth geometries.

7 Elastic Similarity

The elastic-similarity principle is said to apply to similar structures of different sizes if they have a constant ratio between the critical buckling force, F_b, and the actual gravitational, deforming, force, F_g,

$$F_b/F_g = \text{constant} \tag{17}$$

(Hokkanen 1986); that is, the stability safety factor against buckling is the same irrespective of the size of the structure. This safety factor dictates the bulk of supporting structures and thereby also the local elevation of data-points for volume versus population density.

The elastic-similarity principle is an obvious alternative to geometric similarity (isometry) for the design and growth of individual plants. Elastic similarity can be expected in especial when structural strength against the force of gravity is of overriding importance, as in large organisms like trees. The geometry of many tree species also fits better with the elastic-similarity principle than with geometric similarity (McMahon and Kronauer 1976; review in Norberg 1988).

When the top of a vertical column, or a tree trunk, is deflected from the vertical, gravity forces tend to augment the bending, whereas the elastic, structural, deformation sets up a restoring moment, tending to straighten the structure back to the vertical again (Fig. 6). The critical restoring moment is that which exactly equals the moment due to gravity that tends to increase the bending after the top has been deflected a small angle from the vertical. The elastic-similarity principle dictates how the cross-sectional diameter of a vertical column, or a tree trunk, should increase with increasing height to set up a restoring righting moment just large enough to counteract further gravitational bending. Greenhill (1881) showed that the critical diameter below which a vertical column with uniform cross-section along its length, or a tapered tree trunk, buckles under its own weight in-

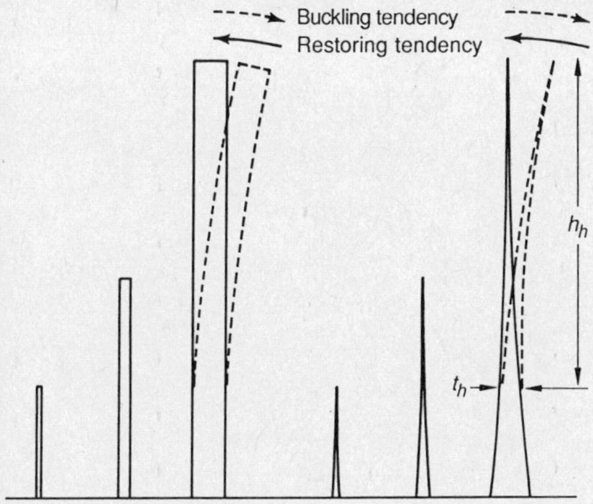

Fig. 6

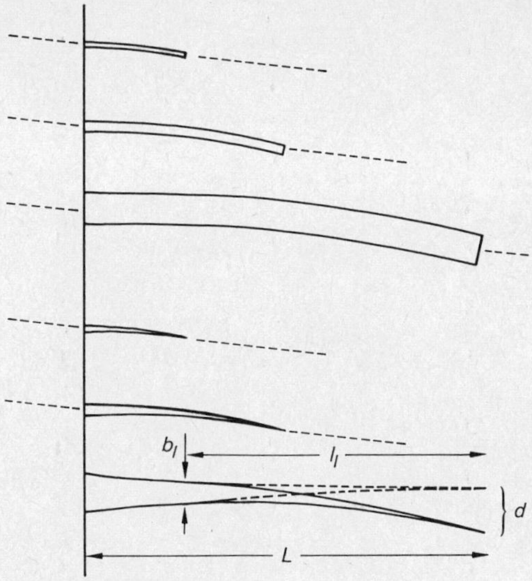

Fig. 7

Fig. 6. Elastically similar vertical columns of a uniform cross-section (*left*), and tapered tree trunks (*right*), of different sizes. The greatest heights that vertical columns and trunks can attain without buckling under their own weight are proportional to the 2/3 power of the cross-sectional diameter at the base. The maintenance of the same safety factor against buckling under gravity forces also requires that the diameter at the base increases as the 3/2 power of the height. In both cases, the different-sized structures are shaped according to the elastic-similarity criterion. Tapered structures (*right*) are elastically self-similar if the local diameter at any height, t_h, is always proportional to the 3/2 power of the remaining distance to the top, h_h

Fig. 7. Elastically similar cantilever beams of a uniform cross-section (*top*), and tapered branches (*bottom*), of different sizes, fixed and supported at the base only. Provided that the beams are set at the same angle to the horizontal, and given that the deflection of the tip, d, under gravity forces must not exceed a constant fraction of the length, L, the greatest lengths that the beams can attain are proportional to the 2/3 power of the cross-sectional diameter at the base. This is an elastic-similarity criterion for branches. Tapered structures (*right*) are elastically self-similar if any local diameter, b_l, is proportional to the 3/2 power of the remaining distance to the tip, l_l

creases with the 3/2 power of the height, a relation later elaborated by McMahon and Kronauer (1976, p. 446) and McMahon and Bonner (1983).

When the elastic-similarity principle applies to different-sized cantilever beams of uniform thickness, or with a tapered shape as well, fixed and supported at the basal end only, and set at some angle to the horizontal, just as branches are, a characteristic feature is that the distance any one of them deflects under its own weight is always a constant fraction of its length (Fig. 7). Therefore, the chord of the curvature forms the same angle with the horizontal, regardless of the beam or branch size. This state is achieved when the basal diameter increases as the 3/2 power of the length of the beam or branch (McMahon and Kronauer 1976).

An "elastically self-similar" trunk or branch (term from McMahon and Kronauer 1976, p. 446) should have a taper ratio such that all local diameters along its length are proportional to the 3/2 power of the remaining distance to the free end. With elastic similarity, there are thus two length scales, one longitudinal, l, and one transverse, t, related as

$$t \propto l^{3/2} , \tag{18}$$

and with the volume of the structure, V, varying as

$$V \propto t^2 l \propto t^{8/3} \propto l^4 . \tag{19}$$

The elastic similarity criterion requires all diameters of an organism, t, to change with total size in the same way,

$$t \propto V^{3/8} , \tag{20}$$

and similarly for all lengths, l,

$$l \propto V^{1/4} . \tag{21}$$

By applying this disproportionate increase in diameter as stem height and branch length increase, a plant maintains a constant safety factor against buckling under gravity forces. This is in contrast with geometric similarity, for which $t \propto l$, leading to a decrease of the stability safety factor as size increases.

When elastic similarity applies, a plant's height-to-breadth ratio h decreases with increasing plant size (albeit very slowly), because the trunk diameter grows faster than do trunk height and branch length, and crown width is determined jointly by branch length and trunk diameter [see Eq. (22)]. With geometric similarity h remains constant.

8 Elastic Similarity and Self-Thinning of Plant Populations

Norberg (1988) derived the self-thinning characteristics for plants conforming to the elastic-similarity principle. The slope of the limiting thinning line, resulting from elastic similarity, would ideally be defined by finding a value for a in Eq. (1). But self-thinning with elastic similarity is described by a curve, rather than by a line, in a log-log diagram. Therefore, it is, unfortunately, not possible to use the convenient power function [Eq. (1)], applicable to geometric similarity, to describe self-thinning under elastic similarity in a theoretically correct way. Nevertheless, because the curvature of the limiting thinning curve turns out to be moderate within the most realistic range of ratios of trunk diameter to crown width (Fig. 9), the convenient power function is eminently useful for testing the elastic-similarity theory by fitting curves to empirical data (Norberg 1988).

Empirical values of tree mass are usually measured only as stem-wood biomass, which is the primary interest in forestry. But in a theoretical model, branch size must also be considered (even if its volume is disregarded), because crown width affects tree spacing and therefore the thinning function. Norberg (1988) took branch diameter and length to be proportional to stem diameter and height, respectively, as dictated by the elastic-similarity principle (see above) and as observed empirically for trees. Radius R of a plant's exclusive space (the hexagonal prism in Fig. 3) is determined by stem radius (at the level of the longest branches) and the horizontal projection of the longest branches. With elastic similarity, these two measures increase at different rates with respect to stem diameter, t,

$$R = 0.5t + k_7 t^{2/3} , \tag{22}$$

which leads to the following expression for the gradient d ln V/d ln N of the thinning curve in terms of stem radius t,

$$\frac{d \ln V}{d \ln N} = \frac{(-4/3)(t^{1/3} + 2k_7)}{t^{1/3} + (4/3)k_7} \tag{23}$$

[from Eq. (22) in Norberg 1988, where all steps in the derivation are explained]. The constant k_7 is a coefficient in a power function relating the horizontal projection of branch length to stem radius.

Since k_7 is usually unknown, a convenient way of predicting the thinning gradient at any stage is to measure the proportion θ that stem diameter (at the height of the longest branches) makes up of crown width,

$$\theta = \frac{t}{t + 2k_7 t^{2/3}} \tag{24}$$

[using Eq. (22)]. Solving for k_7 and inserting into Eq. (23),

$$\frac{d \ln V}{d \ln N} = \frac{-4}{\theta + 2} . \tag{25}$$

This is a simple, fascinating result, permitting the thinning gradient to be easily predicted at any growth stage by knowing only the ratio of stem diameter to crown width, θ (Fig. 8). Norberg (1988, Figs. 4, 5; see Fig. 8 in this chapter) instead used the ratio trunk radius to horizontal projection of the longest branches, ζ, which is related to the ratio between trunk diameter and crown width, θ, as

$$\theta = \zeta/(\zeta + 1) . \tag{26}$$

If instead the thinning gradient is sought for the volume of plant material per unit of ground area, VN, versus population density, N (as in Fig. 2), Eq. (25) changes into

$$\frac{d \ln (VN)}{d \ln N} = \frac{-4}{\theta + 2} + 1 . \tag{27}$$

As θ approaches 0,

$$d \ln V/d \ln N \rightarrow -2 , \tag{28}$$

and

$$d \ln (VN)/d \ln N \rightarrow -1 . \tag{29}$$

With increasing values of θ the gradients increase monotonically (θ can only take on values between 0 and 1 as it is a proportion), and as θ approaches unity,

$$d \ln V/d \ln N \rightarrow -4/3 , \tag{30}$$

and

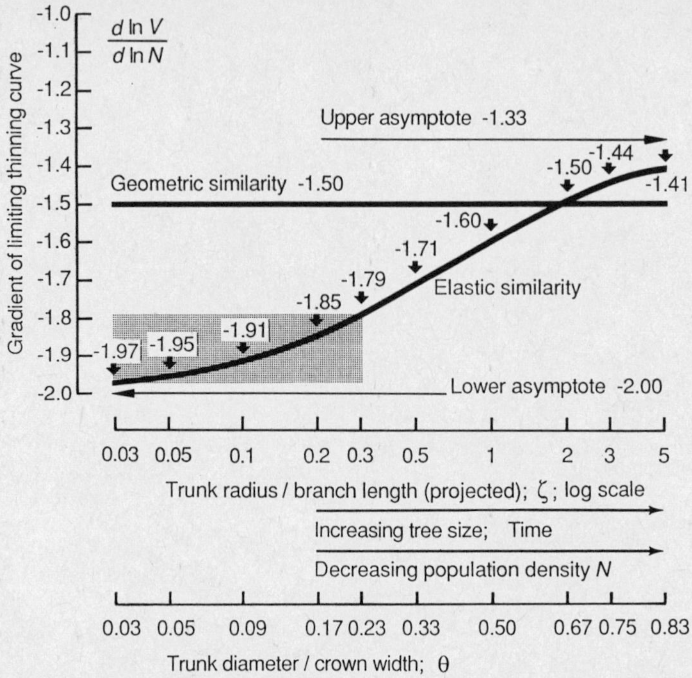

Fig. 8. Predictions of how the gradient of limiting thinning curves in a log-log diagram, showing average plant volume (or mass) against population density (as in Figs. 1, 9) varies with the ratio of trunk diameter to crown width of trees growing according to either the geometric-similarity principle or the elastic-similarity principle. With geometric similarity, the ratio of stem diameter to crown width remains constant throughout plant growth, as does the -1.5 slope of the thinning line. Depending on their characteristic geometry, different plants therefore occupy different stationary positions along the horizontal -1.5 line, and any given population maintains the same position throughout the growth of its plants. As a tree grows taller with elastic similarity maintained, the trunk contributes an increasing proportion of crown width. Along with this individual growth (toward the *right* in the diagram) the population density decreases and the gradient of the limiting thinning curve increases monotonically as shown. *Shaded part of curve* encompasses the most realistic ratios of trunk diameter to crown width for trees and the corresponding predicted gradients of the thinning curves. The curve for elastic similarity is from Eq. (25). If $+1$ is added to the elastic-similarity curve above (all gradient values), the graph instead predicts how the gradient (d ln (VN)/d ln N) of limiting thinning curves in a log-log diagram, showing average plant volume per unit of ground area against population density (as in Fig. 2), varies with the ratio of trunk diameter to crown width [Eq. (27)]. (Modified from Fig. 4 in Norberg 1988)

$$\mathrm{d}\ln(VN)/\mathrm{d}\ln N \rightarrow -1/3\ . \tag{31}$$

But for realistic ratios of stem diameter to crown width, θ, the range of gradients is much less. So, for instance, when the ratio θ increases from 1/30 to 1/4, the gradient d ln V/d ln N changes from -1.97 to -1.78 (Fig. 8), and the gradient d ln (VN)/d ln N changes from -0.97 to -0.78. It is only if the stem diameter were to make up 2/3, or more, of crown width that the thinning gradient would take the geometric-similarity value of -1.5, or larger values (Fig. 8). For plants of a common shape, therefore, the gradients for elastic similarity are significantly

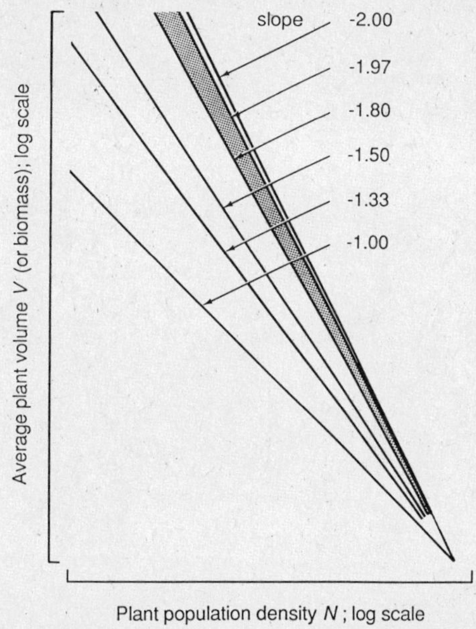

Fig. 9. Generalized diagram of average plant volume versus population density, contrasting different thinning lines. The −1.50 line is for geometric similarity, and the −2 and −1.33 lines show the two extreme, asymptotic values for elastic similarity. With elastic similarity, population thinning proceeds along a curve, rather than a line, in a log-log diagram. For trees growing according to the elastic-similarity principle, the most realistic range of gradients is between −1.97 and −1.80 (*shaded region*, corresponding to shaded zone in Fig. 8). This allows only a very moderate curvature of the thinning curve, which is steepest (gradient closest to −2) when plants are small (*at bottom right*); the steepness then decreases progressively as plants grow larger and populations thin (as shown in Fig. 8). Prostrate plants may take values between −1.00 and −2.00, with the most realistic range from −1.25 to −1.75. The −1.33 curve applies to certain animal populations and is dictated by the scaling exponent 0.75, relating basal metabolic rate to body mass. If instead plant (or animal) volume per unit of ground area is plotted against population density, as in Fig. 2, the new slopes are obtained by adding +1 to those above. (Modified from Fig. 6 in Norberg 1988)

steeper than the −1.5 slope following from geometric similarity (Figs. 8, 9). That is, the biomass volume grows faster with decreasing population density when elastic similarity obtains then when plants remain geometrically similar.

When a power function is fitted to thinning data for elastically similar plants, the slope in a log-log diagram is simply some average of the moderately different gradients predicted for the various stages of the growth and thinning process over the range observed (Norberg 1988). Gradients in thinning functions reflect size-related design principles that apply regardless of whether the plants are different species in different populations (in across-species regressions) or the same individuals at different growth stages, in which case the gradient reflects the growth mode of individual plants (Norberg 1988). The various factors affecting the local elevation, and the y-intercept k, of thinning lines considered in Sections 4.1, 4.2, and 4.3, and in Figs. 4 and 5, apply also to elastically similar plants.

The above gradients apply to stem volume, and as long as branch number remains constant throughout growth they apply also to branch volume and to the combined volume of branches and stem (cf. similar considerations with geometric similarity in Sect. 5). But when the number of branches increases with plant growth, and elastic similarity obtains, the thinning gradients for the combined volume of stem and branches tend to be lower (steeper) than for the stem volume alone (details in Norberg 1988).

In general, the allometry of trees conforms very well with the elastic-similarity principle (review in Norberg 1988). And several empirically obtained thinning slopes for trees (stem-wood data) are as predicted from the elastic-similarity principle (about −1.90). But there are also many tree populations that come near the geometric-similarity slope of −1.5. It is unclear what causes this discrepancy between allometry and thinning slopes. A possible explanation could be that roots and above-ground structures sometimes obey different design principles. If the roots maintain geometric similarity, but above-ground parts maintain elastic similarity, then the lateral spread of the roots is likely eventually to outgrow crown width, whereupon increasingly more open space might open up between crowns. Root competition would then drive population thinning, and the roots would dictate the thinning slope also for above-ground plant parts so that their thinning slope would be at variance with their allometry (see further explanation in Norberg 1988).

9 Packing Density Versus Size of Elastically Similar Plants

Finally, I will consider how the packing density of plant material in a plant's exclusive spatial volume (the idealized hexagonal prism) changes with increasing stem diameter t as plants grow according to the elastic-similarity principle. The volume of plant biomass, $V_{biomass}$, is

$$V_{biomass} = k_c t^2 H = k_c t^2 k_t t^{2/3} , \tag{32}$$

using the elastic-similarity criterion in expression (18),

$$H = k_t t^{2/3} . \tag{33}$$

The volume of the hexagonal prism-shaped volume, V_{space}, (Fig. 3) is

$$V_{space} = AH = 1.5 \cdot 3^{1/2} R^2 H = 2.6(0.5t + k_7 t^{2/3})^2 k_t t^{2/3} , \tag{34}$$

using Eqs. (5) and (22). Now, the packing density with elastic similarity, p_{es}, is

$$p_{es} = \frac{V_{biomass}}{V_{space}} = \frac{k_c t^2 k_t t^{2/3}}{2.6(0.5t + k_7 t^{2/3})^2 k_t t^{2/3}} = \frac{k_c}{2.6(0.5 + k_7 t^{-1/3})^2} , \tag{35}$$

where k_c is a coefficient in the power function (32), relating the combined volume of stem and branches to t^2H, and k_7 [from Eq. (22)] is a coefficient in another power function, relating the horizontal projection of branch length to stem radius. Equation (35) demonstrates that as a plant grows larger (increasing t), while maintaining elastic similarity, the packing density of stem and branches increases within the exclusive spatial volume occupied by the plant. This is in contrast with geometric similarity, where packing density remains constant.

10 Self-Thinning of Plant Populations with Different Age, Size, and Species Composition

Here I will briefly explore whether the self-thinning rule applies to stands showing a bimodal or multimodal size distribution among individuals, to mixed-age stands, and to mixed-species stands.

Because the growth mode dictates the thinning gradient, a thinning exponent that is characteristic for a particular growth mode may apply to such mixed populations, provided that the following two conditions are met: (1) the plants in the constituent size, age, or species classes of any one stand do all adhere to the same growth mode, and (2) inter-class competition is symmetrical throughout the growth and thinning process, such that the number of plants in each class makes up a constant proportion of the total number of plants in the mixed population. If these conditions are met, such a mixed population should exhibit the thinning exponent that is typical for the shared growth mode of the plants in the constituent classes, even if each plant category has a different characteristic k-value (y-intercept) when growing in a pure stand, that is, even if the plants in the different classes differ in shape and structure. The k-value of the mixed population should be a compromise between the k-values of the constituent plant types, weighted by the relative numerical contribution of each plant type to the mixed stand.

By contrast, when competition is asymmetrical, different size-classes vary continuously in their relative numerical contribution to the mixed population, and, therefore, also in their relative contribution to the volume of the average plant in it. The gradient for the entire mixed population may then fail to conform to the gradient associated with the prevailing growth mode, even if that mode is shared by all plants in all constituent classes.

11 Self-Thinning of Animal Populations

11.1 The $-3/2$ Slope in Aquatic Sessile Animals?

The $-3/2$ self-thinning law described for plants might apply to sessile, bottom-living organisms like mussels and barnacles, which are packed closely together. For sessile animals in a densely packed population, where each animal is in direct physical contact with its nearest neighbours, the primary concern is to claim a bottom area large enough for the animal to attach to or to sit on. Under such circumstances, and when the animals grow according to the geometric-similarity principle, the $-3/2$ exponent should apply to their thinning function.

Aquatic, sessile organisms (plants and animals) may not be expected to obey the elastic-similarity principle, since structural strength against the force of gravity is of little or no concern to them because of their buoyancy. If the elastic-similarity principle were to apply, however, densely packed animals might be likened to densely packed plants without branches, and the self-thinning gradient would take the value $-4/3$ [when θ equals 1 in Eq. (25)]. Alternatively, if the projected bottom area covered by tentacles dictates the packing density of sessile animals, and if the elastic-similarity principle were to apply to the body and tentacles alike, then the arguments for plants in Section 8 would apply; possible gradients would rańge from −2 to −4/3 (Fig. 8).

11.2 The −4/3 Slope Dictated by Basal Metabolic Rate

Mobile animals forage over some larger area to cover their energy needs, and for them their metabolic needs are likely to dictate the size-density relationship. The basal metabolic rate of an animal, BMR, varies with its mass M as

$$\mathrm{BMR} \propto M^{0.75} \tag{36}$$

(e.g., Calder 1984; Damuth 1987). Consider animals of different sizes, such as mammalian herbivores, eating similar foods. When food limits population density, the number of animals per unit of ground area, N_a, should be inversely proportional to the energy needs per animal,

$$N_a \propto M^{-0.75} \tag{37}$$

(Peters and Wassenberg 1983; Begon et al. 1986). Exponents very close to −0.75 emerge from empirical across-species generalizations (Damuth 1987), so this seems to be a general rule among animals. Rewriting expression (37) into the plant self-thinning format, i.e., as mass, M, (size) per individual against population density, N_a,

$$M = k_a N_a^{-1/0.75} = k_a N_a^{-1.33} \ . \tag{38}$$

This −4/3 rule may apply to regressions across different growth stages of individuals in one and the same population (Begon et al. 1986), as well as to regressions across different single-species populations, each one of which is composed of different species of different sizes. The exponent −1.33 is larger (shallower line) than those for most plants (Fig. 9), indicating that the size of animals increases more slowly with decreasing population density than the size of plants increases with decreasing population density. It must be stressed, however, that other exponents may be expected for certain animals under other premises (Norberg MS).

12 Conclusion

The apparently universal prevalence of self-thinning according to packing theory may be the best evidence available of the pervasive importance of size- and density-dependence in the regulation of population density among plants and animals, with competition due to crowding driving population self-thinning. The thinning gradients are dictated by demands on space − ultimately energy and nutrients − in both plants and animals. The dominant, self-thinning power "laws" for plants may be −3/2 (with geometric similarity) and about −1.90 (with elastic similarity), whereas for animals the most common thinning gradient may be −4/3 (dictated by the basal metabolic rate).

Acknowledgments. I thank Bo Ebenman and Henry Wilbur for comments on the manuscript. Support was obtained from the Swedish Natural Science Research Council (grant B 4450).

13 References

Begon M, Firbank L, Wall, R (1986) Is there a self-thinning rule for animal populations? Oikos 46:122–124

Calder WA III (1984) Size, function and life history. Harvard Univ Press, Cambridge

Damuth J (1987) Interspecific allometry of population density in mammals and other animals: the independence of body mass and population energy-use. Biol J Linnean Soc 31:193–246

Gorham E (1979) Shoot height, weight and standing crop in relation to density of monospecific plant stands. Nature (London) 279:148–150

Greenhill AG (1881) Determination of the greatest height consistent with stability that a vertical pole or mast can be made, and of the greatest height to which a tree of given proportions can grow. Proc Cambridge Philos Soc 4:65–73

Hokkanen JEI (1986) Notes concerning elastic similarity. J Theor Biol 120:499–501

Hutchings M (1983) Ecology's law in search of a theory. New Sci 98:765–767

Kays S, Harper JL (1974) The regulation of plant and tiller density in a grass sward. J Ecol 62:97–106

Lonsdale WM, Watkinson AR (1983) Plant geometry and self-thinning. J Ecol 71:285–297

McMahon TA, Bonner JT (1983) On size and life. Sci Am Libr, New York

McMahon TA, Kronauer RE (1976) Tree structure: deducing the principle of mechanical design. J Theor Biol 59:443–466

Miyanishi K, Hoy AR, Cavers PB (1979) A generalized law of self-thinning in plant populations (Self-thinning in plant populations). J Theor Biol 78:439–442

Norberg RÅ (1988) Theory of growth geometry of plants and self-thinning of plant populations: geometric similarity, elastic similarity, and different growth modes of plant parts. Am Nat 131:220–256

Peters RH, Wassenberg K (1983) The effect of body size on animal abundance. Oecologia (Berlin) 60:89–96

Weller DE (1987) A reevaluation of the −3/2 power rule of plant self-thinning. Ecol Monogr 57:23–43

Westoby M (1976) Self-thinning in *Trifolium subterraneum* not affected by cultivar shape. Aust J Ecol 1:245–247

Westoby M (1977) Self-thinning driven by leaf area not by weight. Nature (London) 265:330–331

Westoby M (1981) The place of the self-thinning rule in population dynamics. Am Nat 118:581–587

Westoby M (1984) The self-thinning rule. Adv Ecol Res 14:167–225

White J (1977) Generalization of self-thinning of plant populations. Nature (London) 268:373

White J (1980) Demographic factors in populations of plants. In: Solbrig OT (ed) Demography and evolution in plant populations. Blackwell, Oxford, pp 21–48

White J (1981) The allometric interpretation of the self-thinning rule. J Theor Biol 89:475–500

White J, Harper JL (1970) Correlated changes in plant size and number in plant populations. J Ecol 58:467–485

Whittington R (1984) Laying down the −3/2 power law. Nature (London) 311:217

Yoda K, Kira T, Ogawa H, Hozumi H (1963) Self-thinning in overcrowded pure stands under cultivated and natural conditions. J Biol Osaka City Univ 14:107–129

Subject Index